Write-In Text

Sixth Edition

Machine Trades Print Reading

Michael A. Barsamian / Richard A. Gizelbach

Publisher

The Goodheart-Willcox Company, Inc.

Tinley Park, IL

www.g-w.com

The Goodheart-Willcox Company, Inc. Brand Disclaimer: Brand names, company names, and illustrations for products and services included in this text are provided for educational purposes only and do not represent or imply endorsement or recommendation by the author or the publisher.

The Goodheart-Willcox Company, Inc. Safety Notice: The reader is expressly advised to carefully read, understand, and apply all safety precautions and warnings described in this book or that might also be indicated in undertaking the activities and exercises described herein to minimize risk of personal injury or injury to others. Common sense and good judgment should also be exercised and applied to help avoid all potential hazards. The reader should always refer to the appropriate manufacturer's technical information, directions, and recommendations; then proceed with care to follow specific equipment operating instructions. The reader should understand these notices and cautions are not exhaustive.

The publisher makes no warranty or representation whatsoever, either expressed or implied, including but not limited to equipment, procedures, and applications described or referred to herein, their quality, performance, merchantability, or fitness for a particular purpose. The publisher assumes no responsibility for any changes, errors, or omissions in this book. The publisher specifically disclaims any liability whatsoever, including any direct, indirect, incidental, consequential, special, or exemplary damages resulting, in whole or in part, from the reader's use or reliance upon the information, instructions, procedures, warnings, cautions, applications, or other matter contained in this book. The publisher assumes no responsibility for the activities of the reader.

The Goodheart-Willcox Company, Inc. Internet Disclaimer: The Internet resources and listings in this Goodheart-Willcox Publisher product are provided solely as a convenience to you. These resources and listings were reviewed at the time of publication to provide you with accurate, safe, and appropriate information. Goodheart-Willcox Publisher has no control over the referenced websites and, due to the dynamic nature of the Internet, is not responsible or liable for the content, products, or performance of links to other websites or resources. Goodheart-Willcox Publisher makes no representation, either expressed or implied, regarding the content of these websites, and such references do not constitute an endorsement or recommendation of the information or content presented. It is your responsibility to take all protective measures to guard against inappropriate content, viruses, or other destructive elements.

Cover image: Kosoff/Shutterstock.com

Library of Congress Cataloging-in-Publication Data

Barsamian, Michael.
    Machine trades print reading / by Michael Barsamian, Richard Gizelbach.
        p. cm.
    Includes bibliographical references and index.
    ISBN 978-1-63126-105-3
    1. Blueprints. 2. Machinery--Drawings. I. Gizelbach, Richard. II. Title.
T379.B355   2016
604.2'5--dc23
                                    2014037226

Preface

Machine Trades Print Reading is designed to help student's develop the basic skills required for visualizing and interpreting industrial prints. Each unit begins with outcome-based *Learning Objectives* the student must achieve and *Key Terms* the student must learn to demonstrate satisfactory completion of the unit.

These competencies are statements describing the knowledge, skills, and attitudes students should exhibit and language they should be familiar with to succeed in a career in industry. Competencies define the learning outcome, but need not include the conditions and criteria for acceptable performance.

A heavy emphasis has been placed on the use of illustrations to develop the basic skills of visualization and to simplify the learning process. Actual industrial prints are featured to prepare students for the real conditions found in industry. Some of the industrial prints have certain changes or modifications from the original drawings to provide a variety of information that will aid students' learning experiences.

The text consists of 18 units. The first four units give the basics of print reading. They present the important concepts of applications of drawings and prints in industry, visualizing shapes, line usage on drawings, and basic title block format.

Units 5 and 6 provide a review of basic math and measurement skills that are necessary to successfully read and interpret prints and work in industry.

Units 7 through 17 deal with industrial prints. Each unit deals with a specific technical topic, which is first introduced in the text and then shown on the actual prints. A unit on geometric dimensioning and tolerancing (GD&T) has been included due to the widespread use of this system by the manufacturing industry.

The instructional materials, illustrations, and industrial prints in *Machine Trades Print Reading* are presented in a step-by-step sequence. The written material and illustrations in each unit contain the technical information needed to complete the industrial print reading activities. Each succeeding unit contains additional new material. Review questions pertaining to the industrial prints in each unit use the new material as well as material learned from earlier units.

Unit 18 is a collection of prints and activities which review all of the material presented in the text. The reference section includes various charts, tables, and other information that will prove useful to novice and expert print readers alike. The glossary includes definitions of every *Key Term* presented in the text. The reference section and glossary will also serve as valuable resources for students even after completion of this text.

Michael Barsamian
Richard Gizelbach

About the Authors

Michael A. Barsamian is a graduate of the University of Wisconsin-Stout with a Bachelor of Science Degree in Education and advanced graduate credits. He entered the field of career and technical education to prepare his students for their careers in industry. Mr. Barsamian has taught hand drafting, residential drafting, CAD drafting, and print reading for machine trades. In addition, he has provided consultation and customized training to machine shops and other related industries.

Mr. Barsamian has incorporated his educational experiences throughout this text. *Machine Trades Print Reading* is a joint effort with Mr. Gizelbach to present a concise approach to print reading that makes the topic easy to understand.

Richard Gizelbach completed an apprenticeship and became a Journeyman Machinist at Nordberg Mfg. in Milwaukee, Wisconsin. Mr. Gizelbach holds a Bachelor of Science and Master of Vocational Education degrees from the University of Wisconsin-Stout. Mr. Gizelbach also has a Specialist Certificate in Administrative Leadership from the University of Wisconsin-Milwaukee.

Richard Gizelbach is a former instructor of Machine Shop and Computer Numerical Control Programs at Gateway Colleges in Kenosha, and Racine, Wisconsin. His extensive machining knowledge allowed him to develop and teach custom in-plant courses for Gateway and Milwaukee Technical College for several companies throughout southeastern Wisconsin.

Acknowledgments

The authors and publisher would like to thank the following companies and organizations for their contribution of resource material, images, or other support in the development of *Machine Trades Print Reading*.

A&E Manufacturing

Allen-Bradley Co.

CAD Mania

Dessault Systèmes SolidWorks Corp.

Dobby Engineering

L. S. Starrett Co.

Milwaukee Electric Tool Corporation

The American Society of Mechanical Engineers

Reviewers

The authors and publisher would like to thank the following individuals for their valuable input in the development of *Machine Trades Print Reading*.

Rich Ambacher
Technology Education Teacher
Washington Twp. High School
Sewell, NJ

Terry Anselmo
Precision Machining Instructor
Sun Area Technical Institute
New Berlin, PA

Rob Ataman
Instructor/Coordinator
Red River College
Winnipeg, MB

Ronnie Brittain
Computer-Integrated Machining
Program Coordinator
Catawba Valley Community
College School of Business,
Industry, and Technology
Hickory, NC

George Cable
Industrial and Continuing
Education Instructor
Hawkeye Community College
Cedar Falls, IA

Rick Calverley
CNC Manufacturing Supervisor
Lincoln College of Technology
Grand Prairie, TX

Dr. Chesley Chambers
Metals Instructor
Floyd County Schools College &
Career Academy
Rome, GA

Vernon K. Chandler
Teacher
Sevier County High School
Sevierville, TN

Clair W. Cornish
Professor
Durham College
Whitby, ON

Don Dean
Instructor/Coordinator
Manufacturing Technologies
C-TEC of Licking County
Newark, OH

Ron Delagrange
Engineering Technology Instructor
Jefferson County High School
Louisville, GA

Frank D. Gulluni
Director, Manufacturing
Technology Center
Asnuntuck Community College
Enfield, CT

Leonard Hall
CTE Teacher/Engineering and
Metal Fabrication Academy Teacher
WEMOCO Career and Technical
Education Center
Spencerport, NY

Michael M. Hilber
Engineering/Manufacturing
Teacher
Secondary Technical Education
Program—Anoka Hennepin District 11
Anoka, MN

Bruce Jacobs
Industrial Arts Technology/STEM/
CTE, Visual Communications &
Yearbook Teacher
Mt. Bauer Middle School
Auburn, WA

Terry L. Jamison
Precision Metal Machine
Technologies Instructor
York County School of Technology
York, PA

James R. Knapp Sr.
Assistant Professor
Thaddeus Stevens College of
Technology
Lancaster, PA

Michael Koppy
Machine Technology Instructor
Lake Superior College
Duluth, MN

Robert Lee
Instructor
Alexander Hamilton High School
Milwaukee, WI

Oygar Lindskog
Lead Instructor, Machine
Technology
Simi Valley Adult School and
Career Institute
Simi Valley, CA

Frank Mueck
Professor
Cambrian College of Applied Arts
and Technology
Sudbury, ON

Edward Richmer
Instructor, Machine Trades
Apprenticeship Mathematics
Prosser Career Education Center
New Albany, IN

Danny Roberts
Instructor Computerized
Manufacturing & Machining
Bluegrass Community & Technical
College
Lexington, KY

Larry Sisk
Precision Machines Instructor
Princeton Community High School
Princeton, IN

Bruce Smith
Instructor
Tulsa Technology Center
Tulsa, OK

Paul Sorenson
Computerized (CNC) Precision
Machining Technology Instructor
Minnesota State College—Southeast
Technical
Winona, MN

Jon Stenerson
Instructor
Fox Valley Technical College
Appleton, WI

Joseph VandenEnden
Professor
Conestoga College ITAL
Kitchener, ON

Daniel W. Wagner P.E.
Managing Director of Workforce
Training
HACC—Central Pennsylvania's
Community College
Harrisburg, PA

Mark Wildenberg
Instructor—Apprentice Millwright
Fox Valley Technical College
Appleton, WI

Andrew V. Zwanch
Precision Machining Technology
Department Chair
Johnson College
Scranton, PA

Features of the Textbook

Learning Objectives clearly identify the knowledge and skills to be obtained when the unit is completed.

Key Terms list the technical terminology to be learned in the unit. Review this list after completing the unit to be sure you know the definition of each term.

Illustrations have been designed to clearly and simply communicate the specific topic.

Pro Tips provide students with advice and guidance that is especially helpful to beginning print readers.

UNIT 3 — Line Usage

Learning Objectives

After studying this unit, you will be able to:
✓ Describe the purposes of lines on prints.
✓ Identify various types of lines found on prints.
✓ Locate corresponding lines or surfaces in various views.

Key Terms

alphabet of lines	dimension line	section
American Society of Mechanical Engineers (ASME)	dimensioning	section line
	drawing standards	stitch line
break line	extension line	symmetry line
centerline	hidden line	viewing-plane line
chain line	leader line	visible line
cutting-plane line	phantom line	

As new ideas develop into useful products, they require precise drawings of parts. Drawings must be consistent so they are readable across the industry.

The *American Society of Mechanical Engineers (ASME)* develops drawing standards used by the industry for the development of engineering drawings. *Drawing standards* are documented practices used to develop drawings. One of the standards is ASME Y14.2, titled *Line Conventions and Lettering.* This standard, often referred to as the *alphabet of lines*, includes guidelines for specific types of lines, each with an intended purpose, as shown in **Figure 3-1**.

Types of Lines

Drawings contain various types of lines. Understanding the alphabet of lines allows the reader to visualize the intent of a drawing.

A line can have different characteristics. A line can be continuous (solid with no breaks), or dashed (short lines separated by spaces), or a combination of both. Each type has a specific meaning.

Line width, or line weight, is another characteristic of a line. There are two line widths used in mechanical drawing. A thick line is twice the width of a thin line. The difference between the two helps identify their purpose. In the following sections, you will learn about the different types of lines and their importance in use.

Visible Lines

Visible lines, also referred to as object lines or outlines, define the shape and surfaces of an object. They show all edges of an object that are visible in a view. Visible lines define the outside border of a part as well as the surfaces within the borders. Visible lines are thick continuous lines that visually stand out on a drawing. See **Figure 3-2**.

Copyright Goodheart-Willcox Co., Inc.   25

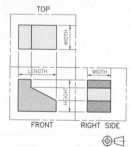

Goodheart-Willcox Publisher
Figure 2-7 Standard third-angle, multiview drawing layout.

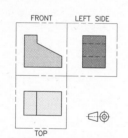

Goodheart-Willcox Publisher
Figure 2-8 Standard first-angle, multiview drawing layout.

Basic Dimensions

The top view contains the length and width of the part. The front view shows the length and height. The right side view gives the height and width. Note that each view contains only two required dimensions of the object.

In most cases, any single view of a part does not have sufficient information to describe the total size of the part. However, the missing view can be determined when any two of the three primary views—top, front, or right side—are given. Refer back to **Figure 2-7** and carefully study the dimensions for each view.

Surface Representation

Each view contains two basic measurements needed to complete the part. These two measurements are the basis for developing an individual surface. **Figure 2-9**, for example, shows an object with front view A projected. In **Figure 2-9**, the front projected surface A has measurements for true size and shape. Any surface that is parallel to a viewing plane is drawn to true size and shape. *True size and shape* occurs when lines and surfaces on the projected

view are identical to the corresponding lines and surfaces on the object. Note that projected surface A is identical to front view A of the object.

In **Figure 2-10**, the top view represents surface C differently from how it appears on the object. On the object, surface C is angular, while the projected

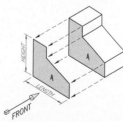

Goodheart-Willcox Publisher
Figure 2-9 Projecting the image to show the front view including the basic dimensions needed to describe the surface.

Copyright Goodheart-Willcox Co., Inc.

PRO TIP A drawing revision is often ordered by an Engineering Change Order (ECO) or an Engineering Change Request (ECR). The ECO or ECR contain the data, notes, and sketches required for the changes on the drawing. Changes include anything that affects the finished product such as dimensions, materials, processes, notes, features, and finishes. The ECR or ECO is usually a separate document in digital form. The document's reference number (ECO NO.) can be included in the revision history block.

located at the top of a sheet next to the revision block as shown in **Figure 4-2(7)**, or on top of the title block as shown in **Figure 4-4(S)**. The revision status of sheets can also be a separate document.

	HISTORY		
		DATE	APPROVED
	...CTION	3–23	P.H.
	...–DELETED	5–3	C.B.
	...DED	5–3	C.B.
	...DDED	5–3	C.B.

Ⓐ Ⓑ   Ⓒ   Ⓓ   Ⓔ

Goodheart-Willcox Publisher
Figure 4-9 The revision history block shows a list of revisions made to the original drawing.

Copyright Goodheart-Willcox Co., Inc.

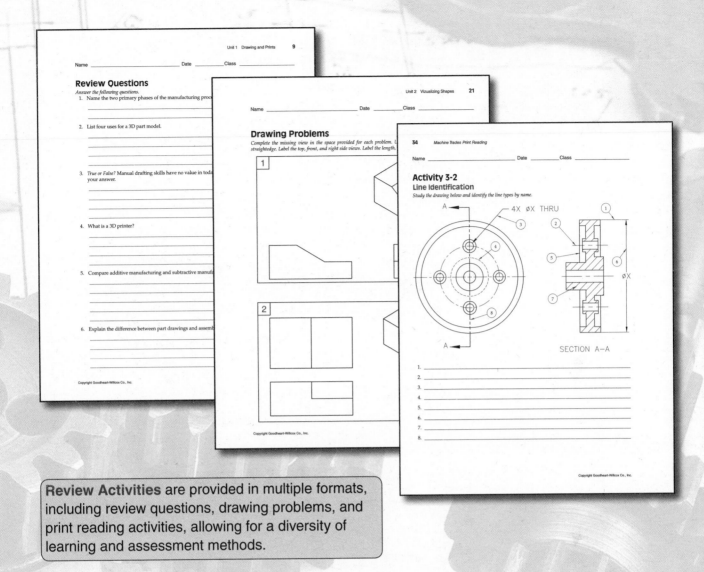

Review Activities are provided in multiple formats, including review questions, drawing problems, and print reading activities, allowing for a diversity of learning and assessment methods.

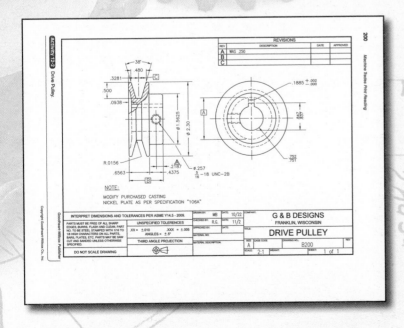

Student Materials

Online Textbook

This online version of the printed textbook gives students access anytime, anywhere whether using an iPad, netbook, PC, or Mac. Using the Online Textbook, students can easily navigate from a linked table of contents, search specific topics, quickly jump to specific pages, zoom in to enlarge text, and print selected pages for offline reading. The Online Textbook is available at www.g-wonlinetextbooks.com.

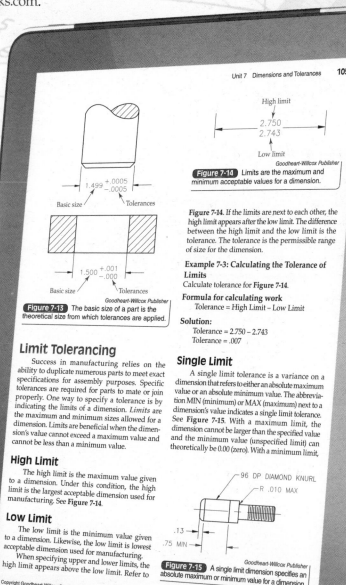

1.499 +.0005 / −.0005

Basic size   Tolerances

1.500 +.001 / −.000

Basic size   Tolerances

Goodheart-Willcox Publisher

Figure 7-13 The basic size of a part is the theoretical size from which tolerances are applied.

High limit

2.750
2.743

Low limit

Goodheart-Willcox Publisher

Figure 7-14 Limits are the maximum and minimum acceptable values for a dimension.

Figure 7-14. If the limits are next to each other, the high limit appears after the low limit. The difference between the high limit and the low limit is the tolerance. The tolerance is the permissible range of size for the dimension.

Example 7-3: Calculating the Tolerance of Limits

Calculate tolerance for **Figure 7-14.**

Formula for calculating work

Tolerance = High Limit – Low Limit

Solution:

Tolerance = 2.750 – 2.743
Tolerance = .007

Limit Tolerancing

Success in manufacturing relies on the ability to duplicate numerous parts to meet exact specifications for assembly purposes. Specific tolerances are required for parts to mate or join properly. One way to specify a tolerance is by indicating the limits of a dimension. *Limits* are the maximum and minimum sizes allowed for a dimension. Limits are beneficial when the dimension's value cannot exceed a maximum value and cannot be less than a minimum value.

High Limit

The high limit is the maximum value given to a dimension. Under this condition, the high limit is the largest acceptable dimension used for manufacturing. See **Figure 7-14.**

Low Limit

The low limit is the minimum value given to a dimension. Likewise, the low limit is lowest acceptable dimension used for manufacturing.

When specifying upper and lower limits, the high limit appears above the low limit. Refer to

Single Limit

A single limit tolerance is a variance on a dimension that refers to either an absolute maximum value or an absolute minimum value. The abbreviation MIN (minimum) or MAX (maximum) next to a dimension's value indicates a single limit tolerance. See **Figure 7-15.** With a maximum limit, the dimension cannot be larger than the specified value and the minimum value (unspecified limit) can theoretically be 0.00 (zero). With a minimum limit,

96 DP DIAMOND KNURL

R .010 MAX

.13

.75 MIN

Goodheart-Willcox Publisher

Figure 7-15 A single limit dimension specifies an absolute maximum or minimum value for a dimension.

Instructor Materials

ExamView® Assessment Suite

Quickly and easily prepare, print, and administer tests with the ExamView® Assessment Suite. With hundreds of questions in the test bank corresponding to each chapter, you can choose which questions to include in each test, create multiple versions of a single test, and automatically generate answer keys. Existing questions may be modified and new questions may be added.

Instructor's Presentations for PowerPoint®

Help teach and visually reinforce key concepts with prepared lectures. These presentations are designed to allow for customization to meet daily teaching needs. They include objectives and images from the textbook.

Instructor's Resource CD

One resource provides instructors with time-saving preparation tools such as answer keys, lesson plans, PDFs of prints from the text, and other teaching aids.

Online Instructor Resources

Online Instructor Resources are time-saving teaching materials organized in a convenient, easy-to-use online bookshelf. Lesson plans, answer keys, presentations for PowerPoint®, ExamView® Assessment Suite software with test questions, PDFs of prints from the text, and other teaching aids are available on demand, 24/7. Accessible from home or school, Online Instructor Resources provide convenient access for instructors with busy schedules.

G-W Online

G-W Online provides robust instructional resources in an easily customizable course management system. This online solution contains an Online Textbook, e-flash cards, activities, and assessments for a deeply engaging, interactive learning experience. G-W Online enhances your course with powerful course management and assessment tools that accurately monitor and track student learning. The ultimate in convenient and quick grading, G-W Online allows you to spend more time teaching and less time administering.

Brief Contents

Contents

Learning Objectives

After studying this unit, you will be able to:

✓ List the two primary phases of the manufacturing process.
✓ Explain the roles of various members of a design team.
✓ Describe potential uses for a 3D part model.
✓ Identify potential benefits of manual drafting skills for workers in the manufacturing industry.
✓ Differentiate part drawings and assembly drawings.
✓ Identify occupations that require print reading skills.

Key Terms

additive manufacturing
computer-aided drafting (CAD)

print reading
three-dimensional (3D) model

three-dimensional (3D) printer

Have you ever wondered how the automobiles, robots, or computers we use are made? These complex machines are assembled from thousands of parts. In order to produce any type of part, component, or machine, a drawing is needed. To produce a machine as complex as an automobile, robot, or computer, thousands of drawings are required.

This unit introduces the manufacturing process and highlights the role of 3D models, drawings, prints, and print reading in the creation of a product.

Design and Manufacturing Process

The manufacture of a machine or product includes two primary phases: a design phase and a production phase. Drawings and prints are created during the design phase and applied during the production phase.

Design Phase

Once the need for a machine or product is identified, the design phase for the product begins. A design team is formed. A design team may include members from various departments of the company:

- Engineers, designers, and CAD technicians from the design department create the design documents and ensure that the product meets the design criteria or objective.
- Machinists and inspectors from the production department review the design and look for ways to make the product easier to manufacture.
- Marketing representatives from the sales department ensure that the product will meet the needs of the customer.
- Accountants from the finance department monitor costs with the goal of creating a profitable product.

Most machines or products are composed of many parts, **Figure 1-1**. The engineers, designers, and CAD technicians on the design team must design each individual part included in the product. Each part is manufactured individually, **Figure 1-2**. Often, the parts in a single product are manufactured by several different companies, then shipped to a single facility for assembly.

3D Modeling

Individual parts are normally designed using *computer-aided drafting (CAD)* software. In many cases, parts are designed as 3D models. A *3D model* is a computer-based representation of a part or assembly in which the model includes the actual physical geometry and material characteristics.

A 3D model of a part has several uses, **Figure 1-3**:

■ Simulation/analysis. A 3D model of a part can be analyzed by computer software to predict how the part will behave in actual situations. Simulation allows engineers to test their designs without actually producing a part, which reduces cost.

■ Direct export for prototyping or production. Prototyping is the process of preparing a sample of a part or design for testing purposes. A prototype may be a scaled model, a sample produced using a material different from the final part, or an actual final part. 3D models can be saved in electronic file formats that can be loaded into 3D printers and CNC machines to produce prototypes and parts.

■ Drawing creation. Two-dimensional (2D) drawings are often required to manufacture a part. CAD software can automatically generate these 2D drawings from the 3D model. See **Figure 1-4**.

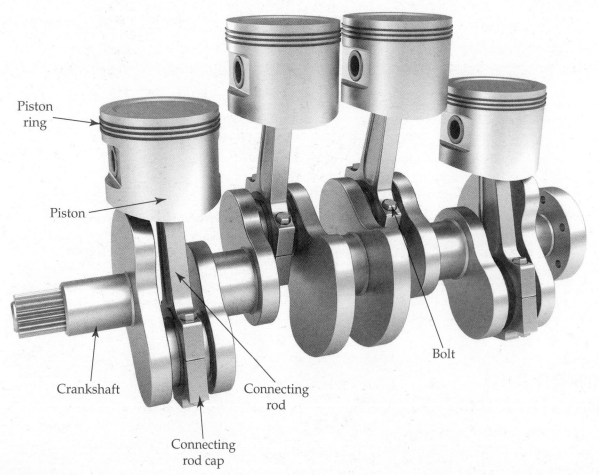

Piston ring

Piston

Crankshaft

Connecting rod cap

Connecting rod

Bolt

Umberto Shtanzman/Shutterstock.com

Figure 1-1 This 3D assembly model is composed of several parts. Each individual part is created as a 3D model.

Figure 1-2 Parts are individual pieces that cannot be separated into smaller items. Parts are assembled together to create a product or machine.

Potential Uses of a 3D Part Model

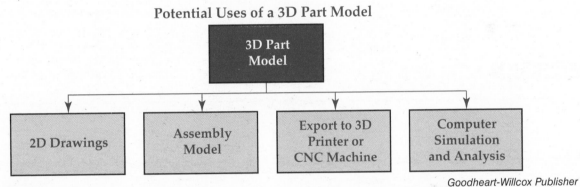

Figure 1-3 A 3D part model has many uses. The specific needs and requirements of the project determine which of these uses are applied.

■ Assembly modeling. 3D models of all the parts in a product can be combined in a single electronic file, with each part positioned as it would be in the final product. This assembly model can be used to ensure that the parts fit together correctly and can be analyzed as a finished product.

Prototyping

Many design projects include a prototyping stage. This stage begins when the initial part design is complete. A sample of the part, or prototype, is manufactured. This prototype can be tested and assembled with other part prototypes to create a prototype of the product.

Three-dimensional (3D) printers are often used to create prototypes. A *3D printer* is a manufacturing device that uses an electronic file of a 3D model to create a part. Most 3D printers produce parts by

PRO TIP Before computers became common in the workplace, drawings were created by hand using a process called manual drafting. A draftsperson would use a pencil or ink pen to create the drawing on a paper or plastic sheet. You may still encounter some hand drawings if you work in an industry with old equipment, machines, or facilities.

Manual drafting skills are still extremely useful in today's computer-focused workplace. Being able to sketch quickly and accurately and being able to letter clearly will allow you to communicate more effectively with coworkers in the manufacturing industry. In addition, much of the knowledge applied in drafting is identical to the knowledge required in print reading.

CAD software user interface   Isometric pictorial view

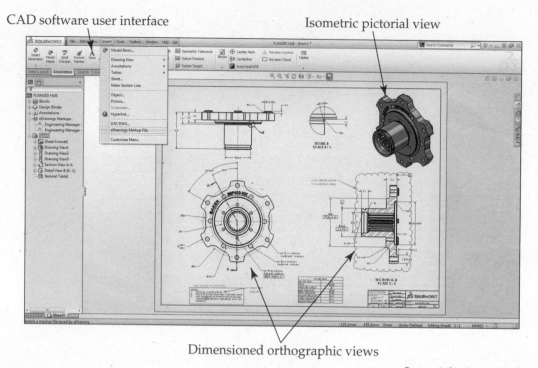

Dimensioned orthographic views

Figure 1-4   A CAD system can generate drawings containing dimensioned orthographic views and pictorial views from a 3D model.

hardening a soft material in very thin horizontal layers. The cross section of each layer matches the corresponding cross section of the part. Each layer is slightly different, creating the shape of the part. Most parts manufactured by 3D printers are made of plastics, but some 3D printers can produce metal parts.

The process of creating a part with a 3D printer is also referred to as *additive manufacturing*. In additive manufacturing, parts are created by adding material to build the part. Most traditional machining processes, such as drilling and milling, are applications of "subtractive manufacturing." With these traditional processes, material is removed from an initial piece to create the finished part.

Drawings

The production of part drawings and assembly drawings is the key output of the design phase. A part drawing is a 2D drawing that provides all of the information needed to manufacture the part. See **Figure 1-5**. A part drawing typically includes scaled, dimensioned views describing the size and shape of the part and the locations of part features (such as holes). A part drawing may also include information about the allowable tolerances (variation in size, shape, and feature position for the part), material, and required surface finish. Basically, a part drawing includes all of the information needed to successfully produce the part.

As mentioned earlier, part drawings are typically created using CAD software. The part drawing can be created from a 3D model. In cases where a 3D part model is not created, the 2D part drawing is created using the CAD software.

Part drawings are often shared with other companies. For example, a company wishes to have a particular part manufactured by a subcontractor. In order to select a subcontractor, the company gives the part drawing to several potential subcontractors and asks those companies to prepare a bid for the project. Often, these types of potential subcontractors may be located in foreign countries.

In order to ensure that the part specifications can be clearly defined by the part drawing, manufacturing companies use drawing and manufacturing standards when designing and making parts. Many of these standards are created by international organizations, such as ANSI and ISO. The use of standards ensures effective communication in the design and production of manufactured products.

Dimensioned views

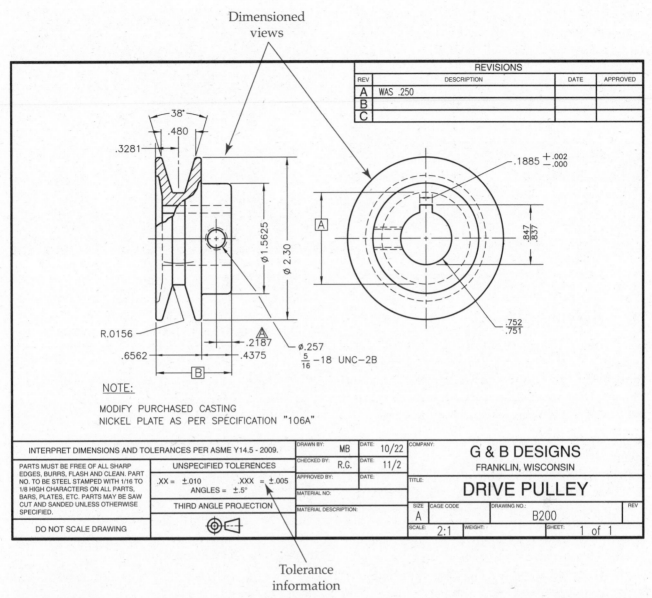

REVISIONS			
REV	DESCRIPTION	DATE	APPROVED
A	WAS .250		
B			
C			

Tolerance information

Goodheart-Willcox Publisher

Figure 1-5 A part drawing contains all the information needed to produce the part, including dimensioned views, tolerances, material specification, and finishing requirements.

PRO TIP Today, most "physical" drawings are created using a plotter. A plotter is basically a computer printer designed specifically for printing large format drawings. However, many drawings are never printed. Instead, an electronic file of the drawing is viewed on a tablet computer, laptop, or mobile device. Electronic drawing files can be stored on company servers or cloud-based applications.

2D assembly drawings, like part drawings, can be created from a 3D assembly model or created as a 2D drawing using the CAD system. Assembly drawings identify each part in the product and show how the various parts fit together.

Prints

Technically, a print is a physical (paper) copy of a drawing. However, the terms *drawing* and *print* are often used interchangeably. *Print reading* refers to the act of interpreting the information shown on a drawing or print. Whether a drawing is viewed on a computer monitor, laptop display, tablet computer, mobile device, or paper, the print reading process remains the same. Print reading is an essential skill for anyone in the fields of manufacturing or engineering.

PRO TIP Prints are often created using an engineering copier. Engineering copiers are similar to standard copy machines and scanners, but they are able to reproduce drawings on large-size paper.

Production Phase

When a complete set of part drawings and assembly drawings is complete, the project moves from the design phase to the production phase. The production phase consists of part production, assembly, and packaging.

Part Production

Part production can be a simple phase or a complex phase, depending on the number and variety of parts in the product. For a simple product, all of the parts may be manufactured, assembled, and packaged at a single facility. For a complex product, such as an automobile, part production may include hundreds of subcontractors (and thousands of parts) in different countries.

Part production often begins by sending the part drawings out to subcontractors to bid. The subcontractors use the drawing to determine how much material will be needed to make the part and the manufacturing processes and operations required. Several types of manufacturing processes—such as forming, separating, conditioning, finishing, and fabricating—may be used to manufacture the part. Using this information, the subcontractor can determine the cost to produce the part and prepare a bid for the job.

As parts are produced, they are inspected to ensure that the parts meet the requirements described in the part drawing. See **Figure 1-6**. For some parts, a basic inspection of random parts may be sufficient. For more critical parts, a

Dmitry Kalinovsky/Shutterstock.com

Figure 1-6 Quality control (QC) inspection is an important part of the manufacturing process. Inspectors measure specific features to ensure that the parts are being produced to specification.

thorough inspection of every part produced may be needed.

Assembly and Packaging

After the parts are produced, they are assembled into the final product. The assembly drawings specify the correct arrangement of parts. After the parts are assembled, the final product is inspected. If the product passes inspection, it is packaged for delivery.

Print Reading Applications

Print reading skills are used by many workers throughout and beyond the production process. The table in **Figure 1-7** shows various careers that use print reading skills.

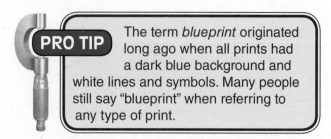

PRO TIP The term *blueprint* originated long ago when all prints had a dark blue background and white lines and symbols. Many people still say "blueprint" when referring to any type of print.

Examples of Print Reading Applications

Phase	Occupation	Use of Print Reading Skills
Design	Engineer, designer, drafter	Interpret drawings created by others that impact the design of a part.
	Marketing team	Visualize completed product based on prints.
	Accountant	Calculate cost estimates for parts.
	Technical writer	Interpret print information to develop technical documentation such as user's manuals and troubleshooting guides.
Production	Estimator	Calculate bids based on prints.
	Machinist, manufacturing technician	Interpret information on prints to determine processes, tools, and operations needed to produce the part.
	QC inspector	Use print to determine which elements of the part to inspect and which inspection methods to use.
Assembly	Manufacturing engineer, manufacturing technician	Use assembly drawings to design assembly process and ensure proper assembly.
Maintenance	Maintenance technician	Use prints to determine appropriate maintenance.
Repair	Maintenance technician	Use prints to troubleshoot issues and perform repairs.

Goodheart-Willcox Publisher

Figure 1-7 Print reading skills are valuable to many occupations. This table lists a small sample of uses of print reading in the manufacturing industry.

Notes

Name _____ Date _____ Class _____

Review Questions

Answer the following questions.

1. Name the two primary phases of the manufacturing process.

   _____

   _____

2. List four uses for a 3D part model.

   _____

   _____

   _____

   _____

3. *True or False?* Manual drafting skills have no value in today's manufacturing industry. Explain your answer.

   _____

   _____

   _____

   _____

4. What is a 3D printer?

   _____

   _____

   _____

   _____

5. Compare additive manufacturing and subtractive manufacturing and name an example of each.

   _____

   _____

   _____

   _____

   _____

6. Explain the difference between part drawings and assembly drawings.

   _____

   _____

   _____

   _____

7. What is the purpose of part inspection?

   _____

   _____

   _____

   _____

   _____

8. List a manufacturing occupation that interests you and explain how print reading skills are used by the occupation.

   _____

   _____

   _____

   _____

   _____

Visualizing Shapes

Learning Objectives

After studying this unit, you will be able to:

✓ List the six principal views of an object.
✓ Understand the basic principles of orthographic projection.
✓ Identify first- and third-angle projection drawings.
✓ Visualize three-dimensional objects from orthographic drawings.
✓ Apply proper dimensioning for projected views.
✓ Draw a missing view of an object from two given views.

Key Terms

first-angle projection
foreshortening

multiview drawing
orthographic projection

third-angle projection
true size and shape

The ability to read and understand prints and technical drawings is an essential skill required in the engineering and manufacturing fields. The key factor in print reading is the ability to visualize separate views of an object in two dimensions.

Multiview Drawing

Manufacturing industries use multiview drawings as a way to visualize and describe a part or assembly. Understanding a multiview drawing is an important aspect of reading prints and technical drawings.

In a *multiview drawing,* two or more views represent the accurate shape of an object or part. A multiview drawing provides a means of visualizing a three-dimensional shape on a two-dimensional surface such as drafting paper or a computer screen. All the necessary details of a part's shape and size are included in a multiview drawing to ensure accurate production of the part.

Orthographic projection is one method used to develop multiview drawings. *Orthographic projection* is a way of showing the shape of an object by projecting two or more views at right angles to each other. See **Figure 2-1**.

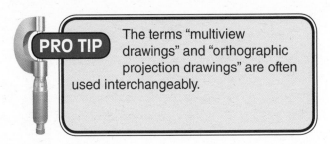

PRO TIP The terms "multiview drawings" and "orthographic projection drawings" are often used interchangeably.

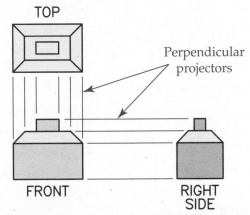

Goodheart-Willcox Publisher

Figure 2-1 A simple multiview drawing is created by orthographic projection.

At first glance, the idea of three-dimensional visualization seems difficult to grasp. Multiview drawings, however, will become easier to understand as your visualization skills improve. The following explanations, figures, and lessons will help improve your print reading skills.

Views in a Glass Box

In a conceptual drawing, two viewing planes—a horizontal plane and a vertical plane—intersect to create quadrants. See **Figure 2-2A**. Each quadrant represents a different viewing angle—first-, second-, third-, and fourth-angle. The viewing angle determines how the views of an object are projected. Only the first- and third-angles are used for drawings because the second- and fourth-angles repeat the same information. The majority of the drawings in this book are third-angle projections.

The United States and Canada primarily use third-angle projections for multiview drawings. A drawing viewed and projected from the third quadrant is a *third-angle projection*. With third-angle projection, the viewer is looking toward the desired view of an object. A viewing plane is located between the viewer and the object. The desired view is transferred onto the viewing plane in front of the viewer. The shape of the object determines how it is shown in a view.

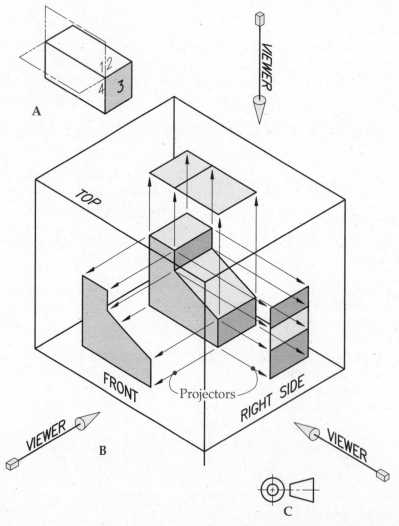

Figure 2-2 Visualizing an object in third-angle projection. A—Viewing planes create four quadrants with the third quadrant highlighted. B—An object in a glass box showing third-angle projection. The three primary views project toward the surfaces of the glass box. C—This symbol indicates a third-angle projection.

To explore third-angle projection further, imagine an object placed in a glass box. Each of the box's six sides represents a viewing plane. Projectors extend the views from the object to the flat surfaces of the glass box. This produces six principal views of the object—top, bottom, front, rear, right side, and left side views. In multiview drawings, however, only three primary views are typically used—the top, front, and right side views. **Figure 2-2B** shows the three primary views projected in the third-angle. Note how the projectors transfer each view to their respective glass surface. On a drawing, the third-angle symbol indicates a third-angle projection, as shown in **Figure 2-2C**.

In **Figure 2-3**, the surfaces of the box are unfolded to show the six principal views. Visualize, then, the unfolding of the glass box:

1. The front surface remains stationary.
2. The other surfaces hinge and rotate toward the front viewing surface.

As the six surfaces are unfolded, the projected views of the top, front, and right side become flat, like a sheet of drafting paper, as shown in **Figure 2-4**.

First-angle projection is the drawing standard used in many international countries. A drawing viewed and projected from the first quadrant is a *first-angle projection*. See **Figure 2-5A**. With first-angle projection, the object is between the viewer and the viewing plane. The viewer transmits the views onto the viewing plane located to the opposite side of the object. See **Figure 2-5B**. On a drawing, a first-angle symbol indicates a first-angle projection, as shown in **Figure 2-5C**.

Like third-angle projection, the front view is stationary, but the surfaces are unfolded away from the front view and flattened to show six principal views. See **Figure 2-6**. Compare **Figure 2-4** and **Figure 2-6**, notice that all the views except for the front view are reversed from each other. The front view is the only common view between the two types of projections.

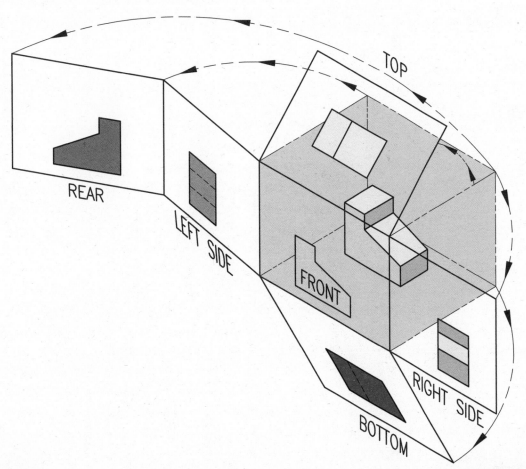

Goodheart-Willcox Publisher

Figure 2-3   The front view surface remains stationary, while the other five view surfaces unfold.

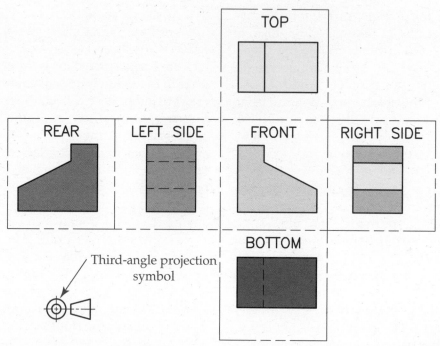

Goodheart-Willcox Publisher

Figure 2-4   Six principal view surfaces of a third-angle drawing are flattened, like a two-dimensional drawing.

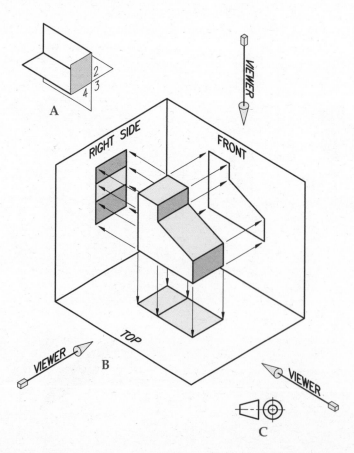

Goodheart-Willcox Publisher

Figure 2-5   Visualizing an object in first-angle projection. A—Viewing planes divided into four quadrants with the first quadrant highlighted. B—An object in a glass box showing first-angle projection. The three primary views project to the opposite sides of the glass surfaces. C—This symbol indicates a first-angle projection.

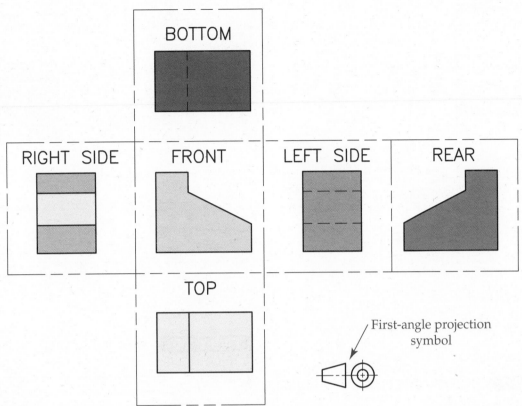

Figure 2-6 Six principal view surfaces of a first-angle drawing are flattened, like a two-dimensional drawing.

Therefore, the first-angle projections are as follows:

- Top view projects to the bottom of the box.
- Bottom view projects to the top of the box.
- Front view projects to the back of the box.
- Rear view projects to the front of the box.
- Right side view projects to the left side of the box.
- Left side view projects to the right side of the box.

As the global market expands, print readers, assemblers, machinists, and fabricators are experiencing more first-angle projection drawings. Both methods of projection are acceptable in modern drawings. However, caution should be taken when drafting and reading drawings. The drafter or designer should never mix different projection methods, which can lead to confusion. It is equally important for the reader to recognize the two types of projections in order to interpret details correctly. Understanding the basic principles of projection will allow you to identify the contours and lines that define a part.

Arrangement of Views

Carefully study the top view in **Figure 2-2**, which is directly above and aligned with the front view by projectors. The right side view also has a direct relationship to the front view through projected lines.

As stated earlier, the standard layout of views is the top, front, and right side. In the design industry, however, there are exceptions. Although the right side view is preferred, using the left side view is acceptable if it describes the object more clearly. Another exception occurs if the left side has fewer obstructing lines than the right side.

Figure 2-7 shows the standard layout of top, front, and right side views for a third-angle projection, along with their basic dimensions. **Figure 2-8** shows the standard layout of top, front, and left side views for a first-angle projection.

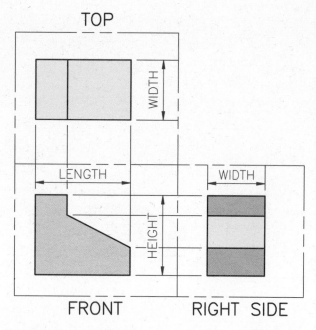

Figure 2-7 Standard third-angle, multiview drawing layout.

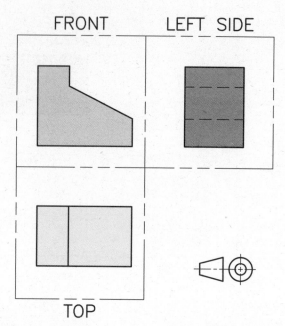

Figure 2-8 Standard first-angle, multiview drawing layout.

Basic Dimensions

The top view contains the length and width of the part. The front view shows the length and height. The right side view gives the height and width. Note that each view contains only two required dimensions of the object.

In most cases, any single view of a part does not have sufficient information to describe the total size of the part. However, the missing view can be determined when any two of the three primary views—top, front, or right side—are given. Refer back to **Figure 2-7** and carefully study the dimensions for each view.

Surface Representation

Each view contains two basic measurements needed to complete the part. These two measurements are the basis for developing an individual surface. **Figure 2-9**, for example, shows an object with front view A projected. In **Figure 2-9**, the front projected surface A has measurements for true size and shape. Any surface that is parallel to a viewing plane is drawn to true size and shape. *True size and shape* occurs when lines and surfaces on the projected

view are identical to the corresponding lines and surfaces on the object. Note that projected surface A is identical to front view A of the object.

In **Figure 2-10**, the top view represents surface C differently from how it appears on the object. On the object, surface C is angular, while the projected

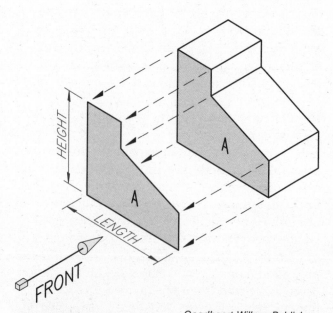

Figure 2-9 Projecting the image to show the front view including the basic dimensions needed to describe the surface.

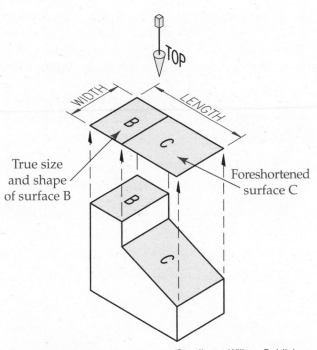

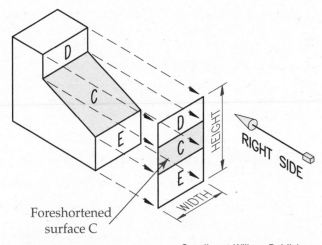

Goodheart-Willcox Publisher

Figure 2-10 In projecting a top view, the angular surface C is foreshortened.

Figure 2-11 The projected right side view also shows the foreshortened angular surface C.

surface C looks rectangular. Any surface on the object that is neither parallel nor perpendicular to a projected view is foreshortened in the view. *Foreshortening* is reducing the apparent size of a surface or line in a view caused by the elimination of depth. The process of projection foreshortens the angular surface as viewed from the top. The projected width and length of the object are true size and shape.

A similar procedure is used in **Figure 2-11** to obtain the right side view. This view shows

surfaces D and E in true size and shape as projected from the pictorial drawing. Since angular surface C is not parallel to the projection plane, it is foreshortened. The height and width of the right side surface do not change as they are projected.

Visualizing Basic Shapes

Multiview drawings can be complex. You must understand the procedures for obtaining top, front, and right side views before attempting to read complicated drawings. Pay particular attention to the various forms or surfaces created by projections and their respective views. Even simple objects may have identical top and right side views, but have different front views, **Figure 2-12**.

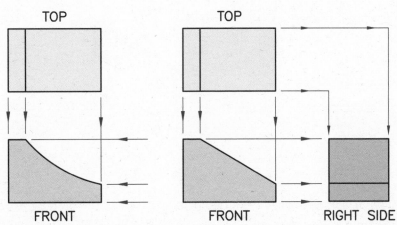

Goodheart-Willcox Publisher

Figure 2-12 Interchangeable surfaces on simple objects use the same top and right side views. Only the front views reveal the difference.

It is important to be able to visualize and understand the basic shape of objects before reading other reference material on the print. If you do not understand the objects shown, you will not understand the reference material. A misunderstanding can cause costly mistakes in the manufacturing process.

A Multiview Application

By applying the procedures of orthographic projection to the L-shaped object shown in **Figure 2-13**, you will create a multiview drawing. First, rotate the object into the front view position (in the same flat plane as paper). Note, as the front view is drawn, the width is not seen. See **Figure 2-14**.

Next, extend the projectors toward the top and right side view positions. Extend the projectors as far

as needed between each view. Then, locate the top view and right side views by transferring the width dimension. Use a 45° miter angle to transfer the width from top to right side view, as shown in **Figure 2-15**.

PRO TIP The L-shape is not a practical machining project because most of the workpiece would be machined away in forming the L. However, the L-shape does lend itself well as a visualization exercise.

This unit covered the basic method of visualizing a multiview drawing. The drawing problems on the following pages allow you to practice your skills while increasing your understanding of multiview drawings.

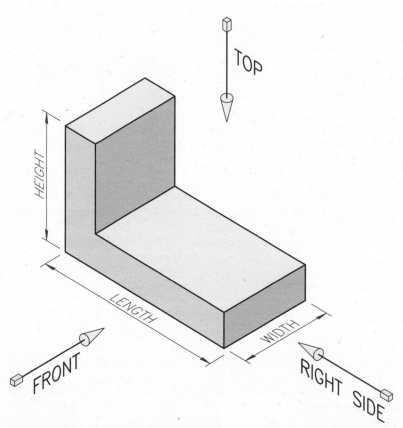

Goodheart-Willcox Publisher

Figure 2-13   A pictorial view of a basic L-shaped object with the three primary views and basic dimensions labeled.

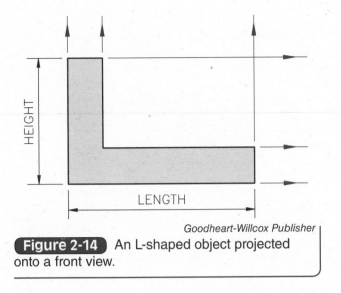

Figure 2-14 An L-shaped object projected onto a front view.

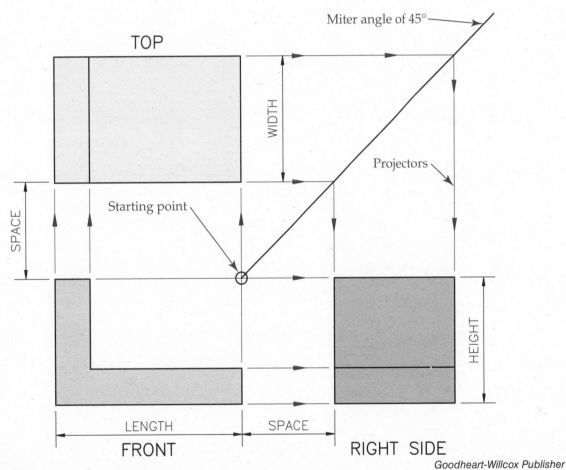

Figure 2-15 When front and top views are drawn, use projectors and a miter angle to transfer the width dimension to the right side view.

Notes

Name _____ Date _____ Class _____

Drawing Problems

Complete the missing view in the space provided for each problem. Use a ruler, machinist's scale, or other straightedge. Label the top, front, and right side views. Label the length, width, and height on each view.

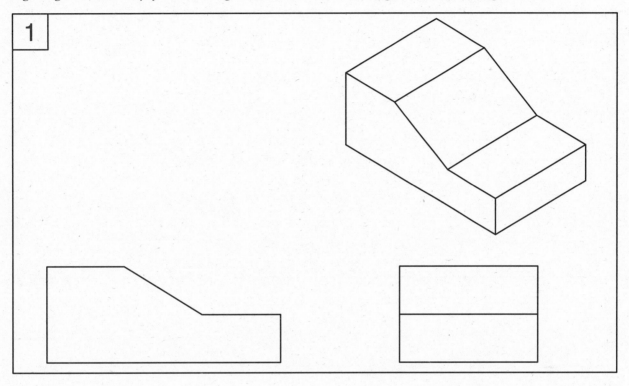

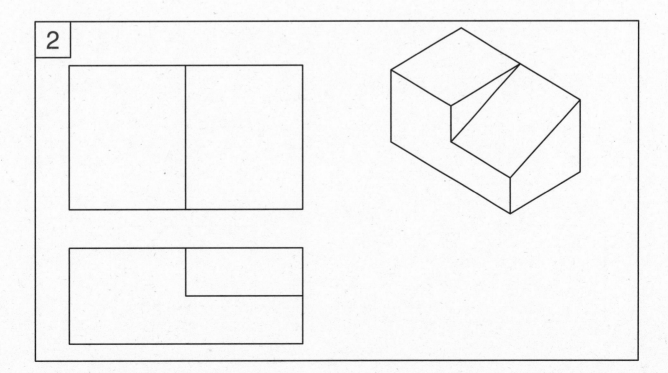

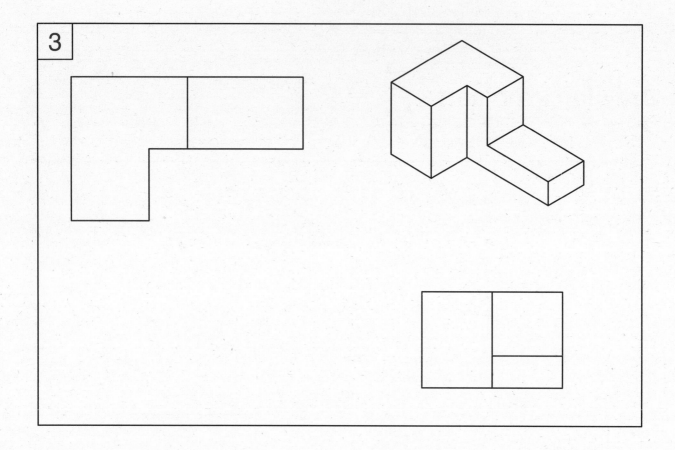

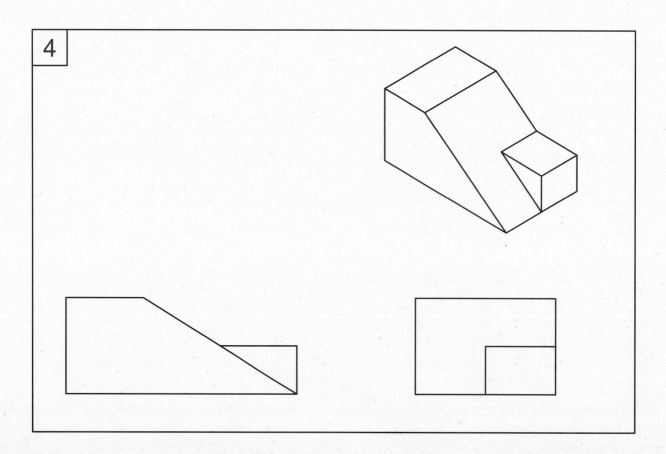

Name _____ Date _____ Class _____

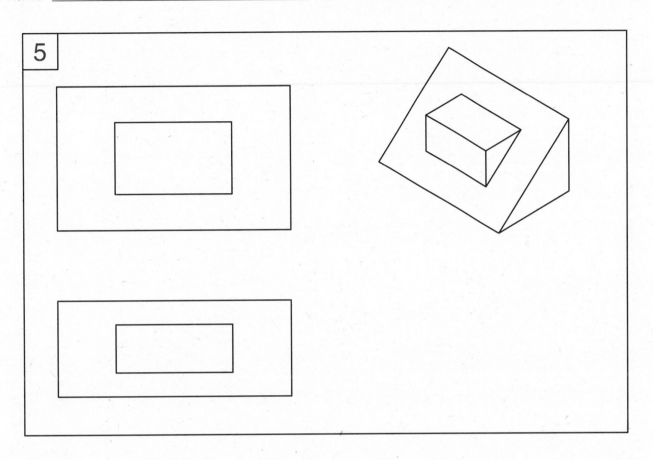

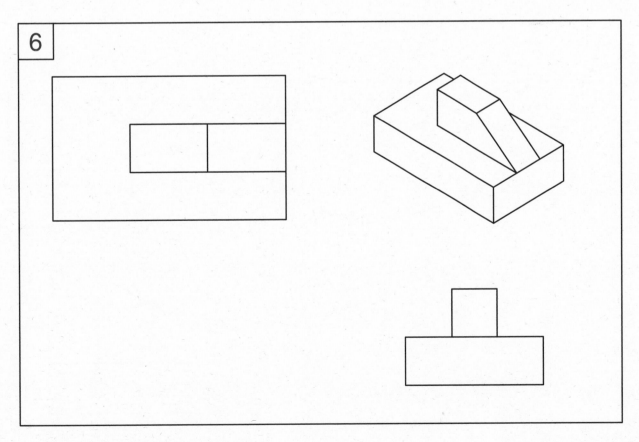

7

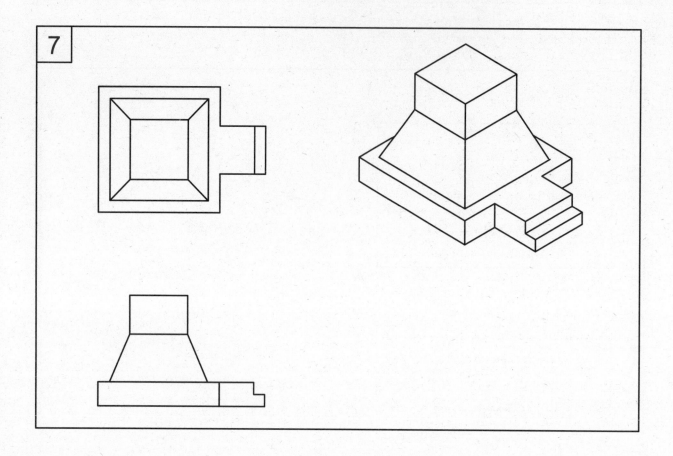

8

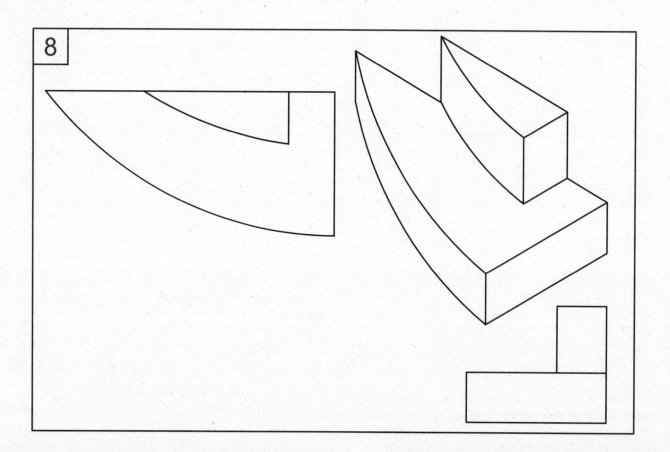

Learning Objectives

After studying this unit, you will be able to:

✓ Describe the purposes of lines on prints.

✓ Identify various types of lines found on prints.

✓ Locate corresponding lines or surfaces in various views.

Key Terms

alphabet of lines
American Society of Mechanical
   Engineers (ASME)
break line
centerline
chain line
cutting-plane line

dimension line
dimensioning
drawing standards
extension line
hidden line
leader line
phantom line

section
section line
stitch line
symmetry line
viewing-plane line
visible line

As new ideas develop into useful products, they require precise drawings of parts. Drawings must be consistent so they are readable across the industry.

The *American Society of Mechanical Engineers (ASME)* develops drawing standards used by the industry for the development of engineering drawings. *Drawing standards* are documented practices used to develop drawings. One of the standards is ASME Y14.2, titled *Line Conventions and Lettering*. This standard, also referred to as the *alphabet of lines*, includes guidelines for specific types of lines, each with an intended purpose, as shown in **Figure 3-1**.

Types of Lines

Drawings contain various types of lines. Understanding the alphabet of lines allows the reader to visualize the intent of a drawing.

A line can have different characteristics. A line can be continuous (solid with no breaks),

or dashed (short lines separated by spaces), or a combination of both. Each type has a specific meaning.

Line width, or line weight, is another characteristic of a line. There are two line widths used in mechanical drawing. A thick line is twice the width of a thin line. The difference between the two helps identify their purpose. In the following sections, you will learn about the different types of lines and their importance in use.

Visible Lines

Visible lines, also referred to as object lines or outlines, define the shape and surfaces of an object. They show all edges of an object that are visible in a view. Visible lines define the outside border of a part as well as the surfaces within the borders. Visible lines are thick continuous lines that visually stand out on a drawing. See **Figure 3-2**.

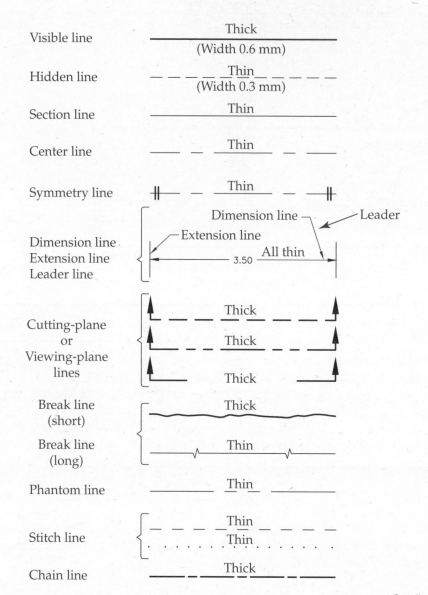

Visible line	Thick (Width 0.6 mm)
Hidden line	Thin (Width 0.3 mm)
Section line	Thin
Center line	Thin
Symmetry line	Thin
Dimension line Extension line Leader line	Dimension line — Leader Extension line 3.50 All thin
Cutting-plane or Viewing-plane lines	Thick Thick Thick
Break line (short)	Thick
Break line (long)	Thin
Phantom line	Thin
Stitch line	Thin Thin
Chain line	Thick

Goodheart-Willcox Publisher

Figure 3-1 The "alphabet of lines" based on the American National Standard, ASME Y14.2.

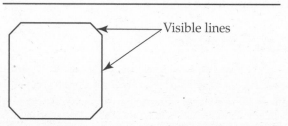

Goodheart-Willcox Publisher

Figure 3-2 Visible lines outline all edges and surfaces of an object seen in a view.

Hidden Lines

Hidden lines show edges or surfaces that are not visible when viewing a part from a specific view. A hidden line, as shown in **Figure 3-3**, is a series of thin line dashes that are evenly spaced. The spaces between the dashes are short.

Centerlines

A *centerline* shows the location of the center point of a hole or an axis of a part. A centerline may also show the center of an arc or a path of motion. Centerlines are thin lines with alternating long and single, short dashes. See **Figure 3-4**.

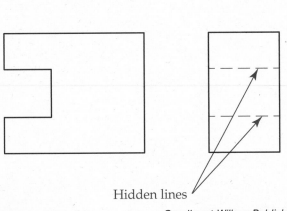

Hidden lines

Figure 3-3 Hidden lines show edges and surfaces not seen in a view.

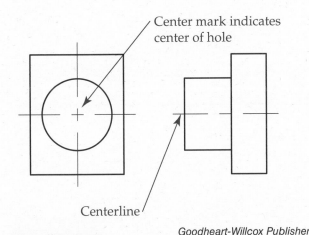

Center mark indicates center of hole

Centerline

Figure 3-4 Centerlines indicate center locations of holes, arcs, and axes of symmetrical parts.

Centerlines are critical to the machinist during layout, machining, and inspection.

Symmetry Lines

A *symmetry line*, common in partial views and sections, shows the center axis of a part where both sides are symmetrical—the same shape and size. It has a centerline with two short thick, parallel lines placed perpendicular at both ends. See **Figure 3-5**.

Section Lines

A *section* is a "cut away" view of a part that shows the hidden interior details of a primary view.

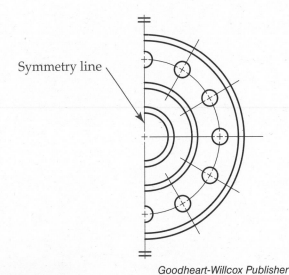

Symmetry line

Figure 3-5 A partial view divided by a symmetry line signifies the part is symmetrically shaped.

Section lines indicate the cut surfaces of a sectional drawing. Section lines are evenly spaced, thin diagonal lines, as shown in **Figure 3-6**. If a section cuts through two or more parts, then the section lines for the additional parts change direction. In addition, different types of section lines identify various types of material. See Unit 13—Sectional Views for information on section lines and sectional views.

Extension and Dimension Lines

Dimensioning is a method of representing measurement on a drawing using extension lines, dimension lines, leader lines, and numerical values. See **Figure 3-7**. Together, they use solid, thin lines and arrowheads to represent the distance between two points. *Extension lines* extend away from, but do not touch, the corners or surfaces of the part to indicate the points of measurement. A *dimension line* spans the distance between the extension lines with arrowheads and a number value. The number value represents the distance between the points of measurement.

Leader Lines

A *leader line* directs the reader to notes, symbols, item numbers, part numbers, dimensions, or specific operations vital to the machining process. The leader line couples a thin angular (slanted) line with an arrowhead. See **Figure 3-8**.

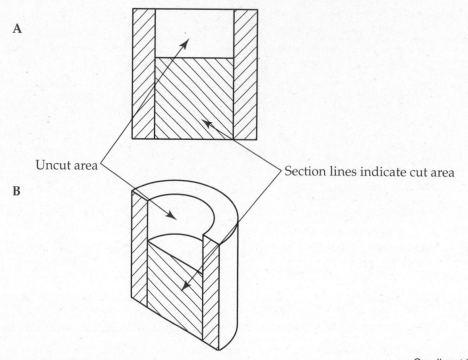

A

B

Uncut area

Section lines indicate cut area

Goodheart-Willcox Publisher

Figure 3-6   Section lines indicate a cut view of a part. A—Sectioned front view of a cylindrical part. B—Sectioned pictorial drawing of the part.

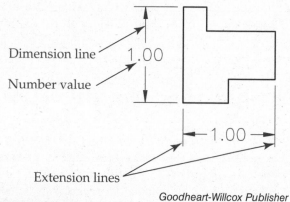

Dimension line

Number value

1.00

1.00

Extension lines

Goodheart-Willcox Publisher

Figure 3-7   Extension and dimension lines specify the size of an object.

Cutting-Plane and Viewing-Plane Lines

Cutting-plane lines indicate the location of a cutting path along a plane, as well as the viewing direction for sectional and removed views. A cutting-plane line is a thick line with oversized arrows, as shown in **Figure 3-9**.

Viewing-plane lines indicate the viewing direction for alternate drawing views. An

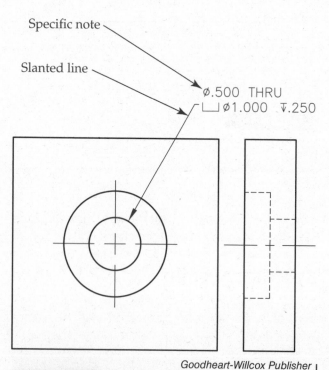

Specific note

Slanted line

ø.500 THRU
⌴ø1.000 ⊽.250

Goodheart-Willcox Publisher

Figure 3-8   A leader line calls out specific information for the referenced part. This note specifies a .500″ diameter hole be drilled through the part and then counterbored 1.000″ in diameter and .250″ deep.

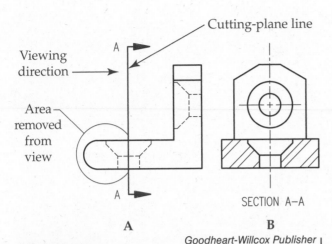

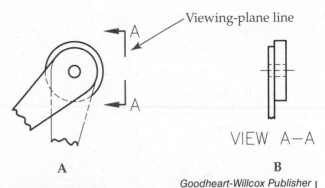

Cutting-plane line

Viewing direction

Area removed from view

SECTION A–A

A

B

Goodheart-Willcox Publisher

Figure 3-9 A cutting-plane line indicates the location and view of the section. A—Arrows indicate the viewing direction, while the cutting-plane line slices through the part. B—The area to the left of the cutting-plane line is removed, while the area to the right is rotated to a front view.

Viewing-plane line

VIEW A–A

A

B

Goodheart-Willcox Publisher

Figure 3-10 Viewing-plane lines identify viewing locations for alternate views. A—Viewing-plane line pointing toward an alternate view. B—Alternate view as noted by the viewing-plane line.

alternate view can explain a part when other views fail. A viewing-plane line is the same line used for cutting planes—a thick line with oversized arrows. See **Figure 3-10A**. A viewing-plane line does not cut through a part; it only references the alternate view. An alternate view is shown in **Figure 3-10B**.

Phantom Lines

Phantom lines indicate movement of parts, repeated details, or extra material on a part

before the machining process. See **Figure 3-11**. They also show filleted and rounded corners on a part. Phantom lines are thin lines made up of one long line followed by two short dashes. Phantom lines always start and end with a long line.

Break Lines

Break lines indicate a removed area on a drawing when it is not necessary to show the complete view. Typically, long parts will not fit on a drawing. Long break lines indicate the shortening of a long part with uniform shape, often with repeating features. Long breaks use thin straight lines with zigzags, as shown in **Figure 3-12**.

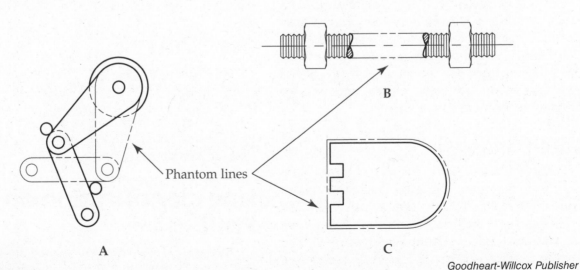

Phantom lines

B

A

C

Goodheart-Willcox Publisher

Figure 3-11 Applications of phantom lines. A—Phantom lines indicating motion. B—Phantom lines signifying repeated details. C—Phantom lines showing extra material before machining.

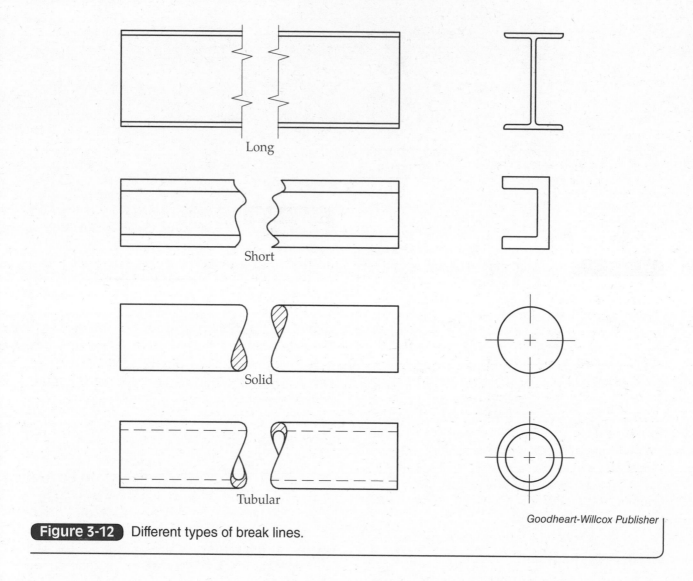

Long

Short

Solid

Tubular

Goodheart-Willcox Publisher

Figure 3-12   Different types of break lines.

Sectional views often use short break lines to provide clearer detail in viewing the part or parts that lie directly below the removed part. Short breaks use thick, freehand lines. Refer to **Figure 3-12**.

Cylindrical and tubular objects call for curved break lines. There are two types of curved break lines—one for solid objects and one for tubular objects. Refer to **Figure 3-12**.

Stitch Lines

Stitch lines indicate a sewing process on a part or an assembly with thin dashed or dotted lines. Thin dashed lines have equal size short dashes and spaces (3 mm). Dotted lines have equal size dots and spaces (3 mm). See **Figure 3-13**. The sewing process can be for function, decoration, or both. As a function, sewing is an excellent way to fasten a material, such as leather or cloth, to

a dissimilar material like metal. As decoration, stitch lines can define an outline and design.

Chain Lines

A *chain line* notes a special treatment or specification about a specific surface of a part, as shown in **Figure 3-13**. A chain line is a thick line with the repeating pattern of a long line broken up by a short dash. Another use of chain lines is to indicate a projected tolerance zone.

Line Identification on a Print

Figure 3-14 shows the application of the alphabet of lines on a two-view drawing. The drawing displays several linetypes as described in this unit.

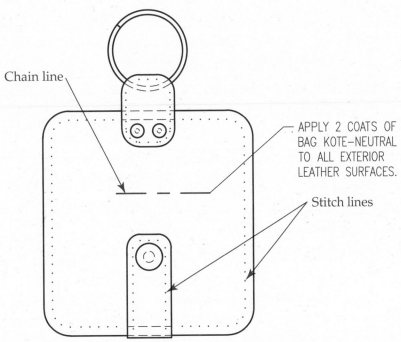

Chain line

APPLY 2 COATS OF
BAG KOTE—NEUTRAL
TO ALL EXTERIOR
LEATHER SURFACES.

Stitch lines

Goodheart-Willcox Publisher

Figure 3-13 Manufacturing of this leather key fob involves a sewing process as indicated by the stitch lines. It also requires a special surface treatment as noted by a chain line.

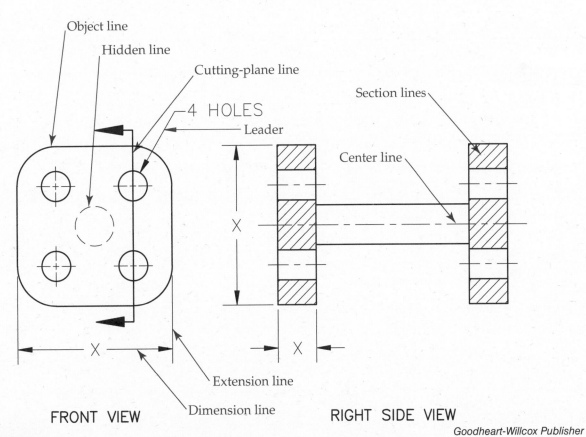

Object line

Hidden line

Cutting-plane line

4 HOLES

Leader

Section lines

Center line

Extension line

Dimension line

FRONT VIEW

X

X

RIGHT SIDE VIEW

X

Goodheart-Willcox Publisher

Figure 3-14 Typical usage of standard lines on a mechanical drawing.

Notes

Name _____ Date _____ Class _____

Activity 3-1
Line Identification
Match the following lines with their names.

_____ 1. Visible line

_____ 2. Hidden line

_____ 3. Section line

_____ 4. Centerline

_____ 5. Symmetry line

_____ 6. Dimension line

_____ 7. Extension line

_____ 8. Leader line

_____ 9. Cutting-plane line

_____ 10. Viewing-plane line

_____ 11. Break line (long)

_____ 12. Phantom line

_____ 13. Stitch line

_____ 14. Chain line

_____ 15. Break line (short)

A — — — — —

B —— — — ——

C ‖— — — — —‖

D   3.50

E   E

F —— — —— — —— —

G ～～～～～～

H — — — — —

I ——————

J ·············

K ——⌐⌐——⌐⌐——

L ▲— — —▲

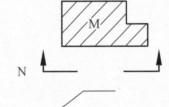

M

N ▲   ▲

O

Name _____ Date _____ Class _____

Activity 3-2
Line Identification

Study the drawing below and identify the line types by name.

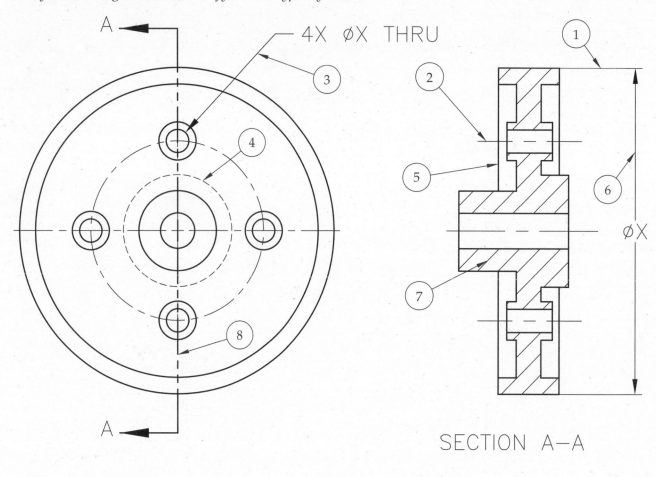

4X ⌀X THRU

SECTION A—A

1. _____

2. _____

3. _____

4. _____

5. _____

6. _____

7. _____

8. _____

Example Activity
Line and Surface Identification

Given pictorial drawings with marked surfaces, identify the corresponding lines and surfaces on the top, front, and right side views. Below is a completed example. Review it before completing Activities 3-2 and 3-3.

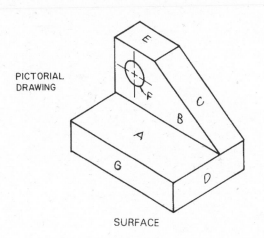

PICTORIAL DRAWING

SURFACE

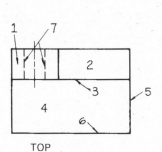

TOP

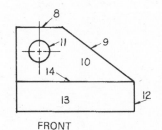

FRONT

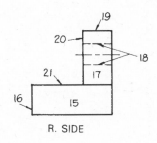

R. SIDE

Surface	Top	Front	Right side
A	4	14	21
B	3	10	20
C	2	9	17
D	5	12	15
E	1	8	19
F	7	11	18
G	6	13	16

Name _____ Date _____ Class _____

Activity 3-3
Line and Surface Identification

Place the correct number in the space provided on the answer grid to identify the corresponding lines and surfaces on pictorial drawings. The circled numbers indicate that the surface is a hidden backside view.

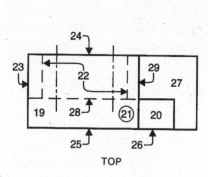

TOP

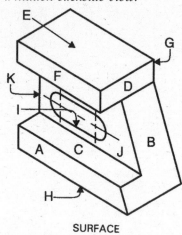

SURFACE

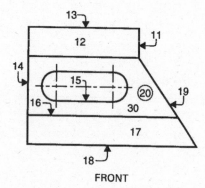

FRONT

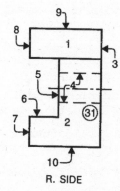

R. SIDE

Surface	Top	Front	Right side
A			
B			
C			
D			
E			
F			
G			
H			
I			
J			
K			

Name _____ Date _____ Class _____

Activity 3-4
Line and Surface Identification

Place the correct number in the space provided on the answer grid to identify the corresponding lines and surfaces on pictorial drawings. The circled numbers indicate that the surface is a hidden backside view.

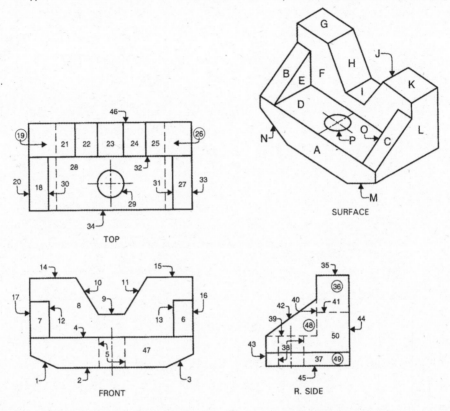

Surface	Top	Front	Right side
A			
B			
C			
D			
E			
F			
G			
H			
I			
J			
K			
L			
M			
N			
O			
P			

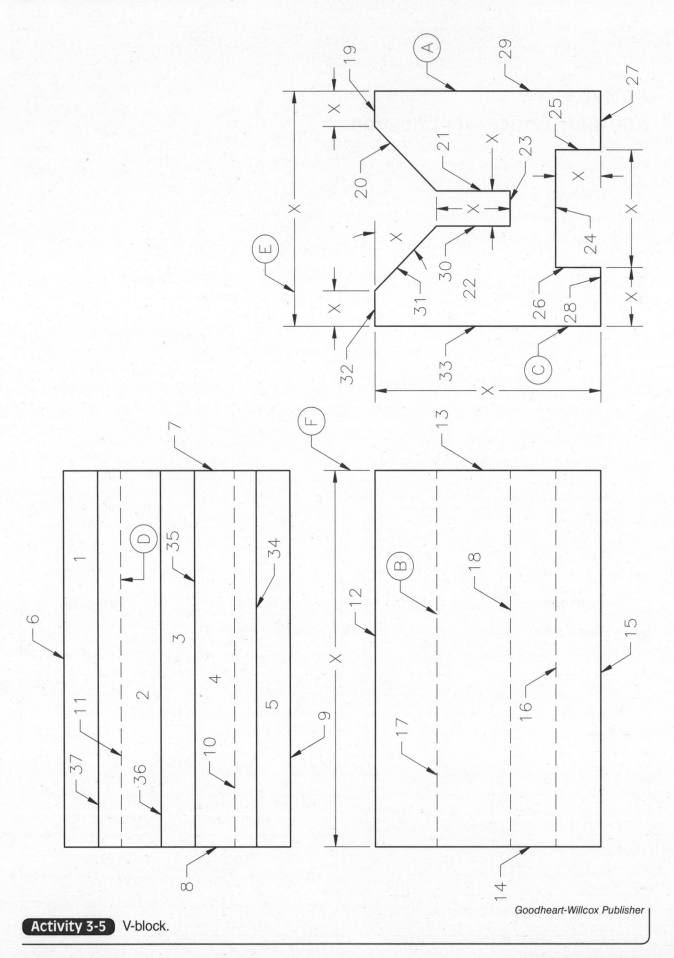

Name _____ Date _____ Class _____

Activity 3-5
V-block

Refer to **Activity 3-5**. *Study the drawings and familiarize yourself with the shapes of the objects. Next, follow the visible lines, hidden lines, and centerlines from view to view. Then answer the questions for each of the drawings.*

1. Line 25 in the side view is what line in the top view?

2. How many surfaces are shown in the top view?

3. Line 33 in the side view is what line in the top view?

4. Line 18 in the front view is what surface in the top view?

5. What type of line is line C?

6. Surface 4 in the top view is what line in the side view?

7. Line 16 in the front view is what line in the side view?

8. Surface 22 in the side view is what line in the top view?

9. Surface 22 in the side view is what line in the front view?

10. Line 20 in the side view is what surface in the top view?

11. Line 27 in the side view is what line in the front view?

12. What type of line is line D?

13. Surface 5 in the top view is what line in the side view?

14. Line 23 in the side view is what surface in the top view?

15. Line 10 in the top view is what line in the side view?

16. Surface 1 in the top view is what line in the side view?

17. What type of line is line A?

18. Line 21 in the side view is what line in the top view?

19. Line 14 in the front view is what line in the top view?

20. Line 35 in the top view is what line in the side view?

21. Line 29 in the side view is what line in the top view?

22. What type of line is line E?

23. What type of line is line B?

24. Surface 1 in the top view is what line in the front view?

25. What type of line is line F?

1. _____

2. _____

3. _____

4. _____

5. _____

6. _____

7. _____

8. _____

9. _____

10. _____

11. _____

12. _____

13. _____

14. _____

15. _____

16. _____

17. _____

18. _____

19. _____

20. _____

21. _____

22. _____

23. _____

24. _____

25. _____

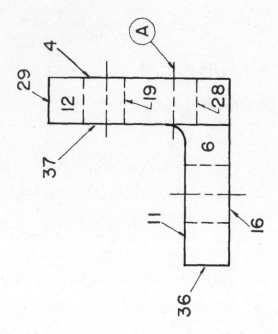

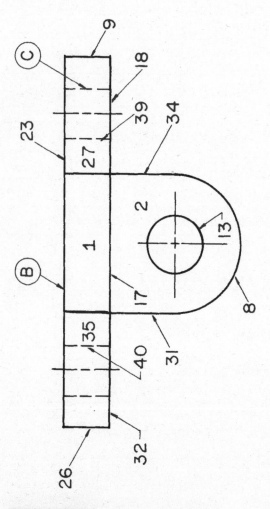

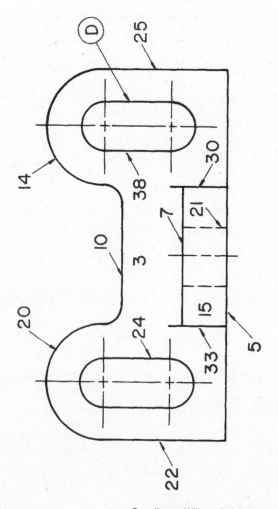

Activity 3-6 Adjusting Bracket.

Name _____ Date _____ Class _____

Activity 3-6
Adjusting Bracket

Refer to **Activity 3-6**. *Study the drawings and familiarize yourself with the shapes of the objects. Next, follow the visible lines, hidden lines, and centerlines from view to view. Then answer the questions for each of the drawings.*

1. Line 7 in the front view is what surface in the top view?

1. _____

2. Line 4 in the side view is what line in the top view?

2. _____

3. Surface 1 in the top view is what line in the side view?

3. _____

4. Line 29 in the side view is what surface in the top view?

4. _____

5. Surface 1 in the top view is what line in the front view?

5. _____

6. Line 9 in the top view is what line in the front view?

6. _____

7. Surface 6 in the side view is what line in the front view?

7. _____

8. Surface 6 in the side view is what line in the top view?

8. _____

9. Line 24 in the front view is what line in the top view?

9. _____

10. What kind of line is line C?

10. _____

11. Surface 3 in the front view is shown as what lines in the top view?

11. _____

12. Line 13 in the top view is what line in the front view?

12. _____

13. Surface 12 in the side view is what line in the top view?

13. _____

14. Line 18 in the top view is what line in the side view?

14. _____

15. What kind of line is line B?

15. _____

16. Line 20 in the front view is what surface in the top view?

16. _____

17. Surface 27 in the top view is what line in the front view?

17. _____

18. Line 38 in the front view is what line in the top view?

18. _____

19. Line 20 in the front view is what line in the side view?

19. _____

20. Line 22 in the front view is what line in the top view?

20. _____

21. What kind of line is line A?

21. _____

22. Line 5 in the front view is what line in the side view?

22. _____

23. Line 33 in the front view is what line in the top view?

23. _____

24. What kind of line is line D?

24. _____

25. Surface 2 in the top view is what line in the side view?

25. _____

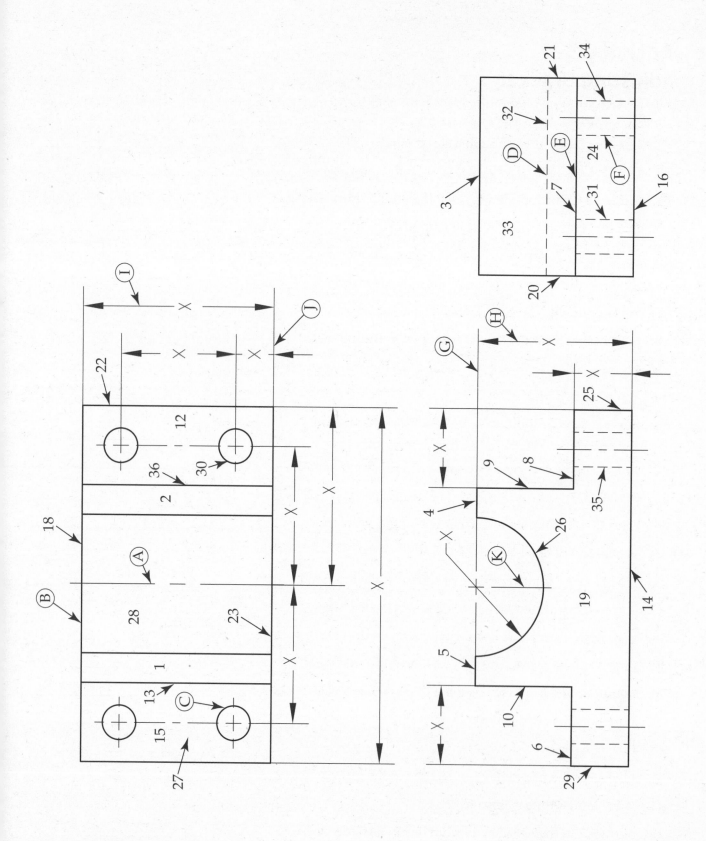

Name _____ Date _____ Class _____

Activity 3-7
Bearing Base

Refer to **Activity 3-7**. *Study the drawings and familiarize yourself with the shapes of the objects. Next, follow the visible lines, hidden lines, and centerlines from view to view. Then answer the questions for each of the drawings.*

1. Line 23 is represented by what surface in the front view? 1. _____

2. Line 9 in the front view is what line in the top view? 2. _____

3. What type of line is line A? 3. _____

4. Surface 28 is represented by what line in the side view? 4. _____

5. Line 31 in the side view is what line in the top view? 5. _____

6. What type of line is line F? 6. _____

7. Line 3 in the side view is which two lines in the front view? 7. _____

8. Surface 19 in the front view is represented by what line in the side view? 8. _____

9. Line 21 in the side view is what line in the top view? 9. _____

10. What type of line is line B? 10. _____

11. Line 22 is what line in the front view? 11. _____

12. Surface 33 is what line in the front view? 12. _____

13. Surface 24 is what line in the top view? 13. _____

14. What type of line is line K? 14. _____

15. Line 6 is what surface in the top view? 15. _____

16. What type of line is line J? 16. _____

17. Line 17 is what line in the side view? 17. _____

18. What type of line is line H? 18. _____

19. What type of line is line C? 19. _____

20. Line 8 is what surface in the top view? 20. _____

21. What type of line is line D? 21. _____

22. Line 5 in the front view is what line in the side view? 22. _____

23. What type of line is line E? 23. _____

24. What type of line is line G? 24. _____

25. What type of line is line I? 25. _____

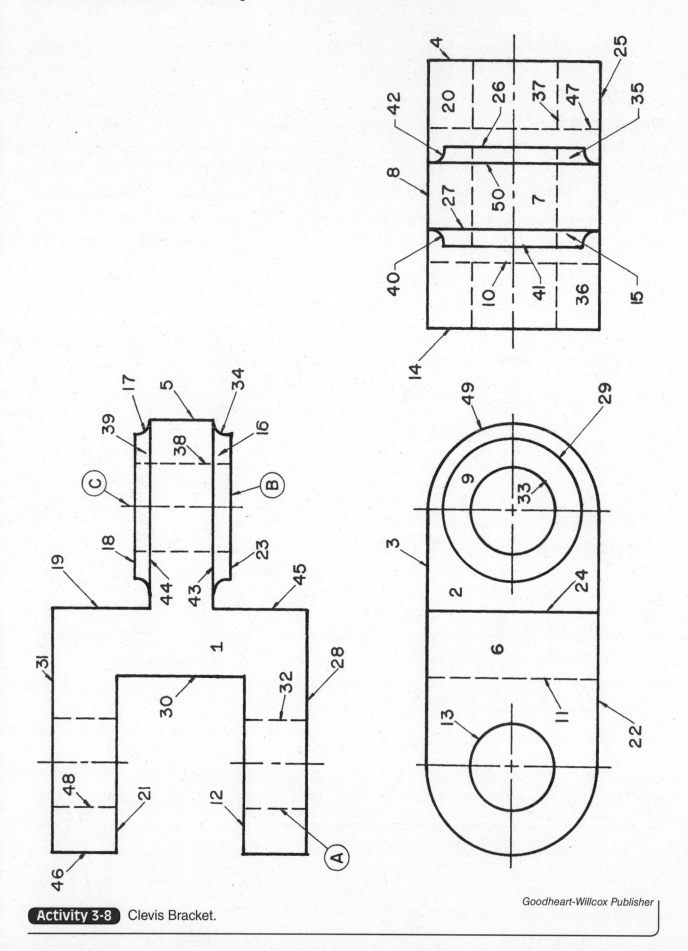

Activity 3-8 Clevis Bracket.

Name _____ Date _____ Class _____

Activity 3-8
Clevis Bracket

Refer to **Activity 3-8**. *Study the drawings and familiarize yourself with the shapes of the objects. Next, follow the visible lines, hidden lines, and centerlines from view to view. Then answer the questions for each of the drawings.*

1. Surface 1 is denoted by what line in the side view?

2. Line 18 in the top view is represented by what line in the side view?

3. Surface 7 is represented by what line in the front view?

4. Line 12 in the top view denotes what line in the side view?

5. Surface 6 is represented by what line in the side view?

6. Surface 20 is denoted by what line in the top view?

7. Line 30 in the top view is represented by what line in the front view?

8. What kind of line is line A?

9. Surface 9 represents what line in the top view?

10. Surface 9 is represented by what line in the side view?

11. Surface 36 represents what line in the front view?

12. Line 14 in the side view is represented by what line in the top view?

13. Line 13 in the front view is which two lines in the top view?

14. Line 37 is represented by what line in the front view?

15. Surface 15 in the side view is what surface in the top view?

16. Surface 7 in the side view represents what line in the top view?

17. What kind of line is line B?

18. Line 37 is what line in the top view?

19. Line 43 in the top view is what line in the side view?

20. Surface 2 in the front view is represented by what line in the top view?

21. What kind of line is line C?

22. Line 21 in the top view is what line in the side view?

23. Line 5 in the top view is what surface in the side view?

24. Line 29 denotes what line in the side view?

25. Surface 35 in the side view is what surface in the top view?

1. _____

2. _____

3. _____

4. _____

5. _____

6. _____

7. _____

8. _____

9. _____

10. _____

11. _____

12. _____

13. _____

14. _____

15. _____

16. _____

17. _____

18. _____

19. _____

20. _____

21. _____

22. _____

23. _____

24. _____

25. _____

Notes

Title Blocks and Notes

Learning Objectives

After studying this unit, you will be able to:

✓ Describe drawing formats according to industry standards.

✓ Identify and define the parts of a title block.

✓ Interpret information found in a title block and its components.

✓ Understand how notes convey important information.

Key Terms

application block
angle of projection block
CAGE code
drawing number
feature
general notes
local notes

part number
parts list
revision
revision history block
revision status of sheets
sheet format
sheet size

surface texture
title block
tolerance
tolerance block
zones

The major topics covered in this unit will include drawing formats, title blocks, tolerance blocks, change blocks, parts lists, and notes.

Drawing Formats

Companies have standard sheet sizes and formats for the layout of their drawings. A *sheet size* is the size of a drawing layout presented in digital form (soft copy) or on paper (hard copy). Letters as shown in **Figure 4-1** easily

Standard Sheet Sizes (US Customary)	
Size Code	Sheet Dimensions (inches)
A	8.5 × 11
B	11 × 17
C	17 × 22
D	22 × 34
E	34 × 44

Standard Sheet Sizes (Metric)	
Size Code	Sheet Dimensions (mm)
A4	210 × 297
A3	297 × 420
A2	420 × 594
A1	594 × 841
A0	841 × 1189

Goodheart-Willcox Publisher

Figure 4-1 ASME standard sheet sizes for US Customary and metric drawings.

identify sheet sizes. The sheet size depends on the area needed to represent the part or assembly and the size and scale of the drawing. A large part with multiple pieces may not fit on a small sheet. Therefore, a large size will be required. Whereas, a small part made of one component may require a smaller size.

A *sheet format* is a standard that controls the layout of information on a drawing. The industry standard for decimal inch drawings is ASME Y14.1-2012. See **Figure 4-2**. For metric drawings, the sheet format standard is ASME Y14.1M-2012. The only difference between the standards is the systems of measurement. Otherwise, the layouts are the same. A sheet format provides a standardized way to arrange information. The following sections cover the basic elements of a drawing format.

Title Block

The *title block* is a boxed area located at the bottom-right of a drawing sheet that contains general information about a project. The purpose of the title block is to identify the project and its components. The rectangular blocks or boxes in a title block allow information to be efficiently organized. Refer to **Figure 4-2(3)**. Some information is specific about the part, while other information is general to the project. The company name, drawing number, part number, and drawing date are some of the information used in a title block.

Many companies have their own title block format. The content and arrangement will vary from one company to another. Some companies will follow the current ASME standard while others will not. Although different in layout, all will contain similar data, as shown in **Figure 4-3**.

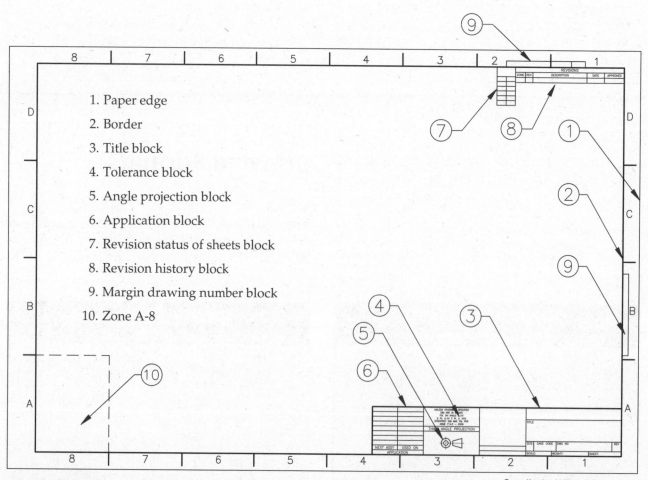

1. Paper edge
2. Border
3. Title block
4. Tolerance block
5. Angle projection block
6. Application block
7. Revision status of sheets block
8. Revision history block
9. Margin drawing number block
10. Zone A-8

Goodheart-Willcox Publisher

Figure 4-2   A typical D-size drawing sheet format, as defined by the ASME Y14.1-2005 standards. The actual size of a D-size drawing is 22″ × 34″.

		−01	A	316 STAINLESS STEEL PLATE	TITANIUM NITRIDE	.25
		#	LTR.	MATERIAL	SURFACE TREATMENT	REMARKS

DESCRIPTION
MOUNTING BRACKET

G & B DESIGNS
SOUTH MILWAUKEE   WISCONSIN
SPECIAL   EQUIPMENT   DIVISION

PART   NO.

DRAWN BY: MB   DATE: 4−12
CHECKED BY: RG   DATE: 4−23
APPROVED BY: BH   DATE: 4−28

TOLERANCES
UNLESS OTHERWISE SPECIFIED
.X = ±.10   .XXX = ±.005
.XX = ±.02   .XXXX = ±.0005
ANGLES =   ±.5°

SCALE: 1:1
SHEET SIZE A
SHEET 1 OF 1

30-415-4589

LTR.	CHANGE	DATE	BY

DESIGN APPROVAL	DATE	UNSPECIFIED TOLERANCES	
DES.ENG.		.XXX ± .005	
MFG.ENG.		.XX ± .015	
Q.C.		.X ± .050	
		Machine Finish √ 125 Max	
SALES		Angles ± 1°	
		Concentricity .005 FIM	
		Squareness .001 per in.	

This Drawing is the property of the Dumore Corporation. It must not be reproduced or copied without written permission.

PART NO.
658-0169

DUMORE CORPORATION
1300 17TH STREET, RACINE, WI 53403

DO NOT SCALE DRAWING	REMOVE ALL BURRS BREAK SHARP CORNERS

TITLE   PLATE−WHEEL GUARD

| MATERIAL SPEC.   STAMPING | Req. per Piece | RECEIVED AS | DWG BY | JKM | DATE | 9/23 | SCALE |
| .0593 C.R. STEEL-SHEET | Pattern No. | RS−658−0169 | CKD. BY | JR | DATE | 9/24 | 1:1 |

REVISIONS

A	ADDED PAINT FINISH: S−3048	10/3

Figure 4-3   Title blocks may vary from one company to another. Even though the layouts are different from each other, they both contain the same types of information.

The drawings in the book show the different formats used in the industry.

Company Information

The company information block includes the name and location of the company that is responsible for the drawing. See **Figure 4-4(A)**. Often, the company's logo or trademark is included with the name.

Drawing Title

The drawing title or title is the descriptive name assigned to the drawing. The drawing title is located in the lower-right corner in a title block. Refer to **Figure 4-4(B)**. The title consists of a single noun or several that describe the drawing. Some companies may use PART NAME or DESCRIPTION in place of TITLE.

Size

The size block identifies the size of the drawing sheet. A letter references the size. Refer to **Figure 4-4(C)**.

CAGE Code

The Commercial and Government Entity code or *CAGE code* block identifies government projects by a five-digit government classification code. The CAGE code refers to a specific government activity. Refer to **Figure 4-4(D)**.

Drawing Number

The *drawing number* is the identification number used for referencing, filing, and archiving a drawing. Most firms assign a drawing number using a designated system based on their needs. Numbering systems use a code of numbers, letters, and dashes to represent departments, drawing dates, project numbers, and part numbers, as shown in **Figure 4-5**. Drawing numbers are typically located at the lower-right corner in a title block. Refer to **Figure 4-4(E)**.

Margin Drawing Number Block

The margin drawing number block is used for placing a drawing number along the sheet border in the margin area. Refer to **Figure 4-2(9)**. The block includes a drawing number, a sheet number, and a revision character, as shown in **Figure 4-6**. A margin drawing number can be located on the border in the upper-right or lower-right margins of a drawing. Placing a drawing number in a margin is optional. However, margin drawing numbers allow the information to be visible on folded or rolled drawings.

Revision Block

Revisions are any changes made to the original drawing. Drawing revisions improve part design, clarify details, change dimensions, correct errors, reduce costs, and change manufacturing procedures. The revision block allows quick reference to the current drawing revision. The revision block is generally located in the lower-right corner of a title block, as shown in **Figure 4-4(F)**. Letters designate a revision, but numbers are also used.

It is becoming more common to exclude the revision block from the drawing sheet and store the revision history digitally. Detailed revision information is stored in a digital document. The revision history is identified by a letter after the drawing number in the title block.

PRO TIP

Storing and identifying large drawings can be challenging. A common storage practice is to fold large drawings, such as a D-size or E-size, down to an A-size. This keeps the name and drawing number accessible from the outside of the folded drawing.

Rolling-up drawings can also achieve the same results. The proper way to roll a drawing is to start at the end opposite of where the drawing title and number are located. Next, with the printed side facing out, roll the drawing along its length. When finished, the identification information will be accessible from the outside of the roll. These practices make for easy identification of stored drawings.

More companies are now storing drawings digitally. A hard copy of the drawing is only printed when needed. This eliminates the need for large amounts of physical storage space. It also simplifies the process of sending out drawings, as they can simply be e-mailed to the recipient.

Goodheart-Willcox Publisher

Figure 4-4 The standard components of an ASME standard title block. See text for letter references.

455–D–1494827

Drawing number series

Sheet size

Organization designation

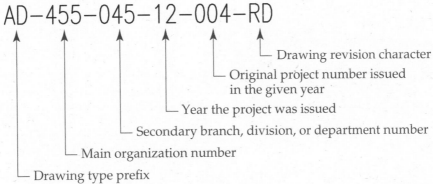

AD–455–045–12–004–RD

Drawing revision character

Original project number issued in the given year

Year the project was issued

Secondary branch, division, or department number

Main organization number

Drawing type prefix

Goodheart-Willcox Publisher

Figure 4-5 Companies use coding systems similar to these for establishing drawing numbers. Drawing numbers are for drawing identification, as well as for archiving purposes.

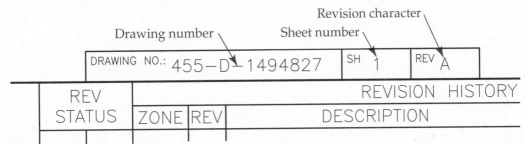

Revision character

Sheet number

Drawing number

DRAWING NO.: 455–D–1494827	SH 1	REV A

REV STATUS	ZONE	REV	REVISION HISTORY
			DESCRIPTION

Goodheart-Willcox Publisher

Figure 4-6 The drawing number and related information can be located in the margin above the revision block at the upper-right corner of a drawing sheet.

PRO TIP
When working on a project, it is important to check the drawing number and revision letter before proceeding with any specific work operations. Prints with similar drawing numbers can easily be mistaken for another. Checking the revision letter prevents you from making a part with old specifications. Double-checking the drawing numbers and revision letters will prevent costly mistakes.

Scale Block

The scale block indicates the scale of a drawing. Refer to **Figure 4-4(G)**. Typically, a drawing has one scale as noted in the scale block. However, drawings with different scales may be required to specify a part or assembly. VARIES in the scale block indicates multiple scales are on a sheet. Each drawing will have a separate scale noted. When a drawing has no specific scale, DRAWING NOT TO SCALE may be on the drawing sheet and NTS (not to scale) in the scale block.

PRO TIP

When reading the dimensions of an object, do not scale (measure) them from the drawing. Scale only reflects the size of the object in relationship to the drawing. It is common for drawings to be printed a different size than the original. When this happens, the scale of the object is no longer accurate. Always use the dimensions listed on the drawing. Dimensions listed on the drawing are always the actual sizes of the object regardless of the scale and size of a drawing.

Weight Block

When required, the weight block indicates the weight of the manufactured product. Refer to **Figure 4-4(H)**. Actual or estimated values specify the weight. The actual weight is the "true" weight of a finished part. The estimated weight is a calculated value that controls the part's weight during the manufacturing process. The first sheet is the only sheet that shows the weight.

Sheet Block

The sheet block tells you how many sheets make up the total number of sheets in a drawing and shows the current sheet number for a drawing or a set of drawings. Refer to **Figure 4-4(I)**.

Part Number

A *part number* identifies a specific item or part. Refer to **Figure 4-4(J)**. Part numbers are located in the main title block, a parts list, an application block, or separately on a drawing. A part number may or may not be included in a title block. A part number can represent the manufactured part or a distributor's number for a supplied part. Some companies also use the part number as the drawing number.

Drawn By and Date

The drawn by area in the title block contains the initials of the drafter who produced the drawing. Refer to **Figure 4-4(K)**. The date area records the starting time or initial completion of the drawing.

Checked By and Date

The checked by area in the title block contains the initials of the drafting department's supervisor or manager who checks the prints for clarity and accuracy. The date area records the time of inspection. Refer to **Figure 4-4(L)**.

Approved By and Date

The approved by area in the title block contains the initials of the drafting department's supervisor or manager who approves the drawing for use. The date area records the time of approval. Refer to **Figure 4-4(M)**.

Material Number

Material number is the number assigned by a company to identify the material required for making a part. The number typically refers to a stock number for a specified material. Refer to **Figure 4-4(N)**.

Material Description

Material description indicates the specific type of material required for manufacturing a part. It may also specify the size of stock. Refer to **Figure 4-4(O)**.

Material Finish

The material finish area contains finish specifications required for manufacturing a part. Refer to **Figure 4-4(P)**.

Angle of Projection Block

The *angle of projection block* identifies whether the drawing is a first-angle or third-angle projection as described in Unit 2—Visualizing Shapes. Refer to **Figure 4-4(Q)**.

Tolerance Block

Tolerance is the amount of variation permitted in the value of a dimension representing a part or any of its features. The tolerance is the difference between the lowest allowable value and the highest allowable value. A *feature* is a physical component of a part identified on a drawing such as a surface, hole, slot, edge, or any other described items. The *tolerance block* indicates the general tolerance limits specified for the drawing. Refer to **Figure 4-4(R)**. Tolerances can be specified for one (.X), two (.XX), three (.XXX), and four (.XXXX) place decimal dimensions, or for fractional, metric, and angular dimensions. The limits in the tolerance block apply to their referenced dimensions unless a dimension has a specific tolerance noted.

PRO TIP Unspecified tolerances may vary from part to part. For example, the tolerance for a two-place (.XX) decimal dimension may be ±.010 inch for one part while another part may use ±.015 inch for two-place decimal limits. In addition, one company may use one set of tolerances while another company's tolerances may differ. Identifying the tolerances before machining a part will allow you to make the part to the required specifications.

Surface texture is the desired surface condition of a manufactured part. Surface texture or surface finish tolerances are noted on the part drawing, but can also be included in the tolerance block. Machined parts usually require certain surface textures. A value system with a designated number describes the surface texture. As the value of the number decreases, the surface condition improves. Surface texture numbers vary anywhere from 2000 to .5. See **Figure 4-7**.

Application Block

The *application block* is used to identify a part's assemblies, systems, and subsystems. Application blocks are typically located at the left side of the title block. Refer to **Figure 4-2(6)**. The NEXT ASSY column identifies a part's next higher assembly with a number (drawing or part) that corresponds to its next higher assembly drawing. See **Figure 4-8**. The USED ON column is used to match a part to its system or subsystem.

TOLERANCES UNLESS OTHERWISE SPECIFIED	
.X = ±.10	.XXX = ±.005
.XX = ±.010	.XXXX = ±.0005
ANGLES = ±1°	

FINISH SPECIFICATIONS:
MACHINED SURFACES: 125/MAX
GROUND SURFACES: 32/MAX
REMOVE BURRS & SHARP EDGES TO .015 MAX.
DO NOT SCALE

Goodheart-Willcox Publisher

Figure 4-7 A tolerance block gives general tolerance limits that apply to dimensions that do not have any specified tolerances.

Revision History Block

The *revision history block*, revision block, or change block is a record of a drawing's changes. It is becoming less common for prints to include a revision history block on the sheet. The revision history is stored digitally, and only the latest revision letter appears on the sheet.

When the revision history block does appear on a print, the block of information is generally located in the upper-right corner of the drawing as shown in **Figure 4-2(8)**. However, it is common to be located in the title block area.

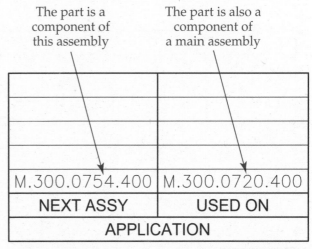

Goodheart-Willcox Publisher

Figure 4-8 The application block is used for recording a part's assemblies and subassemblies.

New revisions appear underneath the previous revisions as they occur. A revision history block contains the following information as shown in **Figure 4-9**:

- ZONE (column A)—identifies the revised area of the drawing. Details on a large drawing can be hard to locate. Separating a large drawing into smaller areas called *zones* makes locating features on a drawing easier. Zones appear in a grid formation using numbers and letters located outside the border for identification. Refer to **Figure 4-2(10)**.
- REV (column B)—identifies each change with a symbol, starting with the letter A. Numbers also identify the changes. In addition, the symbol appears on the drawing to locate and identify the change.
- DESCRIPTION (column C)—explains the revision. REVISION, CHANGE, or REVISION RECORD are alternate terms for this column heading.
- DATE (column D)—records the date of a revision.
- APPROVED (column E)—contains the authorizing initials of the person responsible for approving the revisions.

Some companies may require additional information for their drawing revisions. Additional columns can easily provide the necessary space for the information.

PRO TIP A drawing revision is often ordered by an Engineering Change Order (ECO) or an Engineering Change Request (ECR). The ECO or ECR contain the data, notes, and sketches required for the changes on the drawing. Changes include anything that affects the finished product 'such as dimensions, materials, processes, notes, features, and finishes. The ECR or ECO is usually a separate document in digital form. The document's reference number (ECO NO.) can be included in the revision history block.

Revision Status of Sheets

A project with multiple sheets may require a revision status of sheets block. The *revision status of sheets* block records the revision status for each sheet. See **Figure 4-10**. The block is located at the top of a sheet, next to the revision block as shown in **Figure 4-2(7)**, or on top of the title block as shown in **Figure 4-4(S)**. The revision status of sheets can also be a separate document.

		REVISION HISTORY		
ZONE	REV	DESCRIPTION	DATE	APPROVED
	A	RELEASED FOR PRODUCTION	3–23	P.H.
E2	B	PURCHASE PER PRINT—DELETED	5–3	C.B.
D2	C	.088 & .50 WIDE—ADDED	5–3	C.B.
D2	D	BLUE PAINT STRIPE—ADDED	5–3	C.B.

A   B   C   D   E

Figure 4-9 The revision history block shows a list of revisions made to the original drawing.

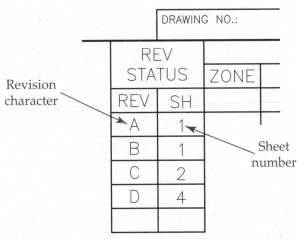

Goodheart-Willcox Publisher

Figure 4-10 The revision status block, shown only on the first sheet of a set of drawings, records all of the revisions of the sheet set.

- PART NUMBER—Part number (PART NO.) or part identification number (PIN) refers to the actual manufacturer's part number, including parts from other sources.
- DESCRIPTION—Description or part name indicates the actual name of that item in the assembly. Usually, the description consists of one or several words that describe the part.
- QUANTITY—Quantity (QTY) or quantity required (QTY REQD) column contains the actual number of parts required for the assembly.
- MATERIAL—Material (MAT) column contains a commercial description of the material specified.

Parts List

The *parts list* is a listing of parts needed for manufacturing a product. See **Figure 4-11**. Parts lists are common with assembly drawings. Other names for a parts list are material list, bill of materials, or parts schedule. The information found in a parts list includes the following:

- ITEM—Item number, find number, or key that directs the print reader to a specific part on the assembly drawing. A number in a circle with a leader line is the common way to note an item on a drawing.

Notes

Notes on a drawing provide additional data and information not found elsewhere on the drawing. There are two types of notes used on a drawing—general notes and local notes. *General notes* provide general information that is relevant to the whole drawing. *Local notes* provide additional information that relates to a specific part or feature. Notes provide information such as heat-treat directions, machining directions, finish requirements, measuring requirements, and other specifications.

It is best to read notes before studying the views of the part because they may advise you of certain requirements or restrictions regarding

ITEM	PART NO.	DESCRIPTION	QTY.	MATERIAL
4	3964	SLIDE BASE	1	SAE 06
3	5718	PLATE	1	1020 C.R.S.
2	6263	ROLLER	4	SAE 06
1	918	PIN	4	SAE 06
ITEM	PART NO.	DESCRIPTION	QTY.	MATERIAL
PARTS LIST				

Goodheart-Willcox Publisher

Figure 4-11 A parts list contains descriptions and quantities of specific parts needed for manufacturing a product.

the part. The word NOTE usually precedes information it is furnishing. Abbreviations often appear in notes and in other parts of drawings. The Reference Section in the back of the book lists the industries' common abbreviations.

Some typical notes found on prints could include:

FINISH: BLACK OXIDE
FINISH THIS SIDE WORKING FROM
    COMPLETED OPPOSITE END

BREAK ALL SHARP CORNERS
NO DIE PARTING MARKS PERMISSIBLE ON
    THIS SURFACE

NOTES:
WHEN GRINDING .3762/.3757 DIA.
USE MICROMETER NOT AIR SNAP GAUGE
HAVE SLIGHT TAPER TOWARD PINION

NOTES:
HEAT TREAT PER ES-12-1001
TEMPER PER ES-12-3000 R 39-43
(ES NUMBER FOR IN-HOUSE USE)

MINIMUM DEPTH OF 5/8 DRILL TO CLEAR
    SPINDLE END

NO SHARP EDGE ON 4 SQUARE CORNERS

#3 GA. (.2391)
HOT ROLLED, PICKLED, AND OILED SHEET
    STEEL

Name _____ Date _____ Class _____

Activity 4-1
Drawing Format and Title Block

Study **Activity 4-1** *and identify the parts of the drawing format and title block. Write your answers in the blanks provided.*

1. _____

2. _____

3. _____

4. _____

5. _____

6. _____

7. _____

8. _____

9. _____

10. _____

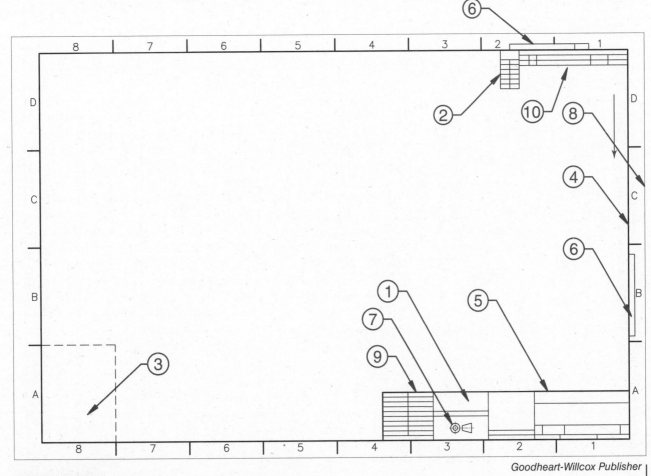

Activity 4-1   Drawing Format and Title Block.

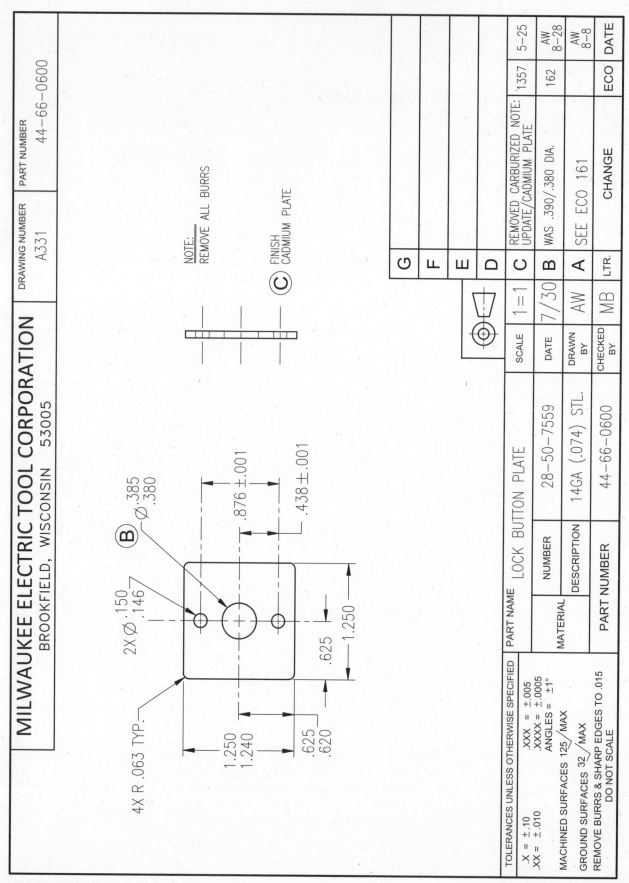

MILWAUKEE ELECTRIC TOOL CORPORATION
BROOKFIELD, WISCONSIN   53005

DRAWING NUMBER	PART NUMBER
A331	44-66-0600

NOTE:
REMOVE ALL BURRS

Ⓒ FINISH
CADMIUM PLATE

2X Ø .150 / .146

Ⓑ Ø .385 / .380

.876 ±.001

.438 ±.001

1.250

.625

4X R .063 TYP.

1.250 / 1.240

.625 / .620

G				
F				
E				
D				
C	REMOVED CARBURIZED NOTE: UPDATE/CADMIUM PLATE	1357	5-25	
B	WAS .390/.380 DIA.	162	AW 8-28	
A	SEE ECO 161		AW 8-8	
LTR.	CHANGE	ECO	DATE	

SCALE	1=1
DATE	7/30
DRAWN BY	AW
CHECKED BY	MB

PART NAME	LOCK BUTTON PLATE	
MATERIAL	NUMBER	28-50-7559
	DESCRIPTION	14GA (.074) STL.
PART NUMBER		44-66-0600

TOLERANCES UNLESS OTHERWISE SPECIFIED
.X = ±.10
.XX = ±.010
.XXX = ±.005
.XXXX = ±.0005
ANGLES = ±1°
MACHINED SURFACES 125/MAX
GROUND SURFACES 32/MAX
REMOVE BURRS & SHARP EDGES TO .015
DO NOT SCALE

Activity 4-2   Lock Button Plate.

Milwaukee Electric Tool Corporation

Name _____ Date _____ Class _____

Activity 4-2
Lock Button Plate

This activity will test your print reading ability. Study the title block, notes, and views in **Activity 4-2**. *Read the questions, refer to the print, and write your answers in the blanks provided.*

1. What is the title or name of this drawing?
2. What is the name of the company responsible for the drawing?
3. Who created the drawing?
4. What is the part number?
5. What type of material is used to make the part?
6. When was revision A issued?
7. What does ECO mean?
8. List the scale of the drawing.
9. The drawings are shown in what angle?
10. What operations must be performed on the part according to the notes in the tolerance block?
11. What is the maximum allowed surface finish value for machined surfaces?
12. Explain revision B.
13. List the material number.
14. What is the drawing number?
15. What is the tolerance for angles?
16. What type of finish is specified for the part?
17. What is the maximum allowed surface finish value for ground surfaces?
18. When was the original drawing issued?
19. List the views shown on the drawing.
20. The removal of a carburizing operation is specified by what ECO number?

1. _____
2. _____
3. _____
4. _____
5. _____
6. _____
7. _____
8. _____
9. _____
10. _____
11. _____
12. _____
13. _____
14. _____
15. _____
16. _____
17. _____
18. _____
19. _____
20. _____

Notes

UNIT 5

Applied Math

Learning Objectives

After studying this unit, you will be able to:

✓ Identify and define whole numbers, fractions, and decimal fractions.
✓ Add, subtract, multiply and divide whole numbers.
✓ Convert between improper fractions and mixed numbers.
✓ Reduce fractions to their lowest terms.
✓ Add, subtract, multiply, and divide fractions.
✓ Add, subtract, multiply, and divide decimal fractions.
✓ Convert fractions to decimal fractions.
✓ Convert decimal fractions to fractions.

Key Terms

decimal
fraction
whole number

A basic understanding of math is critical to success in a machine shop. This unit provides a brief review of some basic math concepts.

Whole Numbers

Whole numbers are simply numbers without fractions or decimal points, numbers such as 1, 2, 3, 4, etc. Whole numbers are sometimes referred to as *counting numbers*. Adding, subtracting, multiplying, and dividing whole numbers primarily requires memorizing a few math facts.

Adding and Subtracting Whole Numbers

Example: Adding this column of whole numbers requires memorizing the *sum* of 3+5 and the sum of 8+2.

$$\begin{array}{r} 3 \\ 5 \\ +2 \\ \hline 10 \end{array}$$

The same type of memorization of math facts is required to subtract whole numbers. We know that

the result of subtracting 12 from 37 is 25, because we know that 2 from 7 is 5 and 1 from 3 is 2.

$$\begin{array}{r} 37 \\ -12 \\ \hline 25 \end{array}$$

The key to both addition and subtraction is to line up the columns of digits correctly. The whole numbers should be aligned on the right.

In subtraction, if the number being subtracted (the number on the bottom) is larger than the number it is being subtracted from (the number on the top), borrow 10 from the next digit to the left and add it to the one on the right. Write small numerals above the column to help you keep track.

Example

$$\begin{array}{r} {}^{3\,1}\!\!\!\!\not{4}2 \\ -17 \\ \hline 25 \end{array}$$

Multiplying Whole Numbers

Multiplication of whole numbers requires memorization of a multiplication table. The only way to get $6 \times 5 = 30$ is to know that multiplication fact or to add 6 + 6 + 6 + 6 + 6. That becomes way too tedious for bigger multiplication problems. To multiply numbers whose values are ten or more (those with more than one digit), align the digits representing 0 through 9 (the ones place) in the right-hand column. Then multiply the top row by the ones digit in the second row:

Example

$$
\begin{array}{r}
31 \\
\times\,15 \\
\hline
155 \quad\longleftarrow 5 \times 31
\end{array}
$$

Tens column — Ones column

Next, multiply the top row by the tens digit in the second row. Write your answer beneath your previous answer. Because you are multiplying by the tens digit, place a zero in the ones column of the *product* (the result of multiplication):

$$
\begin{array}{r}
31 \\
\times\,15 \\
\hline
155 \\
310 \quad\longleftarrow 10 \times 31
\end{array}
$$

If the problem has more digits in the second row, the above steps are repeated for each digit with the products being written in rows beneath one another, with the right-most digit in each row being written in the column for the place it represents: 100s, 1000s, etc.

When all of the multiplication is complete, add the products just as you would for a simple addition problem. The result of this addition is the product (answer) of the multiplication problem.

$$
\begin{array}{r}
31 \\
\times\,15 \\
\hline
155 \\
+\,310 \\
\hline
465 \quad\longleftarrow \text{Product}
\end{array}
$$

Dividing Whole Numbers

Division of whole numbers is the reverse of multiplication, but the problem must be set up differently. The *dividend* (the number being divided) is written inside the division symbol. The *divisor* (the number the dividend will be divided by) is written to the left of the symbol:

Divisor ⟶ $8\overline{)56}$ ⟵ Dividend

By knowing the multiplication table, we know that $8 \times 7 = 56$. So, if 56 is divided into 8 parts, each part will have 7, or $56 \div 8 = 7$. 7 is the *quotient* (the answer to a division problem) and it is written above the division symbol and above the ones place of the 56:

$$
7 \longleftarrow \text{Quotient}
$$
$$
8\overline{)56}
$$

When the divisor is more than 9, the process is divided into steps as follows:

$$
7\overline{)518}
$$

Seven goes into 51 seven times. Write the 7 above the 1 (the right column of the 51). Now multiply 7×7, which is 49. Write the 49 beneath the 51 in the division symbol.

$$
\begin{array}{r}
7 \\
7\overline{)518} \\
49
\end{array}
$$

Subtract the product of your multiplication (49) from the digits above it in the dividend (51). The result of your subtraction is 2.

$$
\begin{array}{r}
7 \\
7\overline{)518} \\
-\,49 \\
\hline
2
\end{array}
$$

Drop the next digit to the right in the dividend (8 in this case) down beside the result of your subtraction. That makes the number at the bottom 28. Determine how many times the divisor can go into this number. In this case, 7 will go into 28 four times, so write 4 above the 8 in the dividend.

$$
\begin{array}{r}
74 \\
7\overline{)518} \\
-\,49 \\
\hline
28
\end{array}
$$

Finally, multiply the 4 in the quotient by the divisor (7) and subtract this product from the number on the bottom line. The difference is zero, so the division is complete.

$$
\begin{array}{r}
74 \\
7\overline{)518} \\
-\,49 \\
\hline
28 \\
-\,28 \\
\hline
0
\end{array}
$$

If there are more places in the dividend, just keep doing the same division, multiplication, subtraction, and drop down for each digit moving to the right.

Example

If the last number produced by the drop-down cannot be divided evenly by the divisor, that number is called the *remainder*. In the following example, the quotient is 115 with a remainder of 7.

```
        115
    8 ) 927
      − 8
      ────
        12
       − 8
      ────
        47
      − 40
      ────
         7  ←──── Remainder
```

The remainder represents a portion less than one whole. The remainder portion is equal to a fraction with the remainder (7) as the numerator and the divisor (8) as the denominator. Thus,

$$927 \div 8 = 115\,\frac{7}{8}$$

Fractions

A *fraction* is part of a whole number. On a machinist's rule, each inch is broken into smaller divisions. These smaller divisions are fractions of an inch. See **Figure 5-1**.

A fraction is made up of two values separated by a fraction bar. The value below the fraction bar is called the *denominator*. The denominator signifies the number of "parts" in a whole. The value above the fraction bar is called the *numerator*. The numerator signifies the number of "parts" in the quantity. See **Figure 5-2**.

A *proper fraction* is a fraction with a numerator smaller than the denominator. Thus, a proper fraction always represents a value between zero and one. An *improper fraction* is a fraction with a numerator greater than the denominator. A *mixed number* is a number that contains both a whole number component and a fraction component. See **Figure 5-3**. Improper fractions and mixed numbers express quantities greater than one.

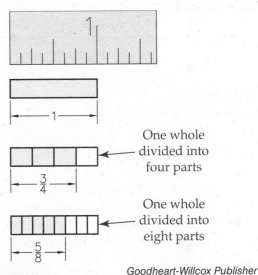

Goodheart-Willcox Publisher

Figure 5-2   Fractions express a quantity as number of parts or pieces that are smaller than a whole.

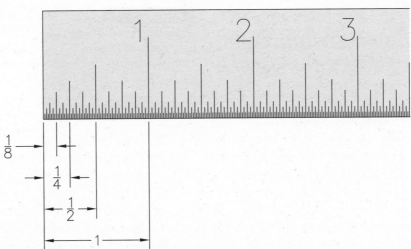

Goodheart-Willcox Publisher

Figure 5-1   A machinist's rule is divided into fractional portions.

Numerator

$$\frac{7}{8}$$

Denominator

$$\frac{13}{8}$$

$$2\frac{3}{16}$$

Proper Fraction
Numerator less than
denominator.
Represents a value
less than one.

Improper Fraction
Numerator greater than
denominator.
Represents a value
greater than one.

Mixed Number
Contains a
whole number and
a proper fraction.

Goodheart-Willcox Publisher

Figure 5-3   A fraction can be classified as a proper fraction or an improper fraction. An improper fraction is best expressed as a mixed number.

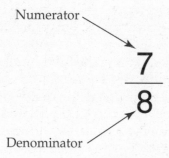

PRO TIP   Avoid expressing quantities as improper fractions.
Instead, express the quantity as a mixed number. Mixed numbers communicate values more clearly than improper fractions do. For example, which of the following is easier to understand: "I worked 33/4 hours today" or "I worked 8 1/4 hours today"?

Properties of Fractions

$$\frac{0}{4} = \frac{0}{16} = 0$$

$$\frac{2}{2} = \frac{16}{16} = 1$$

$$\frac{8}{1} = 8 \qquad \frac{4}{1} = 4$$

Goodheart-Willcox Publisher

Figure 5-4   Understanding the basic rules or properties of fractions will make performing calculations easier.

If a fraction's numerator is zero, the fraction is equal to zero. If a fraction's numerator is the same as its denominator, the fraction is equal to one. Also, a fraction with a denominator of one is equal to the whole number contained in the numerator. See **Figure 5-4**.

Converting between Improper Fractions and Mixed Numbers

To convert an improper fraction to a mixed number:

1. Divide the numerator by the denominator. The whole number value of the quotient is the whole number portion of the mixed number. The remainder is the numerator in the fraction part of the mixed number.
2. The denominator in the mixed number is the same as the denominator in the improper fraction.

This process is illustrated in **Figure 5-5**.

To convert a mixed number to an improper fraction:

1. The denominator in the improper fraction is the denominator in the fraction part of the mixed number.
2. The determine the numerator of the improper fraction, first multiply the denominator by the whole number part of the mixed number, then add the numerator in the fraction part of the mixed number to this product.

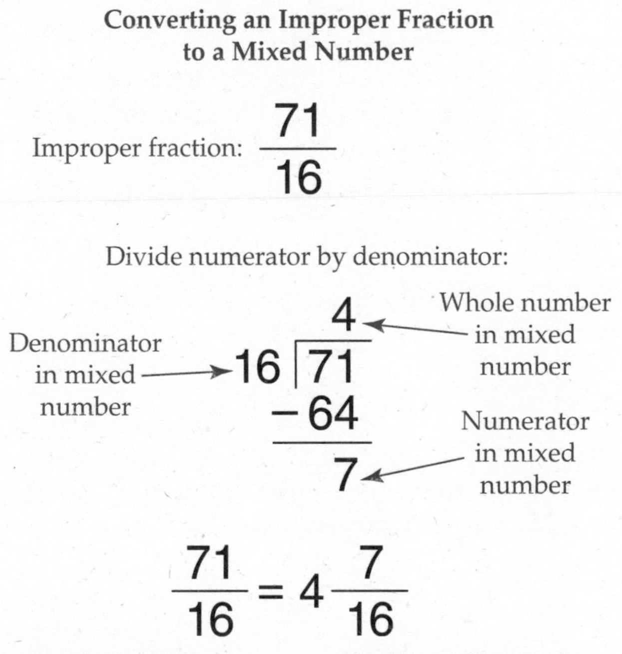

Converting an Improper Fraction to a Mixed Number

Improper fraction: $\dfrac{71}{16}$

Divide numerator by denominator:

Denominator in mixed number $\longrightarrow$ $16\overline{\smash{)}71}$

Whole number in mixed number

-64

7

Numerator in mixed number

$$\dfrac{71}{16} = 4\dfrac{7}{16}$$

Goodheart-Willcox Publisher

Figure 5-5 Improper fractions are best expressed as mixed numbers.

This final sum is the numerator in the improper fraction.

Multiplying Fractions

To multiply fractions:
1. Convert any mixed numbers or whole numbers to improper fractions.
2. Multiply the values in the numerators to obtain the numerator in the product.
3. Multiply the values in the denominators to obtain the denominator in the product.

Example
Multiply.

$$\dfrac{1}{8} \times 3 = \dfrac{1}{8} \times \dfrac{3}{1} = \dfrac{3}{8}$$

Example
Multiply.

$$\dfrac{1}{8} \times 3\dfrac{1}{2} = \dfrac{1}{8} \times \dfrac{7}{2} = \dfrac{7}{16}$$

Dividing Fractions

In a division equation, the number being divided is the *dividend*. The number after the division sign is the *divisor*. The divisor is the number of parts the dividend is being separated into. The rest of the division expression is the *quotient*. See **Figure 5-6**.

To divide fractions, you convert the division problem into a multiplication problem by converting the divisor to its inverse. An inverse of a fraction is obtained by switching a fraction's numerator and denominator. For example, the inverse of 3/8 is 8/3, and the inverse of 2 (or 2/1) is 1/2.

To divide fractions:
1. Convert any mixed numbers or whole numbers to improper fractions.
2. Convert the division problem to a multiplication problem by (1) changing the division sign to a multiplication sign and (2) replacing the divisor with its inverse.
3. Solve the multiplication problem.

Example
Divide.

$$\dfrac{1}{8} \div 3 = \dfrac{1}{8} \times \dfrac{1}{3} = \dfrac{1}{24}$$

Example
Divide.

$$2\dfrac{1}{8} \div 2 = \dfrac{17}{8} \times \dfrac{1}{2} = \dfrac{17}{16} = 1\dfrac{1}{16}$$

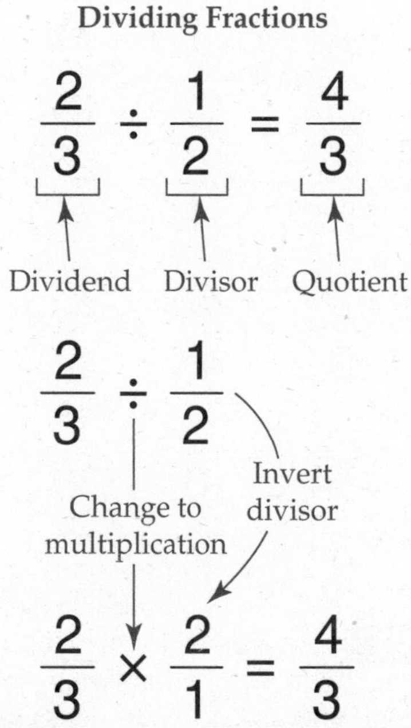

Dividing Fractions

$$\dfrac{2}{3} \div \dfrac{1}{2} = \dfrac{4}{3}$$

Dividend   Divisor   Quotient

$$\dfrac{2}{3} \div \dfrac{1}{2}$$

Change to multiplication

Invert divisor

$$\dfrac{2}{3} \times \dfrac{2}{1} = \dfrac{4}{3}$$

Goodheart-Willcox Publisher

Figure 5-6 To divide by a fraction, multiply by its inverse.

Reducing Fractions to Lowest Terms

As stated earlier, a fraction that has the same number for its numerator and denominator is equal to one. Therefore, when a fraction's numerator and denominator are multiplied or divided by the same number, the value of the fraction remains unchanged. Fractions representing the same value using different numerators and denominators are *equivalent fractions*. See **Figure 5-7**.

A fraction is said to be expressed in lowest terms when its numerator and denominator cannot be divided evenly (that is, without a remainder or fractional portion) by the same whole number.

To reduce a fraction to lowest terms:

1. Determine if there is a single number that both the numerator and denominator can be divided by evenly. If not, the fraction is already expressed in lowest terms.
2. If a number is identified, divide both the numerator and the denominator by this number. The resulting fraction is an equivalent fraction.
3. Repeat step 1 using the equivalent fraction. Continue the process until you cannot evenly divide the numerator and denominator by a single number.

Example

Reduce 4/8 to its lowest terms.

First, both 4 and 8 can be divided evenly by 2. Therefore:

$$\frac{4}{8} = \frac{4 \div 2}{4 \div 8} = \frac{2}{4}$$

Thus, 4/8 and 2/4 are equivalent fractions. Repeating the process, both 2 and 4 can be divided evenly by 2. Therefore:

$$\frac{2}{4} = \frac{2 \div 2}{4 \div 2} = \frac{1}{2}$$

Since 1 and 2 cannot be divided by a number, the fraction is expressed in its lowest terms.

Adding Fractions

To add fractions, the fractions being added must have the same common denominator. When adding fractions with a common denominator, add the numerators of the fractions and use the common denominator in the sum.

Example

Add 3/16, 1/16, and 15/16.

Since the fractions have the same denominator, add the numerators:

$$\frac{3}{16} + \frac{1}{16} + \frac{15}{16} = \frac{19}{16}$$

This improper fraction can be converted to a mixed number:

$$\frac{19}{16} = 1\frac{3}{16}$$

Example

Add 3/8 and 3/4.

The fractions have different denominators, so 3/4 is replaced with its equivalent fraction 6/8. Now that the denominators are the same, the fractions can be added:

$$\frac{3}{8} + \frac{6}{8} = \frac{9}{8} = 1\frac{1}{8}$$

Equivalent Fractions

$$\frac{5}{8} \times 1 = \frac{5}{8}$$

$$\frac{5}{8} \times \frac{2}{2} = \frac{10}{16}$$

$$\frac{5}{8} \times \frac{4}{4} = \frac{20}{32}$$

$$\frac{5}{8} = \frac{10}{16} = \frac{20}{32}$$

Goodheart-Willcox Publisher

Figure 5-7 Equivalent fractions can be determined by multiplying a fraction by one, with one expressed as a fraction with equal numerator and denominator. Every fraction has an infinite number of equivalent fractions.

Adding mixed numbers

Example

Find the sum of 3 1/8, 8 1/4, 5 1/16, 22 1/2.

Add the integers and fractions independently. Whole numbers:

$$3 + 8 + 5 + 22 = 38$$

Convert all of the fractional components to have a common denominator (16) and then add:

$$\frac{2}{16} + \frac{4}{16} + \frac{1}{16} + \frac{8}{16} = \frac{15}{16}$$

Add the whole number sum to the fraction sum: 38 15/16

Subtracting Fractions

Subtracting fractions is similar to adding fractions. In order to subtract fractions, the fractions must have the same common denominator. Once the fractions have the same denominator, simply subtract the numerators and keep the common denominator in the difference.

Example

Subtract 1/8 from 5/8.

Since both fractions have the same denominator, simply subtract 1 from 5 to find the numerator and keep the common denominator (8):

$$\frac{5}{8} - \frac{1}{8} = \frac{4}{8}$$
$$= \frac{1}{2} \text{ (reduced)}$$

Example

Subtract 3/8 from 2/3

When fractions have different denominators, you must first reduce them to fractions having a common denominator. The lowest common denominator of the two fractions is 24. Convert to equivalent fractions:

$$\frac{2}{3} \times \frac{8}{8} = \frac{16}{24}$$
$$\frac{3}{8} \times \frac{3}{3} = \frac{9}{24}$$

Now that the fractions have a common denominator, subtract the fractions by subtracting the numerators:

$$\frac{16}{24} - \frac{9}{24} = \frac{7}{24}$$

Subtract 4 3/4 from 11 3/16

First find the difference of the fractions. Since 3/16 is smaller than 12/16, you must borrow 16/16 from 11 and add it to 3/16 resulting in 19/16. Convert 3/4 to 12/16 to create common denominators. Subtract the fraction components and then subtract the whole number components:

$$11\tfrac{3}{16} \longrightarrow 10\tfrac{19}{16}$$
$$\underline{-\,4\tfrac{3}{4}} \longrightarrow \underline{-\,4\tfrac{12}{16}}$$
$$6\tfrac{7}{16}$$

Decimal Fractions

Decimal fractions are commonly called simply *decimals*. The decimal system uses increments of ten. Decimal fractions are fractions whose denominators are multiples of ten. If the denominator is 10, the fraction is tenths. If the denominator is 100, the fraction is so many hundredths.

Decimal fractions are often written on a single line with a dot separating digits representing one or more from the decimal fraction. The dot between the whole number and the decimal fraction is the *decimal point*. Every place to the left of the decimal point increases the value of the digit in that place tenfold. That is why the second place to the left of the decimal point is called the tens place and the third place to the left is the hundreds place, etc. Moving to the right of the decimal point, the place values decrease tenfold. A decimal fraction of 5/10 can be written as 0.5. A decimal fraction of 12/1000 can be written as 0.012.

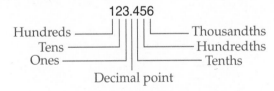

The value of the number in the example above is one hundred twenty-three and four hundred fifty-six thousandths.

Adding and Subtracting Decimals

To add decimals, line up the decimal points in a column, add the numbers, and put the decimal point in the result in the decimal point column.

Example

$$\begin{array}{r} 1.4 \\ 19.2 \\ +\ 31.7 \\ \hline 52.3 \end{array}$$

Subtracting decimals is very similar. Line up the decimal points in the problem and the answer and subtract as usual.

Example

$$\begin{array}{r} 27.74 \\ -\ 2.23 \\ \hline 25.51 \end{array}$$

If there are more decimal places in the number being subtracted than there are in the number it is being subtracted from, zeroes can be added to the right without affecting the value of the number.

Example

$$\begin{array}{r} 5.70 \\ -\ 2.02 \\ \hline 3.68 \end{array}$$ Added zero

Multiplying Decimals

Decimals are multiplied the same as whole numbers, except for the placement of the decimal point in the product (answer). Add the number of decimal places to the right of the decimal point in both the number being multiplied and the number it is being multiplied by. The decimal point should be placed that many places to the left in the product.

Example

$$\begin{array}{r} 12.25 \\ \times\ 3.75 \\ \hline 6125 \\ 85750 \\ 367500 \\ \hline 45.9375 \end{array}$$ Total of four decimal places

Four decimal places in product

Dividing Decimals

Dividing decimals is also much like dividing whole numbers, except for keeping track of the placement of the decimal point. As a reminder, the number being divided is the dividend, the number it is divided by is the divisor, and the answer is the quotient. To start the division problem, move the decimal point in the divisor all the way to the right. Move the decimal point in the dividend the same number of places to the right. Add zeroes to the right of the dividend, if necessary. Divide as you would for whole numbers.

Example

Divide 20 by 0.4.

$$\begin{array}{r} 50. \\ 4.\overline{\smash{)}20.0} \end{array}$$

1. Move decimal point to right of divisor

2. Move same number of places in dividend

Converting Common Fractions to Decimal Fractions and Rounding Off

To change a decimal fraction to a common fraction, divide the numerator by the denominator.

Example

Change 1/4 to a decimal fraction.

$$\begin{array}{r} .25 \\ 4\overline{\smash{)}1.00} \\ -\ 8 \\ \hline 20 \\ -\ 20 \\ \hline 0 \end{array}$$

Converting Decimals to Common Fractions

To convert a decimal to a fraction, drop the decimal point and write the given number as the numerator. The denominator will be 10, 100, 100, or 1 with as many 0s as there were places in the decimal number.

Example

$$.75 = \frac{75}{100} \qquad .625 = \frac{625}{1000}$$

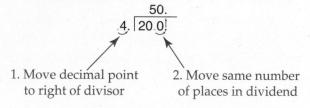

PRO TIP Measured values always include a number and a unit. When you are performing math calculations, always pay attention to the units. In many cases, numbers must have the same unit for the calculation to be true. For example, the following equation is incorrect: 3″ + 1′ = 4″. In order to add 3″ and 1′, you would need to convert so both values have the same units. For example, 3″ + 12″ = 15″.

Name _____ Date _____ Class _____

Activity 5-1
Adding Whole Numbers

Solve the following problems. Show all work in the space provided. Box your answers.

1.
```
   354
 +  81
```

4.
```
   312
   884
   223
 + 507
```

7.
```
   6
   5
   9
   2
   3
 + 8
```

2.
```
   718
 + 239
```

5.
```
 8,416
 2,265
   419
 +  25
```

8.
```
    57
   815
 9,410
     4
    76
 + 843
```

3.
```
 4,663
 + 756
```

6.
```
        89
 1,035,795
       951
 +  10,002
```

9.
```
       803
    24,571
       694
 3,141,592
         7
 +     498
```

10. $4 + 2 + 7 + 1 + 5 =$

11. $87 + 129 =$

12. $56 + 41 + 88 =$

13. $761 + 117 + 523 =$

14. $44 + 762 + 986 + 3 =$

15. $1,119 + 343 + 8,701 + 1,228 =$

16. $7,146 + 45 + 371 + 646 + 932 =$

17. $47,521 + 110,621 + 12 + 965 + 8,461 =$

18. $45,567 + 84,772 + 13,460 + 5,648 + 13 =$

19. Three lengths of aluminum bar stock are needed with the following lengths: 14″, 17″, and 26″. What is the total length of aluminum bar stock needed?

20. G&B Engineering produced 4,651 adjusting bolts in June; 6,975 bolts in July; 5,524 bolts in August; and 8,110 bolts in September. What was the total number of adjusting bolts produced during the four months?

Name _____ Date _____ Class _____

Activity 5-2
Subtracting Whole Numbers

Solve the following problems. Show all work in the space provided. Box your answers.

1.   485
     − 215

4.   9,150
     − 667

7.   13,821
     − 6,624

2.   941
     − 762

5.   7,608
     − 3,349

8.   940,334
     − 512,845

3.   7,759
     − 538

6.   10,945
     − 721

9.   8,457,059
     − 320,372

10.   425 − 56 =

11.   2,401 − 343 =

12.   8,210 minus 4,361 =

13.   Take away 10,562 from 43,851 =

14.   Find the difference between 4,708 and 1,170 =

15.   The difference between 511 and 421 =

16.   912 less 673 =

17.   Subtract 6,183 from 26,845 =

18.   Reduce 71,092 by 1,561 =

19.   Three pieces are cut from a 156″ long bar of mild steel square stock. The pieces are of the following lengths: 60″, 48″, and 36″. What length of square stock remains? Ignore loss of material due to cutting.

20.   A machine shop ordered 11,520 lb of 16 gage steel. The shop used 7,429 lb of the steel to fill a large order. How much 16 gage steel remained after the order was filled?

Name _____ Date _____ Class _____

Activity 5-3
Multiplying Whole Numbers

Solve the following problems. Show all work in the space provided. Box your answers.

1.   52
     $\times$ 13

2.   61
     $\times$ 32

3.   45
     $\times$ 94

4.   117
     $\times$ 25

5.   343
     $\times$ 106

6.   287
     $\times$ 578

7.   3,741
     $\times$ 54

8.   15,187
     $\times$ 14

9.   61,289
     $\times$ 37

10. $89 \times 9 =$

11. $45 \times 18 =$

12. $135 \times 61 =$

13. $74 \times 4{,}573 =$

14. $14 \times 23 \times 35 =$

15. $38 \times 52 \times 24 =$

16. $115 \times 33 \times 79 =$

17. $403 \times 524 \times 16 =$

18. $333 \times 159 \times 284 =$

19. A machine shop has been contracted to produce 156 adapter plates. Each adapter plate requires 14 holes to be drilled. What is the total number of holes that must be drilled for the entire job?

20. Each flow control valve assembly requires 4 gaskets and each pressure valve assembly requires 6 gaskets. A 90 ton hydraulic press needs 2 flow control valves and 2 pressure valves. The pneumatic dead weight press needs 3 flow control valves and 2 pressure valves. If a manufacturer receives an order for 12 hydraulic presses and 15 pneumatic presses, how many gaskets are required to fill the order?

Name _____ Date _____ Class _____

Activity 5-4
Dividing Whole Numbers

Solve the following problems. Show all work in the space provided. Box your answers.

1. $6\overline{)462}$

4. $14\overline{)322}$

7. $24\overline{)8,736}$

2. $7\overline{)343}$

5. $7\overline{)5,038}$

8. $144\overline{)20,836}$

3. $9\overline{)785}$

6. $40\overline{)2,080}$

9. $1,245\overline{)454,425}$

10. $9,856 \div 112 =$

15. 458 divided into 5,038 =

11. $4,532 \div 2,569 =$

16. Divide 23,411,410 by 2,378 =

12. $280,917 / 2,401 =$

17. How many 711's are there in 99,950?

13. 1,123 divided by 58 =

18. Divide 471 into 16,956

14. $314,159 / 2,653 =$

19. A forging process starts with a 7″ long section of 3″ diameter 1018 steel round bar. If the manufacturer has 1200″ of round bar in stock, how many parts can be forged and how much material is left over? Assume no loss of material due to cutting.

20. A part requires 11 evenly spaced holes be drilled in a brass plate 44″ long. If the centers of the first and last holes are located 2″ from the edges of the plate, what is the distance between the centers of the holes? (Hint: Draw a sketch of the plate and holes.)

Name _____ Date _____ Class _____

Activity 5-5
Multiplying Fractions

Solve the following problems. Show all work in the space provided. Convert improper fractions to mixed numbers. Box your answers.

1. $\dfrac{1}{6} \times \dfrac{1}{3} =$

2. $\dfrac{5}{7} \times \dfrac{4}{9} =$

3. $\dfrac{4}{5} \times \dfrac{7}{12} =$

4. $\dfrac{11}{13} \times \dfrac{20}{25} =$

5. $\dfrac{14}{64} \times \dfrac{34}{128} =$

6. $\dfrac{88}{111} \times \dfrac{25}{100} =$

7. $5 \times \dfrac{3}{16} =$

8. $7 \times \dfrac{5}{8} =$

9. $3\dfrac{4}{5} \times 2\dfrac{1}{2} =$

10. $7\dfrac{1}{3} \times \dfrac{24}{96} =$

11. $7\dfrac{5}{8} \times 7\dfrac{5}{8} =$

12. $11 \times \dfrac{11}{12} =$

13. $\dfrac{13}{16} \times \dfrac{5}{8} \times \dfrac{1}{4} =$

14. $2\dfrac{1}{3} \times 6\dfrac{5}{9} \times 3\dfrac{4}{27} =$

15. $8\dfrac{3}{4} \times 3\dfrac{3}{4} \times 1\dfrac{1}{7} =$

16. $\dfrac{25}{120} \times \dfrac{45}{46} \times \dfrac{11}{23} =$

17. $5 \times \dfrac{2}{7} \times \dfrac{2}{5} =$

18. $4 \times \dfrac{7}{16} \times 4\dfrac{7}{16} =$

19. An extrusion machine can produce aluminum extrusions at a rate of 3 1/2" per second. If the extrusion machine runs for 7 1/2 hours per day for 5 3/4 days, how many feet of aluminum extrusion are produced?

20. An order requires 64 lengths of 1 1/4" aluminum square stock to be cut. If the band saw blade creates a 1/8" wide cut and a cubic inch of aluminum weighs 1/10 lb, how much waste (by weight) will be created?

Name _____ Date _____ Class _____

Activity 5-6
Dividing Fractions

Solve the following problems. Show all work in the space provided. Convert improper fractions to mixed numbers. Box your answers.

1. $\dfrac{1}{4} \div \dfrac{5}{12} =$

7. $15 \div \dfrac{6}{7} =$

13. $1\dfrac{1}{2} \div 3\dfrac{5}{8} =$

2. $\dfrac{2}{3} \div \dfrac{9}{11} =$

8. $3\dfrac{8}{9} \div 12 =$

14. $2\dfrac{3}{4} \div 2\dfrac{3}{4} =$

3. $\dfrac{2}{5} \div \dfrac{11}{12} =$

9. $45 \div 3\dfrac{3}{4} =$

15. $8\dfrac{5}{32} \div 8 =$

4. $\dfrac{4}{12} \div \dfrac{4}{7} =$

10. $15 \div \dfrac{1}{15} =$

16. $4\dfrac{7}{16} \div 2\dfrac{1}{2} =$

5. $\dfrac{3}{5} \div \dfrac{5}{3} =$

11. $3\dfrac{5}{9} \div 9\dfrac{3}{5} =$

17. $\dfrac{23}{100} \div \dfrac{33}{60} =$

6. $\dfrac{1}{8} \div \dfrac{1}{16} =$

12. $25 \div \dfrac{1}{3} =$

18. $\dfrac{17}{24} \div \dfrac{45}{365} =$

19. How many 3 1/4″ pieces can be sheared from a length of sheet metal 72″ long?

20. A 103 3/4″ length of 2 1/2″ diameter steel round stock weighs about 167 1/16 lbs. How much would a 12″ section weigh?

Name _____ Date _____ Class _____

Activity 5-7
Adding Fractions

Solve the following problems. Show all work in the space provided. Convert improper fractions to mixed numbers. Box your answers.

1. $\dfrac{1}{16}+\dfrac{5}{16}+\dfrac{7}{16}=$

2. $\dfrac{25}{64}+\dfrac{17}{64}+\dfrac{33}{64}=$

3. $\dfrac{117}{343}+\dfrac{79}{343}+\dfrac{49}{343}=$

4. $\dfrac{3}{8}+\dfrac{7}{16}=$

5. $\dfrac{3}{5}+\dfrac{14}{25}+\dfrac{41}{50}=$

6. $\dfrac{1}{3}+\dfrac{7}{11}+\dfrac{14}{33}=$

7. $\dfrac{5}{6}+\dfrac{2}{3}+\dfrac{1}{4}+\dfrac{1}{12}=$

8. $\dfrac{7}{16}+\dfrac{1}{4}+\dfrac{5}{8}=$

9. $\dfrac{11}{32}+\dfrac{15}{16}=$

10. $\dfrac{9}{32}+\dfrac{47}{64}+\dfrac{3}{16}=$

11. $\dfrac{3}{50}+\dfrac{71}{100}+\dfrac{24}{25}=$

12. $\dfrac{47}{100}+\dfrac{113}{200}+\dfrac{23}{50}=$

13. $5\dfrac{5}{16}+3\dfrac{1}{4}+1\dfrac{5}{8}=$

14. $\dfrac{13}{24}+\dfrac{3}{8}+\dfrac{7}{12}+\dfrac{2}{3}=$

15. $56\dfrac{17}{30}+112\dfrac{1}{3}=$

16. $12\dfrac{1}{2}+4\dfrac{5}{8}+\dfrac{1}{16}+25\dfrac{1}{4}=$

17. $\dfrac{145}{256}+\dfrac{221}{512}=$

18. $2\dfrac{22}{49}+\dfrac{238}{343}=$

19. An internal spur gear has an internal diameter of 6 45/64″ and a total thickness of 1 9/32″. What is the outside diameter?

20. An assembly consists of two 1 5/64″ thick plates, a 6 1/2″ knurled pipe, and a threaded rod inserted through the plates and pipe. Two 3/32″ thick washers and a 3/4″ thick hex nut are used on each end of the assembly. If 1/4″ of threaded rod sticks out past the nut on each side of the assembly, what is the total length?

Name _____ Date _____ Class _____

Activity 5-8
Subtracting Fractions

Solve the following problems. Show all work in the space provided. Convert improper fractions to mixed numbers. Box your answers.

1. $\dfrac{7}{9} - \dfrac{2}{9} =$

2. $\dfrac{11}{15} - \dfrac{8}{15} =$

3. $\dfrac{64}{113} - \dfrac{45}{113} =$

4. $\dfrac{5}{6} - \dfrac{1}{3} =$

5. $\dfrac{23}{32} - \dfrac{11}{16} =$

6. $22\dfrac{13}{24} - 15\dfrac{11}{12} =$

7. Reduce $45\dfrac{3}{64}$ by $7\dfrac{27}{64}$.

8. $1{,}045\dfrac{7}{8}$ minus $456\dfrac{1}{4}$.

9. Find the difference between $1{,}750\dfrac{4}{5}$ and $1{,}212\dfrac{3}{10}$.

10. $\dfrac{117}{128}$ less $\dfrac{34}{64}$.

11. Take $\dfrac{23}{64}$ away from $\dfrac{7}{8}$.

12. Subtract $12\dfrac{5}{8}$ from 22.

13. $8\dfrac{1}{8} - 5\dfrac{5}{8} =$

14. $12\dfrac{7}{32} - \dfrac{3}{16} =$

15. $1\dfrac{5}{12} - \dfrac{2}{3} =$

16. $20\dfrac{3}{16} - 19\dfrac{7}{8} =$

17. $56\dfrac{1}{4} - 16\dfrac{3}{4} =$

18. $81\dfrac{15}{32} - 79 =$

19. A project requires ten 8 3/4" long pieces and six 6 3/16" long pieces of angle iron. If you start with a 137 3/8" length of angle iron and each cut wastes 3/32" of material, how much angle iron is left over?

20. A machine shop starts with 1802 1/2" of sheet metal. The shop uses 360 3/16" of material on Monday, 432" on Tuesday, and 343 3/4" on Thursday. If they have 241 3/16" of material left at the beginning of the day Friday, how much sheet metal was used on Wednesday?

Name _____ Date _____ Class _____

Activity 5-9
Adding and Subtracting Decimals

Solve the following problems. Show all work in the space provided. Box your answers.

1. $0.2 + 1.5 + 6.7 =$

2. $4.07 + 2.14 + 0.55 =$

3. $0.712 + 44.196 + 101.883 =$

4.
$$
\begin{array}{r}
4.18 \\
2.56 \\
7.23 \\
+\ 1.09 \\
\hline
\end{array}
$$

5.
$$
\begin{array}{r}
0.015625 \\
0.0117 \\
0.343 \\
+\ 21.15 \\
\hline
\end{array}
$$

6.
$$
\begin{array}{r}
0.159 \\
49.77 \\
11.23 \\
+\ 0.6 \\
\hline
\end{array}
$$

7.
$$
\begin{array}{r}
23.158 \\
14.78 \\
0.485 \\
8.63 \\
+\ 1.09 \\
\hline
\end{array}
$$

8.
$$
\begin{array}{r}
0.0125 \\
1{,}751.2 \\
45.5 \\
0.67 \\
+\ 8.8 \\
\hline
\end{array}
$$

9.
$$
\begin{array}{r}
1.01101 \\
0.1776 \\
10.66 \\
1.415 \\
-\ 199.9 \\
\hline
\end{array}
$$

10.
$$
\begin{array}{r}
0.6 \\
-\ 0.2 \\
\hline
\end{array}
$$

11.
$$
\begin{array}{r}
14.92 \\
-\ 3.43 \\
\hline
\end{array}
$$

12.
$$
\begin{array}{r}
4.9 \\
-\ 0.117 \\
\hline
\end{array}
$$

13. $4.11 - 1.005 =$

14. $20.0001 - 19.0005 =$

15. $12.7 - 0.1941 =$

16. Subtract 44.999 from 48.753.

17. Reduce 36.125 by 0.0625.

18. Find the difference between 68.741 and 68.525.

19. An assembly consists of four threaded rods screwed into a metal plate. The first rod is 2.2475″ long and each rod is 0.5567″ longer than the next. What is the total length of the four rods?

20. A block of cast iron measures 12.245″ by 8.556″ by 4.975″. What are the dimensions of the block after 0.125″ are machined from each surface? (Hint: Draw the block and imagine milling every surface.)

Name _____ Date _____ Class _____

Activity 5-10
Multiplying Decimals

Solve the following problems. Show all work in the space provided. Box your answers.

1.   0.56
    $\times\, 0.24$

7. $45 \times 0.12 =$

13. $0.4 \times 4.4 \times 44 =$

2.   0.343
    $\times\, 1.17$

8. $16.375 \times 24 =$

14. $100 \times 3.14 \times 0.01 =$

3.   0.641
    $\times\, 196$

9. $0.967 \times 43 =$

15. $0.25 \times 25 \times 2.5 =$

4.   5.008
    $\times\, 0.002$

10. $5,280 \times 0.6213 =$

16. $0.26 \times 7.3 \times 1.9 =$

5.   11.0101
    $\times\, 101.1$

11. $547.12 \times 0.003 =$

17. $10.1 \times 101 \times 1.01 =$

6.   357
    $\times\, 0.44$

12. $8.4521 \times 2.15 =$

18. $1.785 \times 24.25 \times 6.7 =$

19. Your company needs to buy a new lathe. The manufacturer offers two payment options: $21,555.00 in cash or a $4,311.00 down payment with 12 monthly payments of $1,551.96 each. How much money would your company save by paying cash?

20. A turning operation reduces a stainless steel shaft from 51.425 pounds down to 48.137 pounds and can produce 38.71 shafts every hour. At the end of 39.875 hours of production, what is the total weight of steel removed?

Name _____ Date _____ Class _____

Activity 5-11
Dividing Decimals

Solve the following problems. Show all work in the space provided. Box your answers.

1. $115 \div 0.25 =$

2. $1.44 \div 0.002 =$

3. $3.78 \div 15 =$

4. $100.0101 \div 0.0001 =$

5. $0.046 \div 0.008 =$

6. $9.133 \div 1,000 =$

Divide. Round your answer to the nearest hundredth.

7. $2.5/5 =$

8. $3/1.6 =$

9. $3.7/6.4 =$

10. $78,345/100,000 =$

11. $8.8/0.051 =$

12. $91/0.33 =$

Divide. Round your answer to the nearest thousandth.

13. $17.5 \div 2.5 =$

14. $28 \div 10.1 =$

15. $21.3554 \div 1,000 =$

16. $3.43 \div 1.17 =$

17. $2.3 \div 3.2 =$

18. $21.154 \div 101 =$

19. A surface grinder removes 0.006″ of cast iron with each pass. How many passes are required to reduce a cast iron casting from 3.139″ to 2.995″?

20. How many whole 2.377″ squares can be cut from a sheet of metal measuring 35.906″ by 71.875″?

Notes

6 Measurement Basics

Measurement is performed to ensure parts meet print specifications and tolerances. This requires the worker to select the method and instrument to accurately measure the item. All part dimensions have tolerances. The required measurement and tolerance determines the proper measuring instrument to use. For example, "three-place" decimal tolerance (thousandths of an inch) would require that a micrometer that measures in thousands of an inch be used.

This unit introduces measurement systems and measuring devices. Linear measurement, diameter measurement, and angle measurement will be covered. The most common measuring tools will be covered and will include the interpretation, use and care of these tools.

Systems of Measurement

Two systems of measurement are common in the machine trades today. The US Customary system is based on the inch, and the International System of Units (SI) based on the meter.

The *US Customary system* of measuring is the standard format in the United States. In the US Customary system, the most common unit of length measurement is the inch. Larger units of length measurement include the foot (12 inches), yard (3 feet), and mile (5280 feet).

The *SI (International System of Units) system*, or *metric system*, is a system of measurement used in most other countries of the world. Metric measurements are also used in some machining applications in the United States. The standard unit of length in the metric system is the meter.

In the US Customary system, measurements can be expressed in both fractional and decimal units. A type of measure, such as length, may be expressed in several different units. The unit selected is often determined by the magnitude of the measure. For example, a distance could be expressed in inches, yards, or miles.

The metric system is much simpler. Units are expressed as whole numbers or decimals—no fractions are used. Each type of measurement is based on a single unit. For example, length measurement is based on a meter. To accommodate a range of magnitude of measures, prefixes are added to the base unit to create derived units. All of these derived units are created using a base-ten system, and all metric units use the same set of prefixes. See **Figure 6-1**.

Using metric measuring units is relatively easy. A meter (m) can be divided into 100 centimeters (cm) and 1000 millimeters (mm). Millimeters are the most common unit used in machine drawings. Some metric rules include 1/2 mm graduations, Figure 6-4.

In today's global economy, many prints show both metric and US Customary measurements. If a drawing includes only one measurement system, a manufacturer using the other system would need to convert the values. This conversion process requires time (cost) and presents an opportunity to introduce errors (if a conversion is done incorrectly).

Inch and millimeter are the most common units to require conversion in the machine trades. A chart of decimal inch, fractional inch, and metric millimeter equivalents is provided in the Reference Section. You can also convert units using the following equation:

$$1'' = 25.4 \text{ mm}$$

Thus, to convert from inches to millimeters, multiply the value in units by 25.4. To convert from millimeters to inches, divide the value by 25.4.

Example

Convert 5.25″ to millimeters. To convert, multiply by 25.4:

$$5.25'' \times 25.4 = 133.25 \text{ mm}$$

Example

Convert 150 mm to inches. To convert, divide by 25.4:

150 mm ÷ 25.4 = 5.91″ (rounded to two decimal places)

Rules

A *steel rule*, or rule, is one tool used to measure short distances and lengths in a machine shop. The edges of a rule are marked with *graduations*, which are small marks used to determine lengths. A set of graduations typically includes marks with different lengths and some marks labeled with numbers. Graduations are designed to make reading the rule easy.

Most steel rules include graduations in one of three systems: fractional inch, decimal inch, and metric.

Fractional inch rules have graduations in fractions of inches. See **Figure 6-2**. The smallest graduation may be 1/8″, 1/16″, 1/32″, or 1/64″.

Common Metric System Prefixes for Length Measurement				
Prefix	**Symbol**	**Factor**	**Length Unit (Abbreviation)**	**Equivalent Values**
kilo-	k	1000	kilometer (km)	1 km = 1000 m
				0.001 km = 1 m
—	—	1	meter (m)	—
centi-	c	0.01	centimeter (cm)	1 cm = 0.01 m
				100 cm = 1 m
milli-	m	0.001	millimeter (mm)	1 mm = 0.001 m
				1000 mm = 1 m

Goodheart-Willcox Publisher

Figure 6-1   This chart shows common metric prefixes. The millimeter is the most commonly used metric unit on machine drawings.

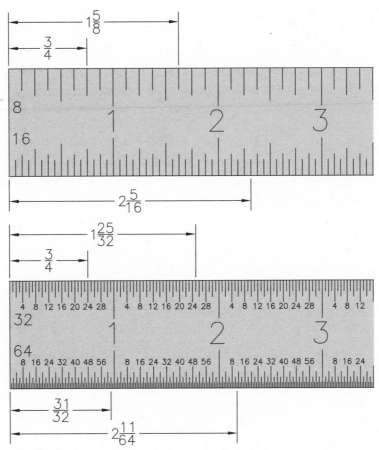

Goodheart-Willcox Publisher

Figure 6-2   Fractional inch rules. The smallest graduation is often marked at the edge of the rule. On the 32nd and 64th rules, the small number marked at each eighth is the number of 32nds (or 64ths) represented by the mark. These small numbers make the rule more easily read.

Decimal inch rules have graduations in tenths or fiftieths of an inch. See **Figure 6-3**. On the fiftieths rule, the distance between each of the four small marks between each tenth mark is 0.02″. Two different sizes of smaller marks are used so it is easy to identify which of the four marks aligns with the measurement. If all four marks were the same length, the rule would be more difficult to read.

On metric rules, graduations of 1 mm and 0.5 mm are common. See **Figure 6-4**. Notice that the one-millimeter graduations are similar to the smaller graduations on the fiftieths of an inch rule: a pattern of one shorter, two longer, and one shorter.

Reading a Rule

To read a rule, align one end of the object to be measured with the end of the rule. Then, read the graduation that aligns with the other end of the object being measured.

PRO TIP   Here are some tips to keep in mind when reading a rule:

- If possible, set the rule on its edge to bring the graduations as close as possible to measured item.
- Be sure the rule coincides with the line to be measured.
- Do not estimate values between graduations. To obtain a more precise measurement, use a more precise rule.
- Sometimes, using the 1″ (or 10 mm) graduation provides a more precise measurement. When using this technique, be sure to subtract 1″ (or 10 mm) from your measurement.

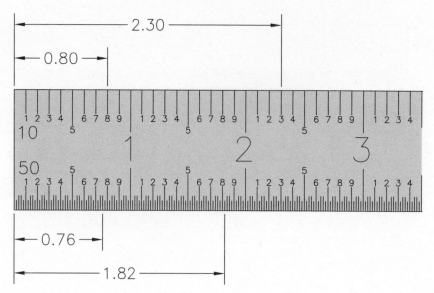

Figure 6-3 Decimal inch rule. The top edge is marked in tenths of an inch. The bottom edge has graduations in fiftieths of an inch. Each mark along the bottom edge of a fiftieths rule is 0.02″ apart.

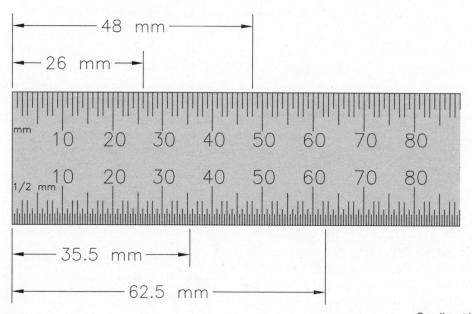

Figure 6-4 Metric rules with 1 mm and 0.5 mm graduations are common in machine shops.

Calipers

Calipers are not used to measure distances directly. Instead, calipers are used to transfer distances to be read more easily with a rule. Outside calipers are used to transfer distance on outside surfaces (surfaces that face away from one another). Inside calipers are used to transfer distances between inside surfaces (surfaces that face one another). Hermaphrodite calipers are used to transfer a distance between an inside face and an outside face (surfaces face the same direction) and for scribing. See **Figure 6-5**.

Firm-joint calipers are adjusted by pulling or pushing the legs to open or close them. Fine adjustment is made by tapping on a leg lightly on a hard surface to close or open the legs.

Outside Caliper

Inside Caliper

Hermaphrodite Caliper

Goodheart-Willcox Publisher

Figure 6-5   This outside caliper is a firm-joint caliper. The inside caliper and hermaphrodite caliper are spring-type calipers.

Spring-type calipers have the legs joined by a strong spring hinge and linked together by a screw and adjusting nut. By turning the adjusting nut, you can open or close the caliper legs in small increments.

Transferring measurements accurately using calipers requires a great deal of skill. See **Figure 6-6**. If not used correctly, it is easy to position a caliper on the work incorrectly, resulting in a faulty measurement.

PRO TIP   When properly positioned, a caliper should fit around or into the workpiece being measured by its own weight. That is, you should not need to "force" a caliper into or over the workpiece. Always make sure the caliper points are at the same vertical location on the workpiece. If one caliper point is higher than the other, the distance between the points is skewed and the resulting measurement will be too large. When measuring a circular feature, always check to make sure that an imaginary line between the caliper points passes through the center of the feature.

Micrometers

A *micrometer* (frequently called a "mike") is a measuring tool that uses a fine pitch screw (40 threads per inch) to take measurements. Micrometers are defined by three main characteristics:

- **System of units.** Micrometers can be inch-based micrometers or metric micrometers, which measure in millimeters.
- **Electronic or manual.** When using a manual micrometer, you read the measured value from the two or scales on the micrometer. Electronic micrometers add a digital readout to the manual scales.
- **Standard or vernier.** Standard micrometers include two scales and measure to an accuracy of 0.001″ (inch-based) or 0.01 mm (metric). A third scale—a vernier scale—is included on some micrometers to increase the accuracy to 0.0001″ or 0.001 mm.

The various parts of an outside micrometer are shown in **Figure 6-7**. The threaded spindle travels inside the sleeve, which is rigid with the frame. Measurements are taken between the anvil and the face (end) of the spindle. The spindle is connected to the thimble.

Outside micrometers measure outer measurements, such as outside diameters or part thicknesses. Inside micrometers and micrometer depth gauges measure interior dimensions and depth dimensions, respectively. See **Figure 6-8**.

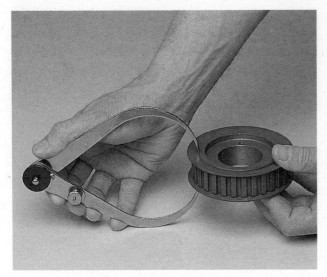

Using an Outside Caliper

Using an Inside Caliper

L. S. Starrett Co.

Figure 6-6 Obtaining accurate measurements with calipers requires much skill and practice. It is easy to use the caliper incorrectly, resulting in an inaccurate measurement.

On an inch-based micrometer, the thimble is divided into 25 divisions, each division represent 1/25th of a turn (.001″). The sleeve has a scale that is one inch long, with ten numbered divisions each equivalent to 0.100″. The 0.100″ numbered divisions are divided into four divisions each equal to .025″. One full turn of the spindle is equal to .025″.

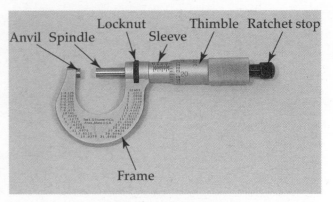

Parts of a Micrometer

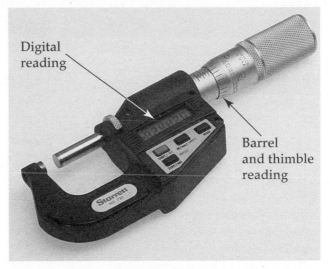

Digital Micrometer

Goodheart-Willcox Publisher

Figure 6-7 The basic parts of a micrometer. Digital micrometers include both digital and manual scales. These micrometers are outside micrometers.

Reading an Inch-Based Micrometer

To read a standard micrometer, you first take a reading from the scale on the sleeve. On an inch-based micrometer, the graduations on the sleeve are in increments of 0.025″. Then, you take a reading from the scale on the thimble. Finally, you add the two readings together. See **Figure 6-9**.

For greater accuracy, some micrometers also include a vernier scale on the sleeve. The vernier scale provides an accuracy of ten-thousandths of an inch (0.0001″). Read a vernier micrometer the same way as a standard micrometer, and then read the vernier scale to determine the ten-thousandths of an inch value. See **Figure 6-10**.

Inside Micrometer

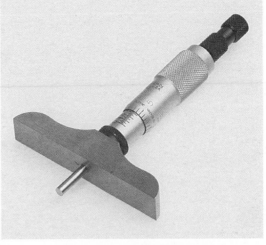

Micrometer Depth Gauge

Goodheart-Willcox Publisher

Figure 6-8 Reading an inside micrometer and a micrometer depth gauge is similar to reading an outside micrometer.

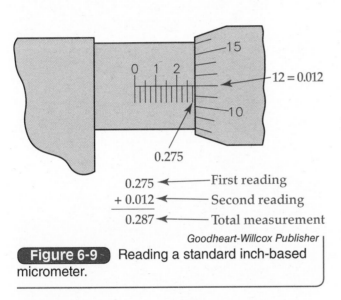

0.275 ← First reading
+ 0.012 ← Second reading
0.287 ← Total measurement

Goodheart-Willcox Publisher

Figure 6-9 Reading a standard inch-based micrometer.

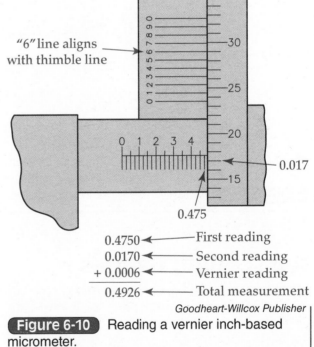

0.4750 ← First reading
0.0170 ← Second reading
+ 0.0006 ← Vernier reading
0.4926 ← Total measurement

Goodheart-Willcox Publisher

Figure 6-10 Reading a vernier inch-based micrometer.

Reading a Metric Micrometer

Reading a metric micrometer is similar to reading an inch-based micrometer. The only differences are the scales. On a metric micrometer, the graduations on the sleeve are in increments of 0.5 mm and the graduations on the thimble are in increments of 0.01 mm. See **Figure 6-11**.

Metric micrometers with vernier scales increase the accuracy to 0.002 mm. The vernier scale is read the same way as the vernier scale on the inch-based micrometer is read. See **Figure 6-12**.

PRO TIP Micrometers are precision measuring instruments and must be handled carefully. Always store a micrometer in its case or box when not in use. Keep micrometers clean and lightly oiled. Follow the manufacturer's directions for calibrating or "zeroing" a micrometer.

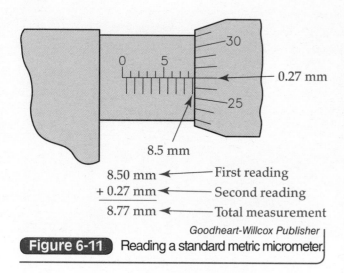

8.5 mm

8.50 mm ← First reading
+ 0.27 mm ← Second reading
8.77 mm ← Total measurement

Goodheart-Willcox Publisher

Figure 6-11 Reading a standard metric micrometer.

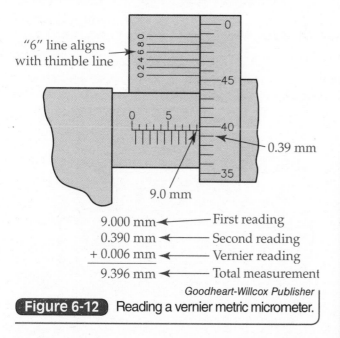

"6" line aligns with thimble line

9.0 mm

9.000 mm ← First reading
0.390 mm ← Second reading
+ 0.006 mm ← Vernier reading
9.396 mm ← Total measurement

Goodheart-Willcox Publisher

Figure 6-12 Reading a vernier metric micrometer.

Telescoping Gauges

A *telescoping gauge* is another type of measurement transferring tool, similar to calipers. Telescoping gauges are used primarily for measuring the diameter of holes or the width of slots. They handle a size range from 5/16" to 6". See **Figure 6-13**.

Telescoping gauges are T-shaped tools in which the shaft of the *T* is used as a handle and the cross-arm is used for measuring. The cross-arms telescope into each other and are pushed out by a light spring.

To use the gauge, the arms are compressed, positioned in the area to be measured, and then allowed to expand. A twist of the locknut on the

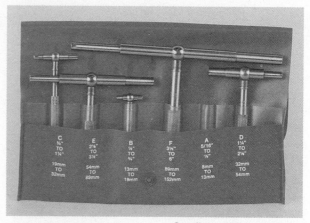

Goodheart-Willcox Publisher

Figure 6-13 This set of telescoping gauges can measure a wide range of hole and slot widths.

end of the handle locks the arms. The gauge is then removed and the distance across the arms are measured with a micrometer. See **Figure 6-14**.

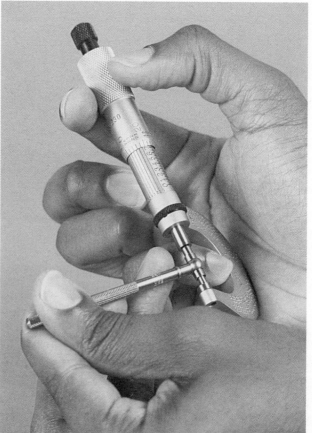

Goodheart-Willcox Publisher

Figure 6-14 An inside micrometer is used to measure a telescoping gauge.

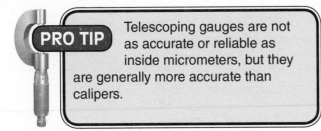

PRO TIP Telescoping gauges are not as accurate or reliable as inside micrometers, but they are generally more accurate than calipers.

Goodheart-Willcox Publisher

Figure 6-15 A set of small hole gauges.

Small Hole Gauges

Small holes, slots, grooves and recesses that are too small for measuring with inside calipers and telescoping gauges can be measured using small hole gauges. Small hole gauges are ordinarily sold in sets of four and cover a measuring range of from 1/8″ to 1/2″ distance. See **Figure 6-15**.

To use a small hole gauge, adjust the knurled knob until the ball at the end of the gauge is touching the sides of the hole. Then, remove the small hole gauge and measure it using an inside micrometer.

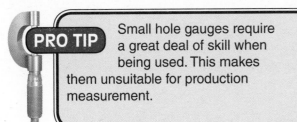

PRO TIP Small hole gauges require a great deal of skill when being used. This makes them unsuitable for production measurement.

Notes

Name _____ Date _____ Class _____

Activity 6-1
Converting Measurements

Convert the following measurements. Round inch values to the nearest thousandth (0.001) and millimeter values to the nearest tenth (0.1).

1. 2.856″ = _____ mm

2. 0.284″ = _____ mm

3. 3.125″ = _____ mm

4. 1.200″ = _____ mm

5. 8.298″ = _____ mm

6. 124 mm = _____ in.

7. 37.5 mm = _____ in.

8. 236 mm = _____ in.

9. 89.5 mm = _____ in.

10. 3 mm = _____ in.

Name _____   Date _____ Class _____

Activity 6-2
Reading Fractional Inch Rules

Write the value identified on the rules.

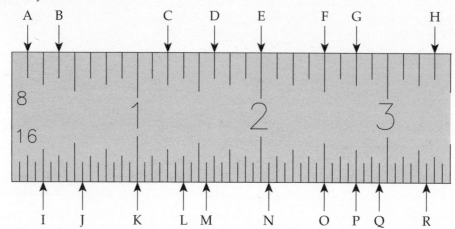

A. _____	F. _____	K. _____	P. _____
B. _____	G. _____	L. _____	Q. _____
C. _____	H. _____	M. _____	R. _____
D. _____	I. _____	N. _____	
E. _____	J. _____	O. _____	

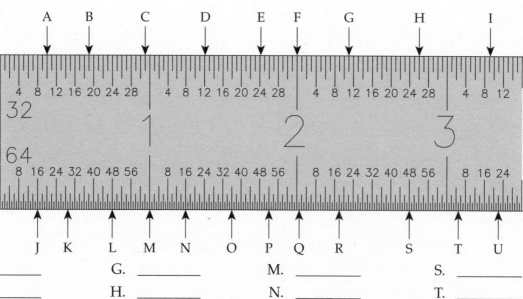

A. _____	G. _____	M. _____	S. _____
B. _____	H. _____	N. _____	T. _____
C. _____	I. _____	O. _____	U. _____
D. _____	J. _____	P. _____	
E. _____	K. _____	Q. _____	
F. _____	L. _____	R. _____	

Name _____ Date _____ Class _____

Activity 6-3
Reading Decimal Inch and Metric Rules

Write the value identified on the rules.

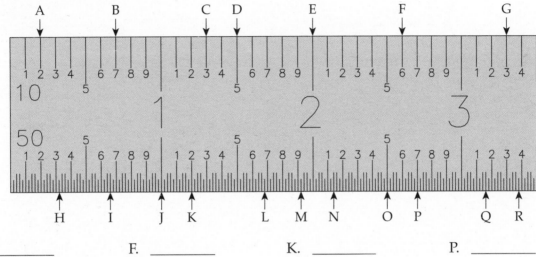

A. _____   F. _____   K. _____   P. _____

B. _____   G. _____   L. _____   Q. _____

C. _____   H. _____   M. _____   R. _____

D. _____   I. _____   N. _____

E. _____   J. _____   O. _____

A. _____   G. _____   M. _____   S. _____

B. _____   H. _____   N. _____   T. _____

C. _____   I. _____   O. _____   U. _____

D. _____   J. _____   P. _____   V. _____

E. _____   K. _____   Q. _____   W. _____

F. _____   L. _____   R. _____   X. _____

Name _____ Date _____ Class _____

Activity 6-4
Reading Standard Inch-Based Micrometers
Write the value displayed on the micrometer.

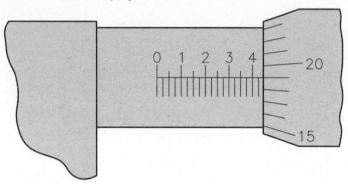

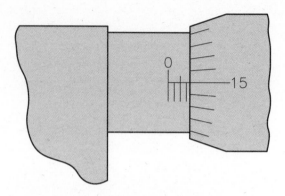

1. _____

2. _____

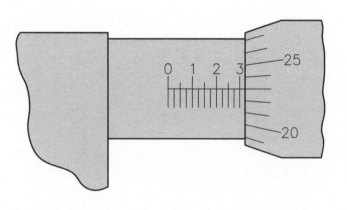

3. _____

4. _____

Name _____ Date _____ Class _____

Activity 6-5
Reading Inch-Based Vernier Micrometers

Write the value displayed on the micrometer.

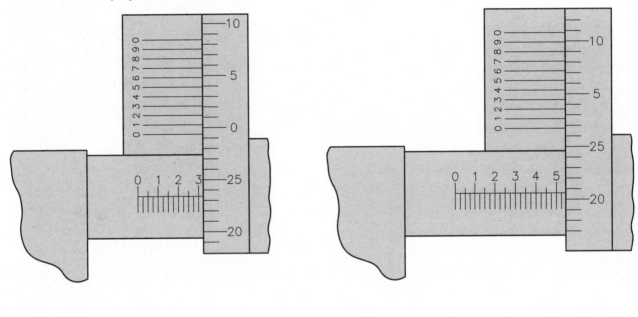

1._____   2. _____

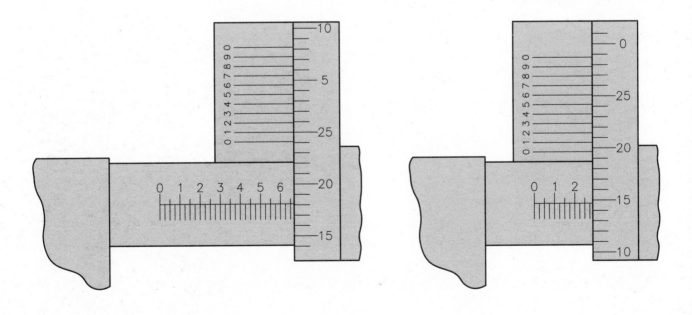

3._____   4. _____

Name _____ Date _____ Class _____

Activity 6-6
Reading Standard Metric Micrometers

Write the value displayed on the micrometer.

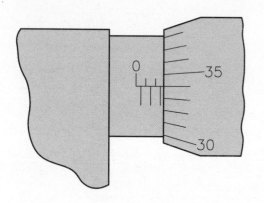

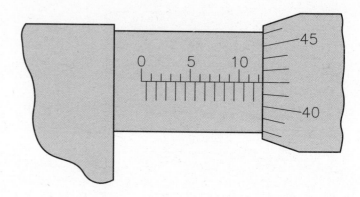

1._____

2._____

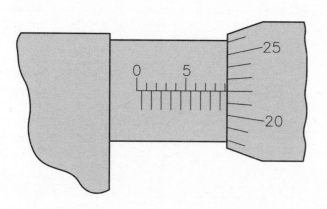

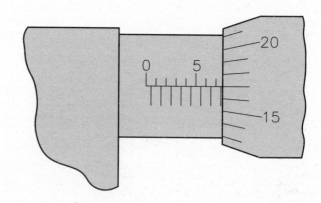

3._____

4._____

Name _____ Date _____ Class _____

Activity 6-7
Reading Metric Vernier Micrometers

Write the value displayed on the micrometer.

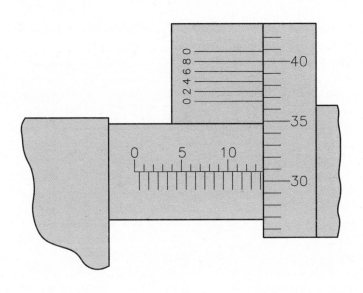

1._____

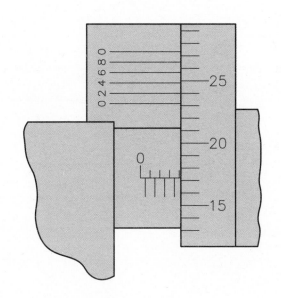

2. _____

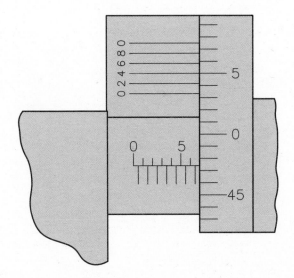

3._____

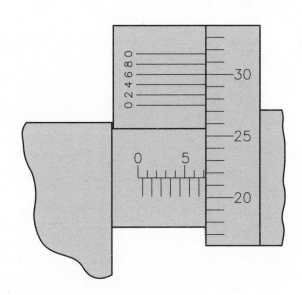

4. _____

Notes

Dimensions and Tolerances

Key Terms

angular dimensioning	direct tolerancing	rectangular coordinate dimensioning
basic size	dual dimensioning	
bilateral tolerance	fractional dimension	reference dimension
datum	limits	specified tolerance
decimal inch	metric dimensioning	tabular dimensioning
decimal inch dimension	nominal size	tolerance
degree	plus and minus tolerancing	unilateral tolerance
dimension	polar coordinate dimensioning	unspecified tolerance

The US customary system is the measurement system used for engineering drawings in the United States. The linear measurement used for length, width, and height is the inch. The measurement used for angles is the degree. Unit 7 explains how this system applies to dimensions, tolerances, and limits. This unit also explains the metric system.

Dimensions

A *dimension* is a definite value of measurement for distance between two given points of an object. Dimensions describe the size and shape of a part, as well as the location of its features. Typical dimensions include length, width, height, angle, radius, and diameter. Dimensions on a drawing can be fractional, decimal, metric, and angular values.

Decimal Inch Dimensions

Decimal inch dimensions are measurements based on decimal inches. *Decimal inch* is a fraction, or decimal equivalent, of an inch. A decimal inch is expressed in bases of ten, such as .1 = 1/10 (tenths), .01 = 1/100 (hundredths), .001 = 1/1000 (thousandths), and so forth. See **Figure 7-1**. Dimensions expressed with more decimal places are more precise than dimensions with fewer decimal places. A part requiring high precision machining will have decimal dimensions expressed in thousandths or ten-thousandths of an inch. See **Figure 7-2**. The decimal inch is the standard unit of linear measurement in the US.

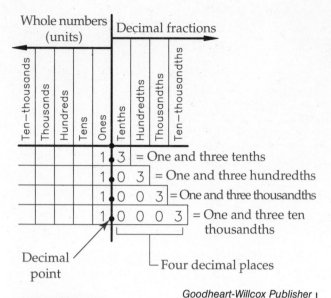

Goodheart-Willcox Publisher

Figure 7-1 Chart showing the placement of whole numbers and decimal fractions in relationship to the decimal point.

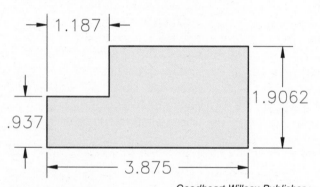

Goodheart-Willcox Publisher

Figure 7-2 Decimal inch dimensioning provides precise measurements in decimal inches.

Fractional Dimensions

Fractional dimensions are measurements based on fractional numbers. The ASME Y14.5-2009, *Dimensioning and Tolerancing* standard does not recommend fractional dimensioning except for specifying nominal sizes. *Nominal size* is a general size or stock size used for the identification of a part. It may not be the actual size of the part. For example, a 3/4-10 UNC bolt has a 3/4″ nominal size diameter. However, the actual diameter size may vary from .7288 to .750.

PRO TIP Certain features on a part may not require high precision measurements, such as a drilled hole for a nominal size fastener. Using a precise machining operation, such as milling, will increase manufacturing costs. Therefore, drilling a standard size hole is preferred.

Fractions are common on welding and casting drawings. Common fractional values found on a drawing include 64ths, 32nds, 8ths, 4ths, and halves. **Figure 7-3** shows typical fractional dimensions.

Metric Dimensions

The International System of Units (SI) is the system of measurement used by most international countries. *Metric dimensioning* uses SI units for its measurement system. The millimeter (mm) is the linear unit of measurement used for metric dimensioning of engineering drawings. The millimeter is equal to 1/1000 of a meter. In addition, decimals express fractions of a millimeter. See **Figure 7-4**.

Due to the growing global economy, the use of metric dimensioning is expanding in the United States. Some drawings will require converting US customary units to SI units. The following example shows how to convert US

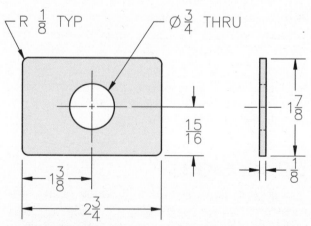

Goodheart-Willcox Publisher

Figure 7-3 Fractional dimensioning provides measurements in fractional inches. Fractions can indicate nominal sizes that require less precision than decimal inches.

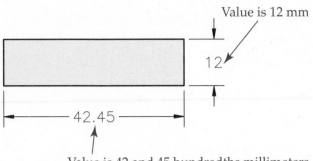

Value is 12 mm

Value is 42 and 45 hundredths millimeters
or 42 $\frac{45}{100}$ mm

Goodheart-Willcox Publisher

Figure 7-4 Metric dimensioning uses millimeters as the base unit of measurement.

customary units to SI units by using the ratio of one inch to 25.4 millimeters.

Example 7-1: Ratio of Inches to Millimeters

Convert .9375 inch to millimeters.

Formulas for Converting Units

1 inch = 25.4 millimeters
Inches = millimeters ÷ 25.4 mm/in
Millimeters = inches × 25.4 mm/in

Solution:

Millimeters = inches × 25.4 mm/in
Millimeters = .9375 in × 25.40 mm/in
Millimeters = 23.81 mm

PRO TIP The abbreviations IN and mm appear on a drawing only when an alternate unit of measurement is used. Only the alternate unit's abbreviation shows on the drawing.

Angular Dimensioning

Angular dimensioning is a measurement of the angle of a line, a surface, or an origin from a given reference point. The reference point can be the vertex of an angle (the point where two lines intersect or meet), or an intersecting line, ray, or plane. The *degree*, represented by the symbol °, is the unit of measurement for an angle based on 360 divisions of a circle's circumference.

Each division, 1/360th of a circle, is a degree. See **Figure 7-5**.

A degree can be broken down into minutes (symbol ′) and seconds (symbol ″). There are 60 minutes to one degree and 60 seconds to one minute.

Circle = 360 degrees (360°)
Degree = 60 minutes (60′)
Minute = 60 seconds (60″)

Figure 7-6A shows an angle measured in degrees, minutes, and seconds.

Decimal units can also represent minutes and seconds. See **Figure 7-6B**. To convert minutes into a decimal, divide the minutes by 60. To convert seconds into decimals, divide the seconds by 3600. The following example shows the procedure.

Example 7-2: Convert Minutes and Seconds to Decimal Degrees

Convert 45° 1′ 30″ to decimal degrees.

Formulas for Converting Units

Degree = 3600 seconds
Minute = 60 seconds
Decimal Degrees° = Degrees° + (minutes ÷ 60)°
        + (seconds ÷ 3600)°

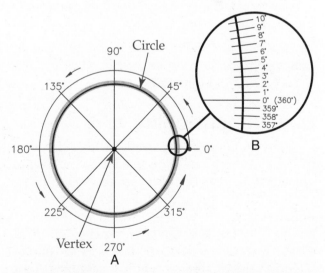

Goodheart-Willcox Publisher

Figure 7-5 The unit of measurement for angles is the degree (°). An angle is measured from a plane, origin, or reference point.
A—Circumference of a circle divided into 360 degrees.
B—An angle is any measurement from 0 degree to 360 degrees.

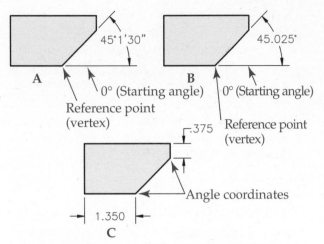

Figure 7-6 Angular dimensions are measured from a plane, origin, or reference point. Examples A, B, and C represent the same angle. A—Angular dimension in degrees, minutes, and seconds. B—Angular dimension in decimal degrees. C—Angular dimension by coordinates.

Solution:

Decimal Degrees° = Degrees° + (minutes ÷ 60)°
  + (seconds ÷ 3600)°
Decimal Degrees° = 45° + (1 ÷ 60)° + (30 ÷ 3600)°
Decimal Degrees° = 45° + .0166° + .0083°
Decimal Degrees° = 45° + .0249°
Decimal Degrees° = 45.0249° (then round-up)
Decimal Degrees° = 45.025°

Coordinates can also locate angles. Linear dimensions locate the coordinates for an angular feature. Refer to **Figure 7-6C**.

Dual Dimensioning

Parts manufactured in global markets may use dual dimensioning on their drawings. *Dual dimensioning* uses both US customary and metric systems for measurement units. When using dual dimensioning, the dominant system appears first, followed by the alternate system. The dominant system is the standard system used by the manufacturer based on their location. Dual dimensions are formatted using brackets [ ] to separate the alternate system from the dominant system, as shown in **Figure 7-7**.

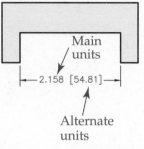

Figure 7-7 Dual dimensioning gives measurements in both US customary and metric values. Brackets indicate the alternate units.

Reference Dimensions

Reference dimensions are dimensions without tolerances that only give basic measurement information. Reference dimensions can identify a part or be a general reference of size. Since reference dimensions are not accurate measurements, they do not have tolerances. For that reason, they are not measurements for machining or inspection purposes. A reference dimension is placed between parentheses ( ) as shown in **Figure 7-8**.

One application for using reference dimensions is for drawings of cast parts. The reference dimensions specify a general size for setting up the molds required for casting a part. However, due to the nature of the process, the size of each cast part will vary from the reference dimensions. In this situation, the reference dimensions are general notes on size and they are not subject to tolerances.

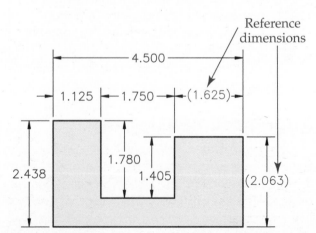

Figure 7-8 A dimension placed between parentheses indicates a reference dimension.

Tabular Dimensions

Tabular dimensioning is a common format for dimensioning parts or assemblies being produced multiple times that have the same shape but different sizes. *Tabular dimensioning* uses letters or numbers referenced to a table instead of dimension lines. A table on the drawing supplies the corresponding values for the lettered parts, as shown in **Figure 7-9**.

Drawings that contain a large amount of repetitious features, such as holes, also benefit from tabular dimensioning. Letters or numbers label the features, and XYZ coordinates specify their locations. Additional information, such as description and quantity, is included as needed. The elimination of excessive extension and dimension lines makes reading dimensions easy. See **Figure 7-10**.

Rectangular Coordinate Dimensions

Rectangular coordinate dimensioning is another method that eliminates excess dimension and extension lines. *Rectangular coordinate dimensioning* uses distances of two or three intersecting planes referenced from a baseline or a datum. A *datum* is an exact point of origin used for dimensioning or locating a feature. A datum can be a line, a surface, or intersecting planes.

A feature such as a hole is located horizontally and vertically from a datum to the feature's center point. Centerlines mark the center point's

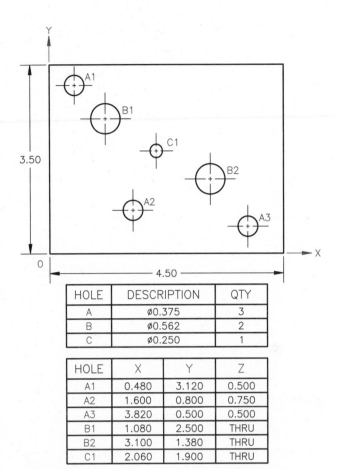

HOLE	DESCRIPTION	QTY
A	⌀0.375	3
B	⌀0.562	2
C	⌀0.250	1

HOLE	X	Y	Z
A1	0.480	3.120	0.500
A2	1.600	0.800	0.750
A3	3.820	0.500	0.500
B1	1.080	2.500	THRU
B2	3.100	1.380	THRU
C1	2.060	1.900	THRU

Goodheart-Willcox Publisher

Figure 7-10 Tabular dimensioning is a beneficial way to dimension repetitive features that would otherwise require many lines. Letters code the holes and a chart correlates them to dimensions and other relevant information.

coordinate directions from the datum. This system helps clarify and avoid difficulty in interpreting the drawing. Refer to **Figure 7-11**. Arrowless dimensioning, or datum dimensioning, are other names for coordinate dimensioning.

Polar Coordinate Dimensions

Polar coordinate dimensioning is a method of locating a point, line, or surface with a linear distance and an angular measurement from a fixed point of two intersecting perpendicular planes. The fixed point of the intersecting planes is the pole. The fixed direction from the pole is the ray. The distance from the pole is the radius. The angle from the ray is the polar angle. See **Figure 7-12**. The common use for polar coordinate dimensioning is for locating features on circular objects.

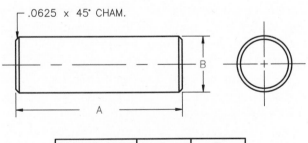

PART NO.	A	B
3204	2	.75
3205	2.5	.875
3206	3	1
3207	3.5	1.125

Goodheart-Willcox Publisher

Figure 7-9 Tabular dimensioning uses letters or numbers instead of numerical values for dimensions. A chart correlates the different measurement values to the coded parts.

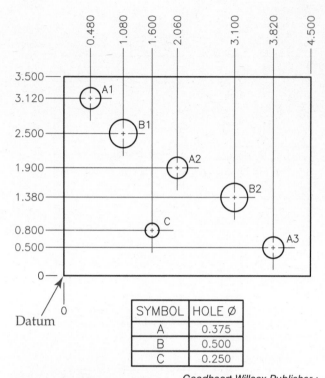

SYMBOL	HOLE Ø
A	0.375
B	0.500
C	0.250

Goodheart-Willcox Publisher

Figure 7-11 Coordinate dimensioning is common on parts that will use computer numerical control (CNC) for machining processes.

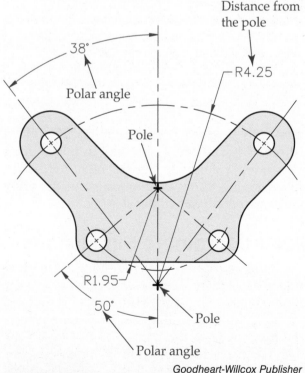

Goodheart-Willcox Publisher

Figure 7-12 Polar coordinates are useful for locating features on circular parts or in a circular position.

Tolerancing

A *tolerance* is an allowable variance of a dimension for a part or feature. Tolerances are extremely important in providing sufficient specifications to allow the proper assembly of units and the interchangeability of parts. Tolerances control the size, form, orientation, location, shape, function, and movement of a part or any of its features. In addition, tolerances compensate for deviations that occur during manufacturing. Since it is impossible to make a part to its exact theoretical size, tolerancing is required to control the quality and precision of a part.

All dimensions have tolerances except for nominal, reference, minimum, and maximum dimensions. All dimensions begin with a basic size. *Basic size* is the specified theoretical value from which limits are applied.

> **PRO TIP** When a dimension is a stock size, the basic size may be the nominal size. Any specified or unspecified tolerances will apply to the given dimension.

Basic size should not be confused with *basic dimensions* as described in Unit 15—Geometric Dimensioning and Tolerancing. A basic dimension is similar, but it differs by relating to the geometric tolerancing of a feature or datum. *Basic size* is used with dimensions, such as a part size, or determining the sizes for hole and shaft systems. For example, a hole with a basic size of 1.500 may require a shaft with a basic size of 1.499, but an allowable maximum variation of .001 on both parts. See **Figure 7-13**.

There are many ways to specify tolerances. Most tolerances are applied to a dimension or a feature. Tolerances can also be included in a note, a table, or a tolerance block. A separate document can also specify tolerances for a specific part, function, or feature.

Direct tolerancing is the method of specifying tolerances directly to dimensions that control location or size. There are two types of direct tolerances—limits and plus and minus tolerances. The following sections explain the direct tolerancing of dimensions. A third type of tolerances, geometric tolerances, are discussed in Unit 15—Geometric Dimensioning and Tolerancing.

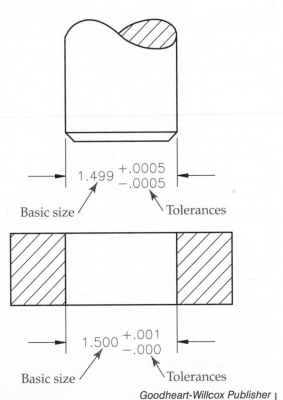

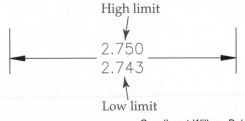

Goodheart-Willcox Publisher

Figure 7-14 Limits are the maximum and minimum acceptable values for a dimension.

Goodheart-Willcox Publisher

Figure 7-13 The basic size of a part is the theoretical size from which tolerances are applied.

Limit Tolerancing

Success in manufacturing relies on the ability to duplicate numerous parts to meet exact specifications for assembly purposes. Specific tolerances are required for parts to mate or join properly. One way to specify a tolerance is by indicating the limits of a dimension. *Limits* are the maximum and minimum sizes allowed for a dimension. Limits are beneficial when the dimension's value cannot exceed a maximum value and cannot be less than a minimum value.

High Limit

The high limit is the maximum value given to a dimension. Under this condition, the high limit is the largest acceptable dimension used for manufacturing. See **Figure 7-14**.

Low Limit

The low limit is the minimum value given to a dimension. Likewise, the low limit is lowest acceptable dimension used for manufacturing.

When specifying upper and lower limits, the high limit appears above the low limit. Refer to

Figure 7-14. If the limits are next to each other, the high limit appears after the low limit. The difference between the high limit and the low limit is the tolerance. The tolerance is the permissible range of size for the dimension.

Example 7-3: Calculating the Tolerance of Limits

Calculate tolerance for **Figure 7-14**.

Formula for calculating work

Tolerance = High Limit − Low Limit

Solution:

Tolerance = 2.750 − 2.743
Tolerance = .007

Single Limit

A single limit tolerance is a variance on a dimension that refers to either an absolute maximum value or an absolute minimum value. The abbreviation MIN (minimum) or MAX (maximum) next to a dimension's value indicates a single limit tolerance. See **Figure 7-15**. With a maximum limit, the dimension cannot be larger than the specified value and the minimum value (unspecified limit) can theoretically be 0.00 (zero). With a minimum limit,

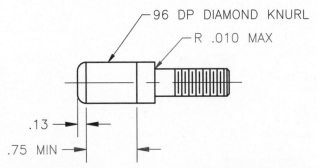

Goodheart-Willcox Publisher

Figure 7-15 A single limit dimension specifies an absolute maximum or minimum value for a dimension.

the dimension cannot be smaller than the specified value and the maximum value (unspecified limit) can theoretically be infinite. However, single limits are only used when other geometry such as dimensions or features control the unspecified limit. Single limits are common for radii, depths of holes, length of threads and knurls, and other features that require a set limit.

Plus and Minus Tolerances

Plus and minus tolerancing is the method of tolerancing that specifies a dimension's variation in a positive and negative direction. The plus symbol (+) indicates the tolerance in the positive direction while a minus symbol (–) represents the tolerance in the negative direction. The plus and minus tolerances are placed next to a basic size. The plus tolerance is above the dimension line while the minus tolerance is below the dimension line. See **Figure 7-16**. The tolerances can be unilateral or bilateral.

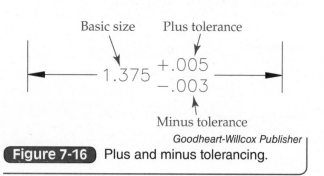

Basic size   Plus tolerance

$$1.375 \begin{array}{c} +.005 \\ -.003 \end{array}$$

Minus tolerance

Goodheart-Willcox Publisher

Figure 7-16 Plus and minus tolerancing.

Unilateral Tolerances

Unilateral tolerance is a variance from a dimension in only one direction, either plus (+) or minus (–). See **Figure 7-17**.

Bilateral Tolerances

Bilateral tolerance is a variance from a dimension in both directions: plus (+) and minus (–) dimensions. The variances can be different values or they can be the same value. See **Figure 7-18**. When the plus and minus variances are the same, the dimension and tolerance will appear on one line with the ± symbol next to them. See **Figure 7-19**. Also, when calculating a plus and minus tolerance with equal value, the tolerance value is doubled because it represents the variance in both directions. For example, if an equal bilateral tolerance is ±.005, then the actual tolerance is .01 (.005 × 2).

Specified Tolerances

A *specified tolerance* is any tolerance directly applied to a dimension. The unit of measurement for a specified tolerance is the same unit used for their related dimension. Tolerances can be specified in fractional, decimal inch, angular, and metric units.

Fractional Tolerances

Fractional dimensions require fractional tolerances. Fractional tolerances appear on a drawing as shown in **Figure 7-20**.

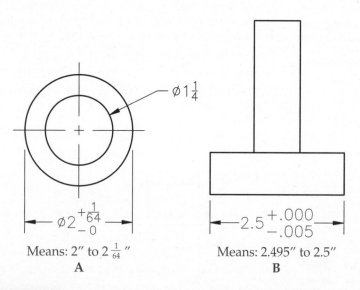

Ø$1\frac{1}{4}$

Ø$2^{+\frac{1}{64}}_{-0}$

Means: 2″ to $2\frac{1}{64}$″

A

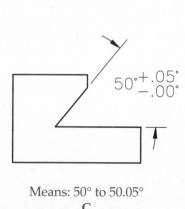

$2.5^{+.000}_{-.005}$

Means: 2.495″ to 2.5″

B

$50°^{+.05°}_{-.00°}$

Means: 50° to 50.05°

C

Goodheart-Willcox Publisher

Figure 7-17 Unilateral tolerances indicate the allowable variation of a basic size from one direction only, either + or –. A—Fractional dimensions with unilateral tolerances. B—Decimal inch dimensions with unilateral tolerances. C—Angular dimensions with unilateral tolerances.

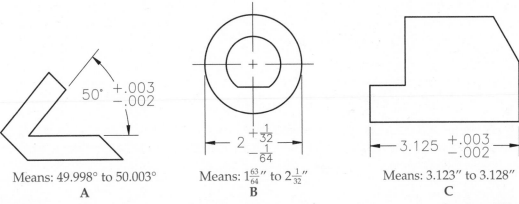

Means: 49.998° to 50.003°
A

Means: $1\frac{63}{64}''$ to $2\frac{1}{32}''$
B

Means: 3.123″ to 3.128″
C

Goodheart-Willcox Publisher

Figure 7-18 Bilateral tolerances indicate variation permitted from basic dimensions in both directions, + and −. A—Angular dimensions with bilateral tolerances. B—Fractional dimensions with bilateral tolerances. C—Decimal inch dimensions with bilateral tolerances.

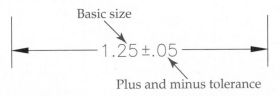

Goodheart-Willcox Publisher

Figure 7-19 Plus and minus tolerancing with an equal tolerance in both directions. The dimensional range is 1.2″–1.3″

Decimal Inch Tolerances

Tolerances for dimensions in decimal inches can be expressed as limits or plus and minus tolerances. See **Figure 7-21**.

Angular Tolerances

Tolerances for angular dimensions are shown in degrees or decimal degrees. Angular dimensions and tolerances are expressed with the same number of decimal places. See **Figure 7-22**.

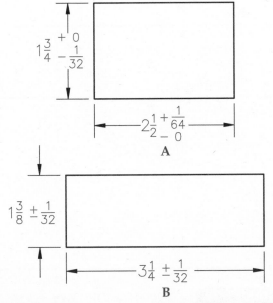

Goodheart-Willcox Publisher

Figure 7-20 Fractional dimensions with specified fractional tolerances. A—Unilateral tolerances. B—Bilateral tolerances.

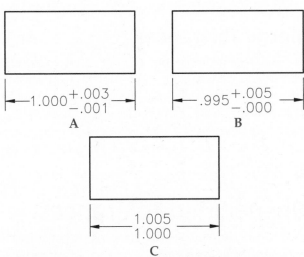

Goodheart-Willcox Publisher

Figure 7-21 Several ways of expressing specified decimal inch tolerances. A—Bilateral tolerance. B—Unilateral tolerance. C—Limits.

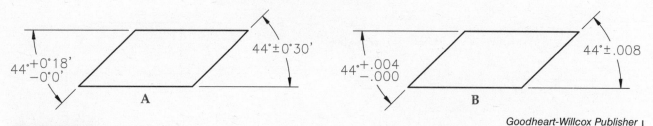

Goodheart-Willcox Publisher

Figure 7-22 Angular dimensions with specified tolerances. A—Unilateral and bilateral, equal tolerances specified in degrees, minutes, and seconds. B—Unilateral and bilateral, equal tolerances specified in decimal degrees.

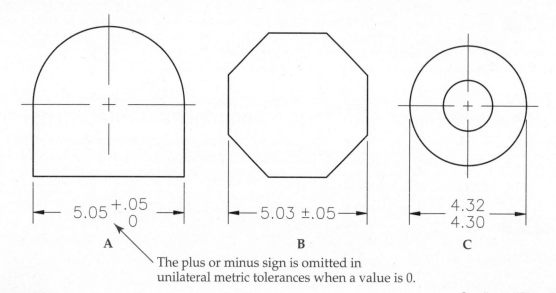

The plus or minus sign is omitted in unilateral metric tolerances when a value is 0.

Goodheart-Willcox Publisher

Figure 7-23 Several ways of expressing specified metric tolerances. A—Unilateral tolerance. B—Equal bilateral tolerance. C—Limits.

Metric Tolerances

Metric dimensions require metric tolerances. The precision of most metric dimensions are two or three decimal places. Higher precision requires four or more decimal places. Also, metric dimensions and tolerances do not use trailing zeros. See **Figure 7-23** for examples of metric tolerances.

Unspecified Tolerances

Not all dimensions will have a direct tolerance. An *unspecified tolerance* is a tolerance that applies to a dimension that does not have a specified tolerance. A general note provides unspecified tolerances for fractional, decimal, angular, and metric dimensions as shown in **Figure 7-24**.

FRACTIONS	$\pm \frac{1}{64}$
2 DIGIT DECIMAL	$\pm.01$
3 DIGIT DECIMAL	$\pm.005$
ANGULAR	$\pm.5°$
METRIC	± 0.5MM

Goodheart-Willcox Publisher

Figure 7-24 Unspecified tolerances appear in a note on a drawing. They apply to all dimensions that do not have any specified tolerances.

Notes

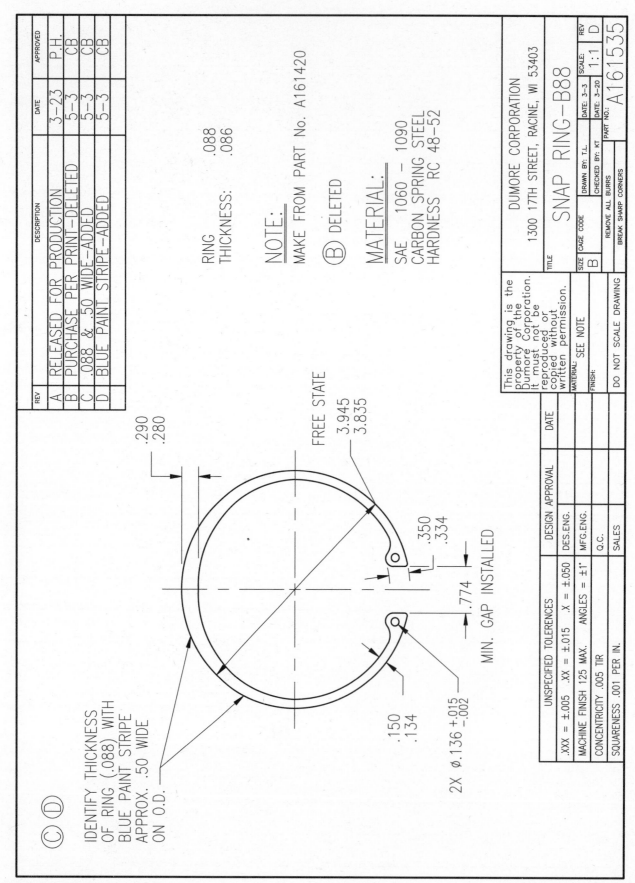

REV	DESCRIPTION	DATE	APPROVED
A	RELEASED FOR PRODUCTION	3-23	P.H.
B	PURCHASE PER PRINT-DELETED	5-3	CB
C	.088 & .50 WIDE-ADDED	5-3	CB
D	BLUE PAINT STRIPE-ADDED	5-3	CB

RING   .088
THICKNESS: .086

NOTE:
MAKE FROM PART No. A161420

Ⓑ DELETED

MATERIAL:
SAE   1060 – 1090
CARBON SPRING STEEL
HARDNESS   RC 48-52

FREE STATE
3.945
3.835

.290
.280

.350
.334

.774

MIN. GAP INSTALLED

.150
.134

2X ⌀.136 +.015/-.002

Ⓒ Ⓓ
IDENTIFY THICKNESS
OF RING (.088) WITH
BLUE PAINT STRIPE
APPROX. .50 WIDE
ON O.D.

DUMORE CORPORATION
1300 17TH STREET, RACINE, WI 53403

TITLE
SNAP RING-B88

		DRAWN BY: T.L.	REV
SIZE	CAGE CODE	CHECKED BY: KT	D
B			

DATE: 3-3   SCALE: 1:1
DATE: 3-20

PART NO.: A161535

REMOVE ALL BURRS
BREAK SHARP CORNERS

MATERIAL: SEE NOTE
FINISH:

DO NOT SCALE DRAWING

	DESIGN APPROVAL	DATE
	DES.ENG.	
	MFG.ENG.	
	Q.C.	
	SALES	

UNSPECIFIED TOLERANCES			
.XXX = ±.005	.XX = ±.015	.X = ±.050	
MACHINE FINISH 125 MAX.		ANGLES = ±1°	
CONCENTRICITY .005 TIR			
SQUARENESS .001 PER IN.			

Activity 7-1   Snap Ring.

Name _____ Date _____ Class _____

Activity 7-1
Snap Ring

Refer to **Activity 7-1**. *Study the drawing and familiarize yourself with the views, dimensions, title block, and notes. Read the questions, refer to the print, and write your answers in the blanks provided.*

1. What is the part number?

2. Name the type of material used to make the part.

3. State what change was made by revision D.

4. What is the scale of the drawing?

5. Who released revision A for production?

6. What size is the drawing?

7. Who created the drawing?

8. What is the ring thickness?

9. What is the required width of the blue paint stripe?

10. What tolerance is specified for angles?

11. Who was responsible for making revision B?

12. What is the name of the part?

13. Dimensions .280, 3.835, .134, and .334 are examples of what type of limits?

14. What tolerance is used for one-place decimal dimensions?

15. What is the maximum value for unspecified surface finishes?

16. What is the minimum allowable size on the .136 diameter dimension?

17. What is the tolerance for the FREE STATE?

18. What is the tolerance for the ring thickness?

19. The ring should be made from what existing part?

20. What is the tolerance for the ring width opposite the gap?

1. _____

2. _____

3. _____

4. _____

5. _____

6. _____

7. _____

8. _____

9. _____

10. _____

11. _____

12. _____

13. _____

14. _____

15. _____

16. _____

17. _____

18. _____

19. _____

20. _____

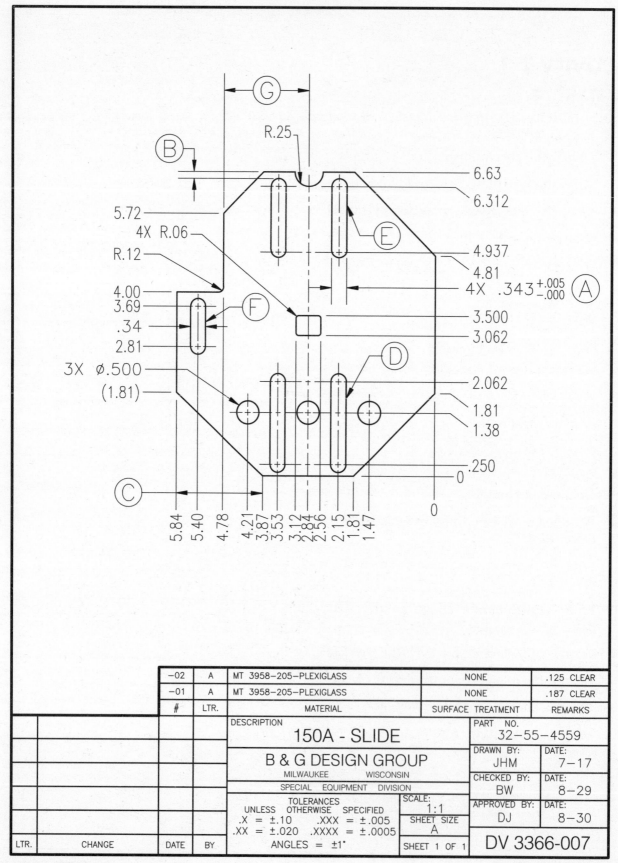

		−02	A	MT 3958−205−PLEXIGLASS	NONE	.125 CLEAR
		−01	A	MT 3958−205−PLEXIGLASS	NONE	.187 CLEAR
		#	LTR.	MATERIAL	SURFACE TREATMENT	REMARKS

		DESCRIPTION **150A - SLIDE**	PART NO. 32−55−4559	
		B & G DESIGN GROUP MILWAUKEE WISCONSIN SPECIAL EQUIPMENT DIVISION	DRAWN BY: JHM	DATE: 7−17
			CHECKED BY: BW	DATE: 8−29
		TOLERANCES UNLESS OTHERWISE SPECIFIED .X = ±.10 .XXX = ±.005 .XX = ±.020 .XXXX = ±.0005	SCALE: 1:1	APPROVED BY: DJ	DATE: 8−30
LTR.	CHANGE	DATE	BY	ANGLES = ±1°	SHEET SIZE A	
					SHEET 1 OF 1	**DV 3366-007**

Name _____ Date _____ Class _____

Activity 7-2
150A-Slide

Refer to **Activity 7-2**. *Study the drawing and familiarize yourself with the views, dimensions, title block, and notes. Read the questions, refer to the print, and write your answers in the blanks provided. When calculating any dimensions or tolerances for a feature with coordinate dimensioning, take into consideration the values are from the datum point.*

1. What is the maximum allowable size for the A dimension?

1. _____

2. List the radii that are dimensioned on the drawing.

2. _____

3. What does the ( ) around the 1.81 dimension signify?

3. _____

4. How much material remains at B?

4. _____

5. What is the maximum allowable size for the Ø.500?

5. _____

6. Calculate the distance for C.

6. _____

7. What is the maximum allowable length of slot D?

7. _____

8. What are the basic size dimensions of the rectangular hole?

8. _____

9. What is the drawing number?

9. _____

10. Is the rectangular hole in the exact center of the part?

10. _____

11. What is the maximum allowable length of slot E?

11. _____

12. Where is the datum origin for the dimensions?

12. _____

13. What is the maximum allowable length of slot F?

13. _____

14. What type of dimensioning was used on the drawing?

14. _____

15. Find the distance for G.

15. _____

16. What is the distance from the bottom of the part to the center of the rectangular hole?

16. _____

17. What material is used to make the 150A-SLIDE?

17. _____

18. What type of tolerance is specified at A?

18. _____

19. Is the center-to-center distance between the three Ø.500 holes the same? What is the distance(s)?

19. _____

20. This part requires what type of surface treatment?

20. _____

Notes

Contours

Learning Objectives

After studying this unit, you will be able to:

✓ Define a circle, arc, and contour.
✓ Know the difference between a fillet and a round.
✓ Identify the size of a fillet, round, or contour.
✓ Explain blend radius and tangent.
✓ Calculate distances using radii dimensions.

Key Terms

arc	contour	round
blend radius	diameter	tangent
circle	fillet	
circumference	radius	

A *contour* is a curved outline of an object that has a shape other than a circle. In order to understand contours, you first need to understand the elements of a circle. A *circle* is an edge that loops 360° around a center point at a fixed distance to form a closed curve. See **Figure 8-1A**. The *diameter* determines the size of a circle. The diameter is the distance of a circle's center axis from outer edge to outer edge. The symbol ∅ means diameter. It is used instead of the word "diameter" before the actual size. It appears as ∅1.00, meaning one inch diameter.

The *radius* is the distance from a circle's or an arc's center point to its outer edge. The radius of a circle is 1/2 the length of its diameter. The symbol R means radius. It is used instead of the word "radius" before the actual size. It would appear as R1.5, meaning 1 1/2" radius. The word for more than one radius is "radii." The *circumference* of a circle is the distance around its closed curve starting and ending at the same point, **Figure 8-1B**.

Arcs

An *arc* is any curved edge with a constant radius that has an angle less than 360°, as shown in **Figure 8-2**. Arcs are dimensioned with leader lines and radii. On a drawing, a leader line identifies an arc while the radius indicates the arc's size. A radius is noted by the radius symbol (R), followed by its size. Most radii have a defined origin at the center point, as shown in **Figure 8-3**. A radius dimensioned with a break line indicates the origin is located outside the drawing area, **Figure 8-4**.

A *blend radius* is a curve that is tangent to other lines or arcs. *Tangent* means the edge of a contour or line touches another curved surface at one point. A blend radius joins curved contours to lines, arcs, or other contours as smoothly as possible to form a surface or closed edge. See **Figure 8-5**.

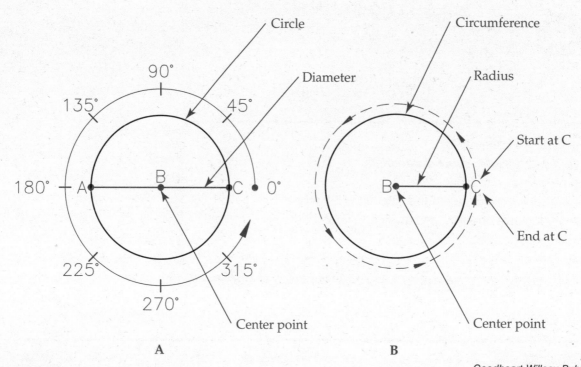

Goodheart-Willcox Publisher

Figure 8-1   Geometry of a circle. A—The diameter is the center axis of a circle, as shown by the line that starts at point A, goes through center point B, and ends at point C. The angles of a circle start at 0° on point C and increase counterclockwise to a full revolution of 360°. B—The radius starts at center point B and ends at point C. The circumference is the length of a circle's edge starting at point C, following the outside of the circle, and ending at point C.

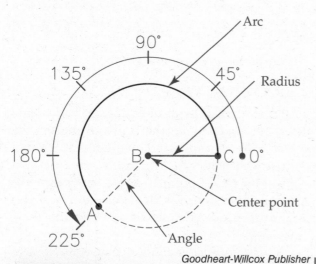

Goodheart-Willcox Publisher

Figure 8-2   An arc is defined by its radius and angle. The radius is the distance between points B and C. The angle determines the arc length, or distance along the arc from A to C.

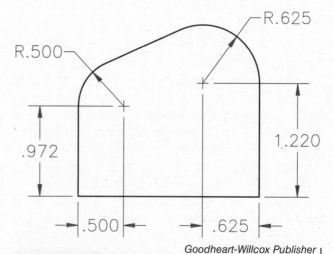

Goodheart-Willcox Publisher

Figure 8-3   The linear dimensions locate the origins of the radii while the radii dimensions define the size of the arcs.

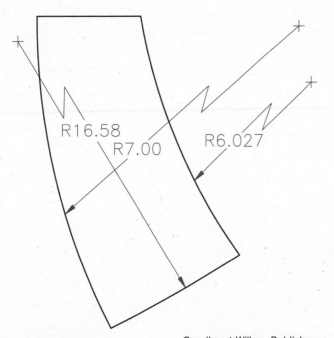

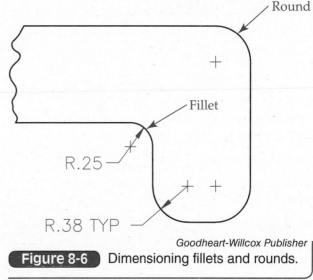

Goodheart-Willcox Publisher

Figure 8-6   Dimensioning fillets and rounds.

Figure 8-6. A general note can also specify a fillet. See **Figure 8-7**.

Rounds

A *round* is a radius applied to the outside edge or corner of a part. Rounding edges or corners improves a part's appearance. Rounding also removes sharp edges that are subject to breakage. Just like a fillet, a radius determines the size. Refer to **Figure 8-6**. A general note can also specify a round. Refer to **Figure 8-7**.

Goodheart-Willcox Publisher

Figure 8-4   The break lines indicate the radii origins are outside of the drawing area.

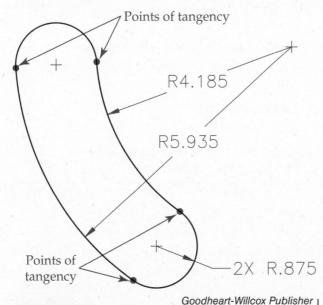

Goodheart-Willcox Publisher

Figure 8-5   With blend radii the contours are tangent to each other, each blending to form a multiple curved edge or surface.

Fillets

A *fillet* is a radius applied to the inside corner of a part. A fillet can increase a part's strength and improve its appearance. A fillet is specified by the size of its radius, as shown in

NOTE:
_____

UNLESS OTHERWISE SPECIFIED ALL ROUNDS AND FILLETS = R 1/8

NOTE:
_____

ALL RADII ARE R.06 UNLESS OTHERWISE NOTED.

Goodheart-Willcox Publisher

Figure 8-7   General notes specify sizes of radii.

Other Contours

The spherical diameter symbol (SØ) is used to indicate a spherical feature such as a ball. The symbol SR precedes the value of a spherical radius. The spherical radius must be tangent with adjacent surfaces and must contain no flats or reversals.

Not all contours are arcs. A contour can be a curved surface without a radius. When a radius does not apply, baselines, datums, points, or coordinates define the curved edges. See **Figure 8-8**.

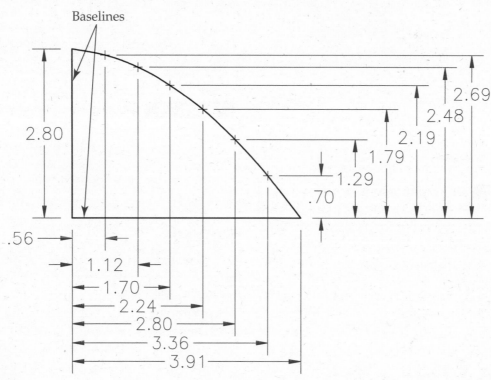

Figure 8-8   The contour of this part does not have a radius. Coordinates measured from the baselines define the contour.

Notes

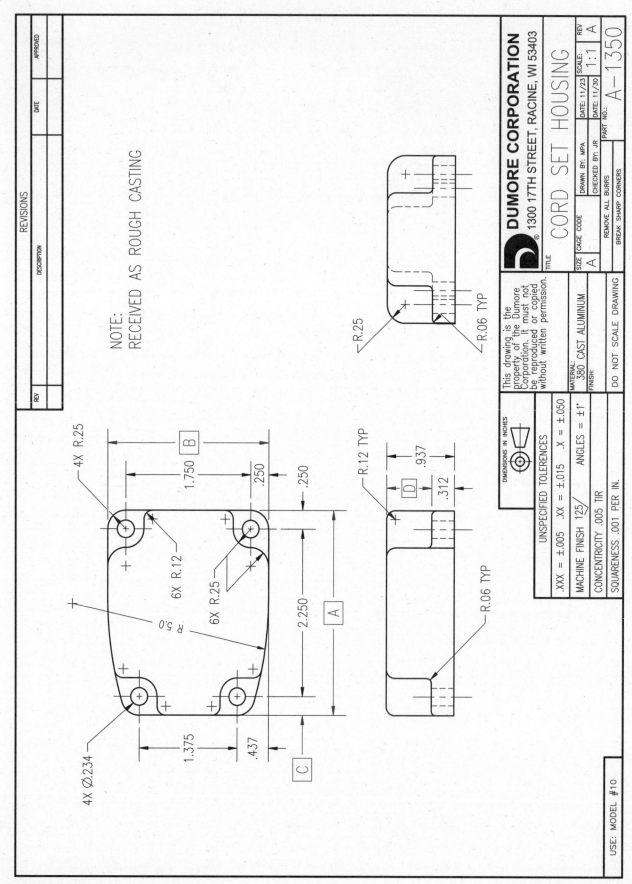

NOTE:
RECEIVED AS ROUGH CASTING

REVISIONS

REV | DESCRIPTION | DATE | APPROVED

4X R.25

B

1.750

.250

.250

6X R.12

6X R.25

R 5.0

A

2.250

1.375

.437

C

4X Ø.234

R.25

R.06 TYP

R.12 TYP

.937

D

.312

R.06 TYP

DIMENSIONS IN INCHES

UNSPECIFIED TOLERENCES

.XXX = ±.005   .XX = ±.015   .X = ±.050

MACHINE FINISH 125/   ANGLES = ±1°

CONCENTRICITY .005 TIR

SQUARENESS .001 PER IN.

This drawing is the property of the Dumore Corporation. It must not be reproduced or copied without written permission.

DUMORE CORPORATION
1300 17TH STREET, RACINE, WI 53403

TITLE
CORD SET HOUSING

CAGE CODE | DRAWN BY: MPA
| CHECKED BY: JR

SIZE
A

DATE: 11/23 | SCALE: | REV
DATE: 11/30 | 1:1 | A

PART NO.: A-1350

REMOVE ALL BURRS
BREAK SHARP CORNERS

MATERIAL:
380 CAST ALUMINUM

FINISH:

DO NOT SCALE DRAWING

USE: MODEL #10

Activity 8-1 Cord Set Housing.

Name _____ Date _____ Class _____

Activity 8-1
Cord Set Housing

Refer to **Activity 8-1**. *Study the drawing and familiarize yourself with the views, dimensions, title block, and notes. Read the questions, refer to the print, and write your answers in the blanks provided.*

1. List the various radii found on the part.

2. Determine dimension A.

3. What type of radius are the R.12 corners?

4. What kind of material is used to produce the part?

5. Determine dimension B.

6. What is the height of the part?

7. What is the largest radius specified for the part?

8. Determine dimension C.

9. How many holes does the casting require?

10. In what condition was the material received?

11. What is the tolerance for two-place decimal dimensions?

12. How many surfaces are shown in the top view?

13. Determine distance D.

14. What is the acceptable maximum height of the part?

15. What scale is the print?

1. _____

2. _____

3. _____

4. _____

5. _____

6. _____

7. _____

8. _____

9. _____

10. _____

11. _____

12. _____

13. _____

14. _____

15. _____

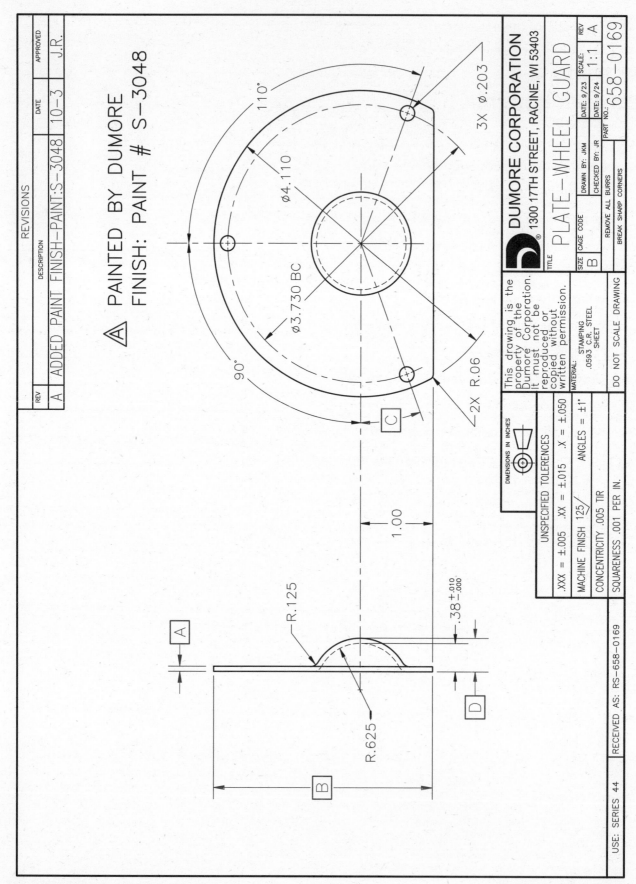

REVISIONS

REV	DESCRIPTION	DATE	APPROVED
A	ADDED PAINT FINISH-PAINT:S-3048	10-3	J.R.

⚠ PAINTED BY DUMORE
FINISH: PAINT # S-3048

110°

⌀4.110

3X ⌀.203

⌀3.730 BC

90°

2X R.06

C

1.00

R.125

A

R.625

.38 +.010 −.000

D

B

This drawing is the property of the Dumore Corporation. It must not be reproduced or copied without written permission.

DIMENSIONS IN INCHES

UNSPECIFIED TOLERENCES
.XXX = ±.005   .XX = ±.015   .X = ±.050

MACHINE FINISH 125/   ANGLES = ±1°

CONCENTRICITY .005 TIR

SQUARENESS .001 PER IN.

MATERIAL:
STAMPING
.0593 C.R. STEEL
SHEET

Ⓓ DUMORE CORPORATION
1300 17TH STREET, RACINE, WI 53403

TITLE PLATE-WHEEL GUARD

| DRAWN BY: JKM | DATE: 9/23 | SCALE: | REV |
| CHECKED BY: JR | DATE: 9/24 | 1:1 | A |

SIZE CAGE CODE
B

REMOVE ALL BURRS
BREAK SHARP CORNERS

PART NO.: 658-0169

DO NOT SCALE DRAWING

USE: SERIES 44   RECEIVED AS: RS-658-0169

Activity 8-2 Plate-Wheel Guard.

Name _____ Date _____ Class _____

Activity 8-2
Plate-Wheel Guard

*Refer to **Activity 8-2**. Study the drawing and familiarize yourself with the views, dimensions, title block, and notes. Read the questions, refer to the print, and write your answers in the blanks provided.*

1. What is the part number?
2. What is revision A?
3. What is dimension A?
4. How many holes are in the part?
5. What type of tolerance is used on the .38 dimension?
6. What is the outside diameter of the part?
7. Name the material specified for the part.
8. What is dimension B?
9. What is radius R.125 called?
10. How many degrees is C?
11. What is the maximum distance for D?
12. What is the maximum acceptable size on the 1.00 dimension?
13. How far are the center points of the holes from the center of the part?
14. What tolerance is required on angles?
15. What process is used for producing the part?
16. List the radii found on the part.
17. Are the three holes equally spaced on a circle?
18. What note is given regarding scale?
19. What information is found on the print regarding corners?
20. What tolerance is required on three-place decimal dimensions?

1. _____
2. _____
3. _____
4. _____
5. _____
6. _____
7. _____
8. _____
9. _____
10. _____
11. _____
12. _____
13. _____
14. _____
15. _____
16. _____
17. _____
18. _____
19. _____
20. _____

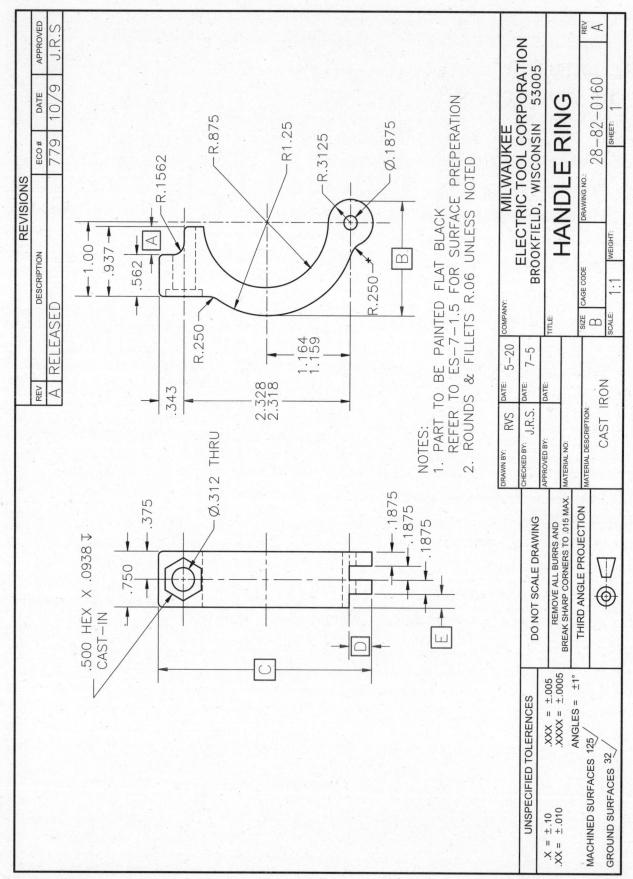

REVISIONS

REV	DESCRIPTION	ECO #	DATE	APPROVED
A	RELEASED	779	10/9	J.R.S

R.1562
R.875
R.1.25
R.3125
Ø.1875
R.250

Ⓐ

Ⓑ

1.00
.937
.562
R.250
.343
1.164 / 1.159
2.328 / 2.318

NOTES:
1. PART TO BE PAINTED FLAT BLACK
   REFER TO ES-7-1.5 FOR SURFACE PREPERATION
2. ROUNDS & FILLETS R.06 UNLESS NOTED

.375
Ø.312 THRU
.750
.500 HEX X .0938 ▼
CAST-IN

.1875
.1875
.1875

Ⓒ
Ⓓ
Ⓔ

MILWAUKEE
ELECTRIC TOOL CORPORATION
BROOKFIELD, WISCONSIN   53005

HANDLE RING

REV A

COMPANY:		
DRAWN BY: RVS	DATE: 5-20	TITLE:
CHECKED BY: J.R.S.	DATE: 7-5	
APPROVED BY:	DATE:	
MATERIAL NO:		
MATERIAL DESCRIPTION:		

CAST IRON

DRAWING NO: 28-82-0160

SHEET: 1

SIZE B   CAGE CODE   SCALE: 1:1   WEIGHT:

UNSPECIFIED TOLERENCES

.X = ±.10
.XX = ±.010

.XXX = ±.005
.XXXX = ±.0005
ANGLES = ±1°

MACHINED SURFACES 125
GROUND SURFACES 32

DO NOT SCALE DRAWING

REMOVE ALL BURRS AND
BREAK SHARP CORNERS TO .015 MAX.

THIRD ANGLE PROJECTION

Activity 8-3 Handle Ring.

Goodheart-Willcox Publisher

Name _____ Date _____ Class _____

Activity 8-3
Handle Ring

Refer to **Activity 8-3**. *Study the drawing and familiarize yourself with the views, dimensions, title block, and notes. Read the questions, refer to the print, and write your answers in the blanks provided.*

1. List the size of the smallest diameter hole.

2. How deep is the hex counterbore?

3. Determine dimension A.

4. What size is the part's largest radius?

5. What is the maximum acceptable distance between the two holes on the part?

6. Determine distance B.

7. What is the part number?

8. List all the radii contained on the part.

9. What is the tolerance on the 2.328/2.318 dimension?

10. What color is the casting painted?

11. Determine the maximum length of dimension C.

12. How thick is the casting?

13. What material is used to make the part?

14. What two views are shown on the print?

15. How far is the .1875 diameter hole from the center of the part?

16. Determine dimension D.

17. What is the maximum acceptable size for the .343 dimension?

18. How is the 1/2″ hex counterbore made?

19. What size are unspecified fillets and rounds?

20. Determine dimension E.

1. _____

2. _____

3. _____

4. _____

5. _____

6. _____

7. _____

8. _____

9. _____

10. _____

11. _____

12. _____

13. _____

14. _____

15. _____

16. _____

17. _____

18. _____

19. _____

20. _____

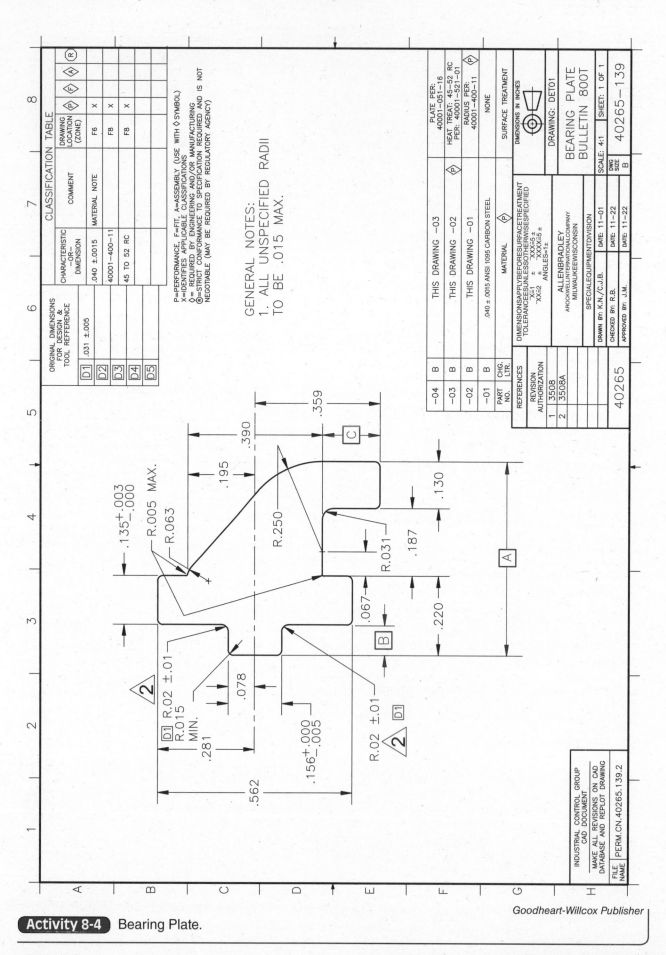

Activity 8-4   Bearing Plate.

Name _____ Date _____ Class _____

Activity 8-4
Bearing Plate

Refer to **Activity 8-4**. *Study the drawing and familiarize yourself with the views, dimensions, title block, and notes. Read the questions, refer to the print, and write your answers in the blanks provided.*

1. Determine dimension A.

2. What was the original radius prior to R.02?

3. What material is used to make the part?

4. What scale is the print?

5. Where is Allen-Bradley located?

6. What is the drawing number?

7. What is the maximum thickness of material allowed to manufacture the part?

8. Determine dimension B.

9. Determine dimension C.

10. What is the size of the smallest radius on the part?

11. What does the diamond symbol with the letter "P" inside represent?

12. What angle projection is this print?

13. What tolerance is used for three-place decimals?

14. What is the size of largest radius shown on the part?

15. What is the maximum dimension for unspecified radii?

16. What specification number relates to the heat treat specifications?

17. What is the overall height of the part?

18. What is the lowest acceptable size for the .156 dimension?

19. What is the file name?

20. List the radii found on the part.

1. _____

2. _____

3. _____

4. _____

5. _____

6. _____

7. _____

8. _____

9. _____

10. _____

11. _____

12. _____

13. _____

14. _____

15. _____

16. _____

17. _____

18. _____

19. _____

20. _____

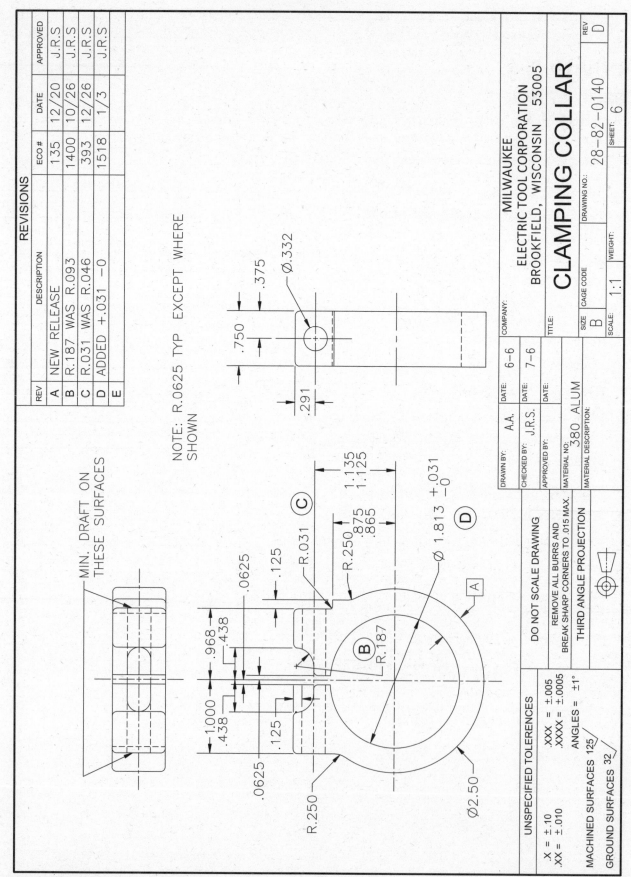

REVISIONS

REV	DESCRIPTION	ECO #	DATE	APPROVED
A	NEW RELEASE	135	12/20	J.R.S
B	R.187 WAS R.093	1400	10/26	J.R.S
C	R.031 WAS R.046	393	12/26	J.R.S
D	ADDED +.031 −0	1518	1/3	J.R.S
E				

NOTE: R.0625 TYP EXCEPT WHERE SHOWN

MIN. DRAFT ON THESE SURFACES

.375
Ø.332
.750
.291

1.135 / 1.125
.875 / .865
R.031 C
R.250
Ø 1.813 +.031 −0
D

.0625
.125
.968
.438
1.000
.438
.125
.0625
R.250
Ø2.50
R.187 B

A

COMPANY:			
DRAWN BY: A.A.	DATE: 6-6		MILWAUKEE
CHECKED BY: J.R.S.	DATE: 7-6		ELECTRIC TOOL CORPORATION
APPROVED BY:	DATE:		BROOKFIELD, WISCONSIN 53005

TITLE: **CLAMPING COLLAR**

MATERIAL NO.: 380 ALUM

MATERIAL DESCRIPTION:

SIZE B	CAGE CODE	DRAWING NO.: 28-82-0140	REV D
SCALE: 1:1		WEIGHT:	SHEET: 6

DO NOT SCALE DRAWING

REMOVE ALL BURRS AND BREAK SHARP CORNERS TO .015 MAX.

THIRD ANGLE PROJECTION

UNSPECIFIED TOLERENCES
.XXX = ±.005
.XXXX = ±.0005
ANGLES = ±1°

.X = ±.10
.XX = ±.010

MACHINED SURFACES 125
GROUND SURFACES 32

Activity 8-5 Clamping Collar.

Name _____ Date _____ Class _____

Activity 8-5
Clamping Collar

*Refer to **Activity 8-5**. Study the drawing and familiarize yourself with the views, dimensions, title block, and notes. Read the questions, refer to the print, and write your answers in the blanks provided.*

1. What material is used to make the part?

2. List all the radii noted for the part.

3. What is the tolerance for two-place decimal dimensions?

4. What size hole goes through both sides of the part?

5. What is the width of the part?

6. What is the center distance between the two holes?

7. What is the size of the largest diameter on the part?

8. What scale is the print?

9. What is the total length of the ⌀.332 hole?

10. What ECO number designated revision B?

11. What is the high limit size of the large diameter hole?

12. Determine dimension A.

13. Which two dimensions give the location of the small diameter hole?

14. What is the width of the slot that splits the clamping collar?

15. What is the drawing number?

16. What distance is the face of the .125 flat to the vertical centerline of the part?

17. What is the tolerance for the .875/.865 dimension?

18. What size are the radii not specified on the print?

19. What specific change happened in revision C?

20. List the three views on the drawing.

1. _____

2. _____

3. _____

4. _____

5. _____

6. _____

7. _____

8. _____

9. _____

10. _____

11. _____

12. _____

13. _____

14. _____

15. _____

16. _____

17. _____

18. _____

19. _____

20. _____

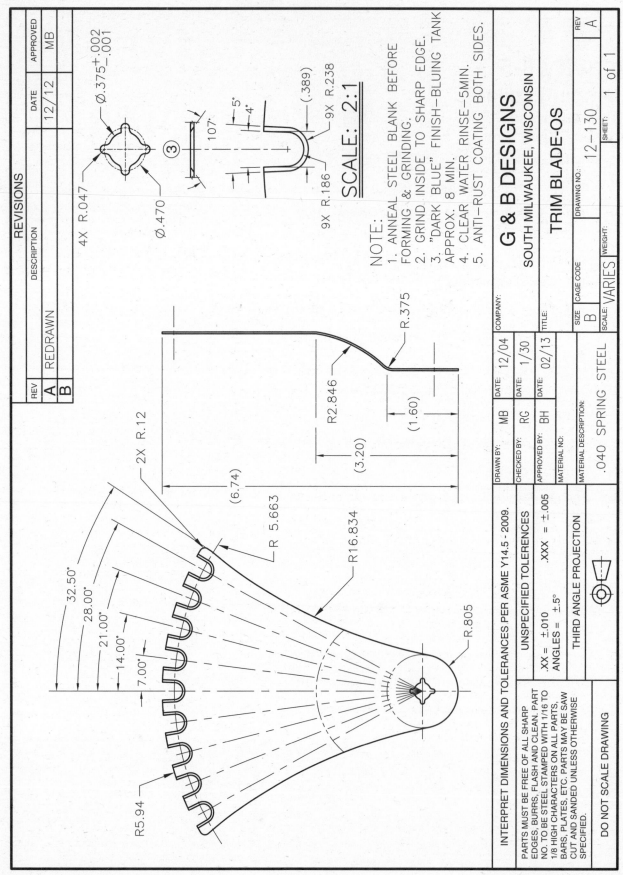

SCALE: 2:1

NOTE:
1. ANNEAL STEEL BLANK BEFORE FORMING & GRINDING.
2. GRIND INSIDE TO SHARP EDGE.
3. "DARK BLUE" FINISH—BLUING TANK APPROX. 8 MIN.
4. CLEAR WATER RINSE—5MIN.
5. ANTI–RUST COATING BOTH SIDES.

REV	DESCRIPTION	DATE	APPROVED
A	REDRAWN	12/12	MB
B			

REVISIONS

Ø.375 +.002 −.001

4X R.047

Ø.470

9X R.238

(.389)

9X R.186

107°

5°

4°

③

2X R.12

R 5.663

R16.834

R2.846

R.375

(6.74)

(3.20)

(1.60)

32.50°
28.00°
21.00°
14.00°
7.00°

R5.94

R.805

G & B DESIGNS
SOUTH MILWAUKEE, WISCONSIN

TRIM BLADE-OS

DRAWN BY: MB	DATE: 12/04	COMPANY:
CHECKED BY: RG	DATE: 1/30	
APPROVED BY: BH	DATE: 02/13	TITLE:
MATERIAL NO:		
MATERIAL DESCRIPTION: .040 SPRING STEEL		

SIZE B	CAGE CODE	DRAWING NO.: 12–130	REV A
SCALE: VARIES	WEIGHT:	SHEET: 1 of 1	

INTERPRET DIMENSIONS AND TOLERANCES PER ASME Y14.5 - 2009.

UNSPECIFIED TOLERENCES	
.XX = ±.010	.XXX = ±.005
ANGLES = ±.5°	

THIRD ANGLE PROJECTION

PARTS MUST BE FREE OF ALL SHARP EDGES, BURRS, FLASH AND CLEAN. PART NO. TO BE STEEL STAMPED WITH 1/16 TO 1/8 HIGH CHARACTERS ON ALL PARTS, BARS, PLATES, ETC. PARTS MAY BE SAW CUT AND SANDED UNLESS OTHERWISE SPECIFIED.

DO NOT SCALE DRAWING

Activity 8-6   Trim Blade-OS.

Name _____ Date _____ Class _____

Activity 8-6
Trim Blade-OS

Refer to **Activity 8-6**. *Study the drawing and familiarize yourself with the views, dimensions, title block, and notes. Read the questions, refer to the print, and write your answers in the blanks provided.*

1. What is the thickness of the trim blade?

2. What is the unspecified tolerance for angles?

3. What is the largest radius listed on the part?

4. At what angle are the cutting edges ground?

5. What is the angle between each tooth?

6. How many teeth are on the blade?

7. List the radii and diameters of the hole.

8. List the radii shown in the right side view.

9. List the radii shown in the front view.

10. What is the scale of the print?

11. What is the approximate depth of each tooth?

12. What is the width at each tooth opening?

13. What process must be done before forming and grinding?

14. How long must the blade be left in the bluing tank?

15. What type of coating is applied to the blade?

1. _____

2. _____

3. _____

4. _____

5. _____

6. _____

7. _____

8. _____

9. _____

10. _____

11. _____

12. _____

13. _____

14. _____

15. _____

Notes

Learning Objectives

After studying this unit, you will be able to:

✓ Identify machining processes for holes including drilling, reaming, countersinking, boring, counterboring, and spotfacing.

✓ Recognize symbols used for specifying holes.

✓ Evaluate specifications for holes.

✓ Determine dimensions for holes.

Key Terms

blind hole	counterdrilling	reamer
bolt circle	countersinking	reaming
boring	drill	spotfacing
counterboring	drilling	through hole

Various machining operations make holes, enlarge holes, finish holes, and prepare surfaces around holes. These operations include drilling, boring, reaming, counterdrilling, countersinking, counterboring, and spotfacing. The ASME Y14.5-2009, *Dimensioning and Tolerancing* standard has various symbols that represent holes on a drawing.

A circle on a drawing can represent a hole, but it also can represent a surface. The size of a hole is defined by its diameter and depth. Basic dimensions include a leader line, a diameter symbol (∅), and a value. THRU placed after the value indicates the hole goes through the part. See **Figure 9-1**.

The depth of a hole is noted by the depth symbol followed by its value. The depth's measurement always follows the diameter's size as shown in **Figure 9-2**. The depth symbol is not required when a drawing clearly shows a hole going through the part.

Multiple holes are specified by placing the number of holes with the "by" symbol (X). The number and symbol are placed before the diameter dimension. For example, nine holes with the same diameter would look like **Figure 9-3**.

Other ways of dimensioning holes include specifying by fractions, abbreviations, machine processes, and drill or tool sizes. See the Reference Section for more information about drill sizes.

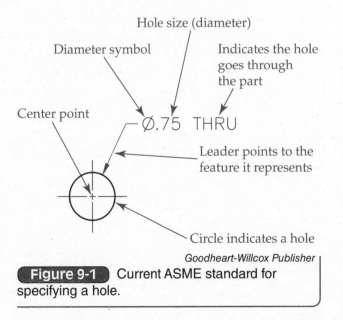

Goodheart-Willcox Publisher

Figure 9-1 Current ASME standard for specifying a hole.

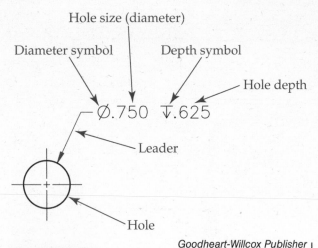

Diameter symbol

Hole size (diameter)

Depth symbol

Hole depth

⌀.750 ▽.625

Leader

Hole

Figure 9-2 Dimension of a hole with a specified depth.

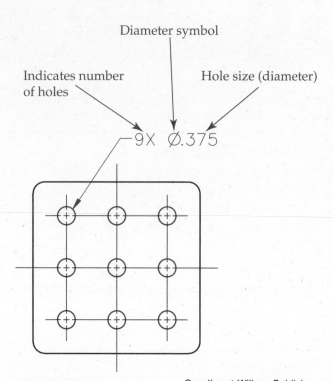

Diameter symbol

Indicates number of holes

Hole size (diameter)

9X ⌀.375

Figure 9-3 Multiple holes of the same size specified once on a drawing.

The ASME Y14.5-2009 standard does not recommend those formats for dimensioning but they are still used. ASME standard recommends using symbols instead of words for specifying dimensions. See **Figure 9-4**. When specific manufacturing processes are required, the ASME recommends placing them as a separate note on the drawing or in a related document.

Drilled Hole

Drilling is the most common machining operation. *Drilling* is the process of cutting a hole in or through a surface with a drill. A *drill* is a cylindrical tool with a sharpened point and edges that cuts a specific size hole. The diameter of the drill determines the size of the hole it makes. The chuck of a rotary machine such as a hand drill,

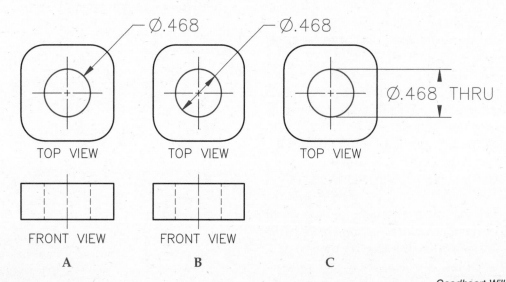

⌀.468

⌀.468

⌀.468 THRU

TOP VIEW   TOP VIEW   TOP VIEW

FRONT VIEW   FRONT VIEW

A   B   C

Figure 9-4 Different ways to dimension the same sized hole. A & B—A front or side view is needed because the top view does not indicate the depth of the hole. C—Only one view is needed because the top view indicates the depth of the hole.

drill press, lathe, or milling machine holds the drill. As the machine spins the chuck, the operator moves the drill toward the part's surface while it rotates and cuts a hole in the material.

There are two classifications of drilled holes—through holes and blind holes. A *through hole* passes completely through the part or material. Through holes appear on a drawing as hidden lines (dashed) that extend to the outside surfaces, as shown in the front view of **Figure 9-5A**. A *blind hole* has a specified depth that does not go completely through the part or material. A drill will produce a conical point at the bottom of the hole, as shown in **Figure 9-5B**. Typical dimensions for drilled holes require information for the quantity, size, and depth of the holes. If no depth is given for a hole, it is assumed to be a through hole.

A circular centerline, known as a *bolt circle* or circle of centers, is used to locate holes in a circular pattern. In specific situations, equal division of centerlines, an equal spacing note, or angular dimensioning specifies the needed information. See **Figure 9-6**. A bolt circle is dimensioned with a leader line, a diameter symbol (∅), a value, and the letters BC.

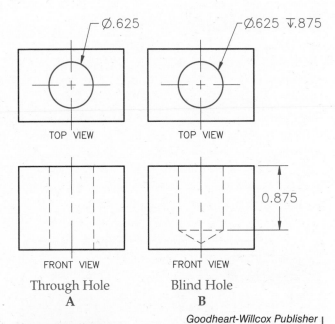

Goodheart-Willcox Publisher

Figure 9-5 Two types of holes. A—A through hole goes completely through a part. B—A blind hole is machined to a specified depth.

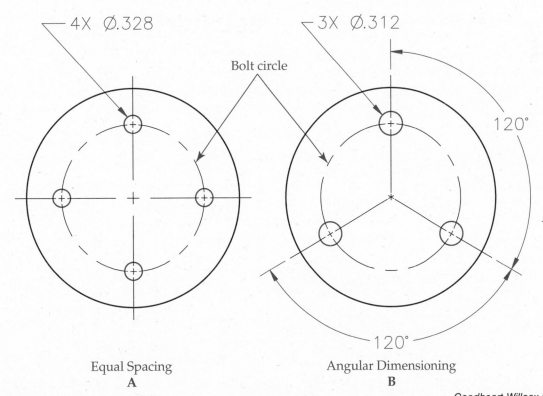

Goodheart-Willcox Publisher

Figure 9-6 Specifying holes in circular patterns. A—The equal division of centerlines indicates equal spacing. B—Angles indicate the spacing along the bolt circle.

Bored Hole

Boring is the operation of enlarging a hole to a close tolerance and fine finish. It produces a straight, round hole more accurately than by drilling. A boring bar and single point tool can produce holes through a wide range of diameters. A bored hole is dimensioned the same as a drilled or reamed hole. The diameter of a bored hole appears in the top view of the part. A tolerance or a specific note may appear after the diameter of the hole to specify the boring process, as shown in **Figure 9-7**.

> **PRO TIP** Extra material around a hole is required to ream an accurate size hole to a fine finish. The basic material allowance is .0156″ (0.3968 mm) for diameters under 1″ (25.4 mm) and .0313″ (0.7938 mm) or more for diameters greater than 1″ (25.4 mm).

Reamed Hole

Reaming is the operation that finishes a hole to a specific size and required finish. Reaming is required for precision fits. Reaming follows drilling or boring to provide a closer tolerance and smoother finish. A *reamer*, a straight or helical multi-fluted rotary cutting tool, is used with a drill press, lathe, or milling machine to remove a small amount of material.

A reamed hole is dimensioned the same as a bored hole. The tolerances and finish specifications designate the machining process. This method allows the machinist to use the proper process that makes the hole to the required specifications. On older drawings, it is common to find a drill and ream size noted for a hole. See **Figure 9-8**.

Counterdrilled Hole

Counterdrilling is a two-step process of drilling a conical shaped hole to a specified depth that allows a fastener's head to sit at or below the part's surface. Counterdrilling does not require a noted angle because the drill's angled point shapes the hole. The typical angle produced by a drill is 120°.

The specifications for a counterdrilled hole appear on two lines. The first line represents the small hole and the second line represents the larger hole. Both lines will have the same information in the following order: the diameter symbol and diameter size, the depth symbol and depth size (if required). See **Figure 9-9**.

Countersunk Hole

Countersinking is the operation of enlarging the entrance of a hole conically to recess the head of a fastener, such as a flat head

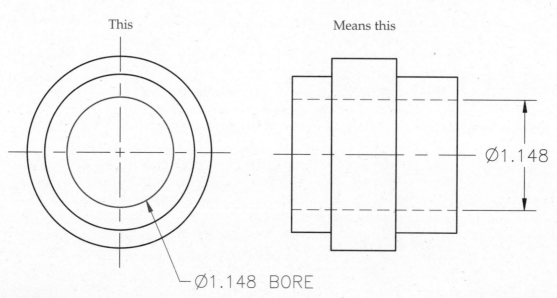

This   Means this

Ø1.148

Ø1.148 BORE

Figure 9-7 Representation of a bored hole.

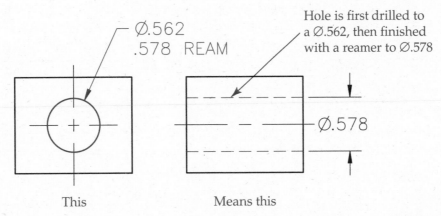

Ø.562
.578 REAM

Hole is first drilled to
a Ø.562, then finished
with a reamer to Ø.578

Ø.578

This   Means this

Figure 9-8   Older drawings included specifications for reamed holes.

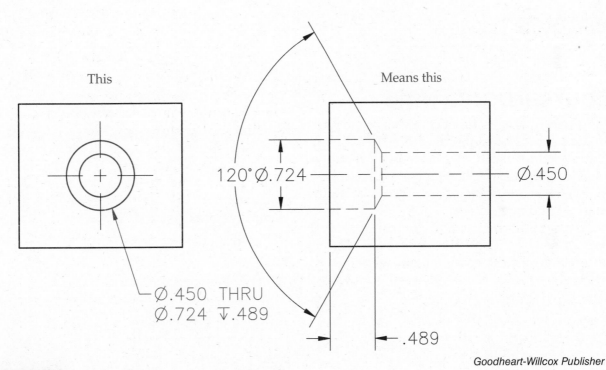

This   Means this

120°Ø.724   Ø.450

Ø.450 THRU
Ø.724 ⊼.489

.489

Figure 9-9   Dimensions for a counterdrilled hole.

screw. Countersinking is the same process as counterdrilling except a countersunk hole has a specified angle to match the angle of the fastener's head. The typical angle used for fasteners is 82°. However, 100° is common for fasteners on thin materials to provide additional surface area for fastening.

The specifications for a countersunk hole appear on two lines. The first line specifies information for the small hole in the following order: the diameter symbol and diameter size, the depth symbol and depth size (if required).

The second line specifies information for the countersunk hole in the following order: the countersink symbol, the diameter symbol and diameter size, the "by" symbol (X) and the included angle of the countersink. See **Figure 9-10**.

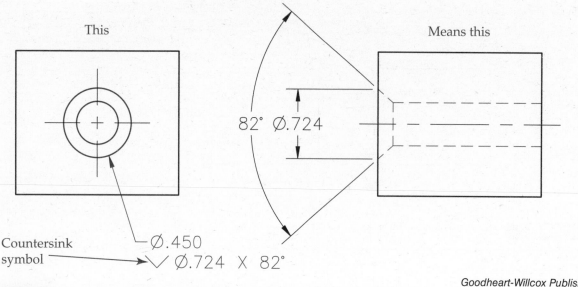

This

Means this

82° Ø.724

Countersink
symbol

Ø.450

Ø.724 X 82°

Goodheart-Willcox Publisher

Figure 9-10   Current ASME standard for specifying a countersunk hole.

Counterbored Hole

Counterboring is the operation of cylindrically enlarging a previously formed hole to a specific diameter and depth. Counterboring produces a flat bottom surface. Its purpose is to provide a recessed hole for fitting fastener heads or seating bearings and pins.

The specifications for a counterbored hole appear on two lines. The first line specifies information for the small hole in the following order: the diameter symbol and diameter size, the depth symbol and depth size. The depth's symbol and size are only included if needed.

The second line specifies information for the counterbore hole in the following order: the counterbore symbol, the diameter symbol and diameter size, the depth symbol and depth size. See **Figure 9-11**.

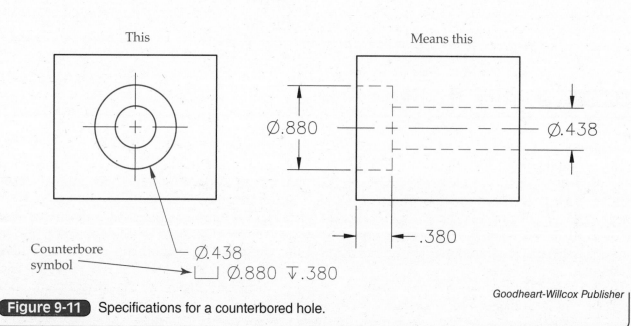

This

Means this

Ø.880

Ø.438

.380

Counterbore
symbol

Ø.438

Ø.880 ▽.380

Goodheart-Willcox Publisher

Figure 9-11   Specifications for a counterbored hole.

Spotface Hole

Spotfacing is the operation of providing a smooth, flat surface around a hole. A rough surface, such as a casting, is spotfaced to accommodate the seating of a washer or bolt head. Spotfacing also provides a flat seat on an inclined surface. The machining process for spotfacing and counterboring are the same, except for the depth of the spotfaced hole. Spotfacing produces a shallow, recessed surface approximately .0625" deep.

According to the current ASME standards, the spotface symbol is the counterbore symbol with the letters SF added. In older editions of the ASME standards, the spotface symbol was the same as the counterbore symbol with no depth specified.

The specifications for a spotface hole appear on two lines. The first line specifies the information for the small hole in the following order: the diameter symbol and diameter size, the depth symbol and depth size (if required).

The second line specifies the information for the spotfaced hole in the following order: the spotface symbol, the diameter symbol and diameter size, the depth symbol and depth size. See **Figure 9-12**.

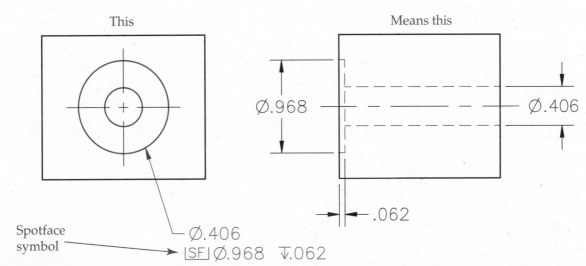

This   Means this

Ø.968   Ø.406

.062

Spotface symbol

Ø.406
SF Ø.968 ⊽.062

Goodheart-Willcox Publisher

Figure 9-12   A spotface hole is similar to a counterbore, except its depth is shallower than a counterbored hole.

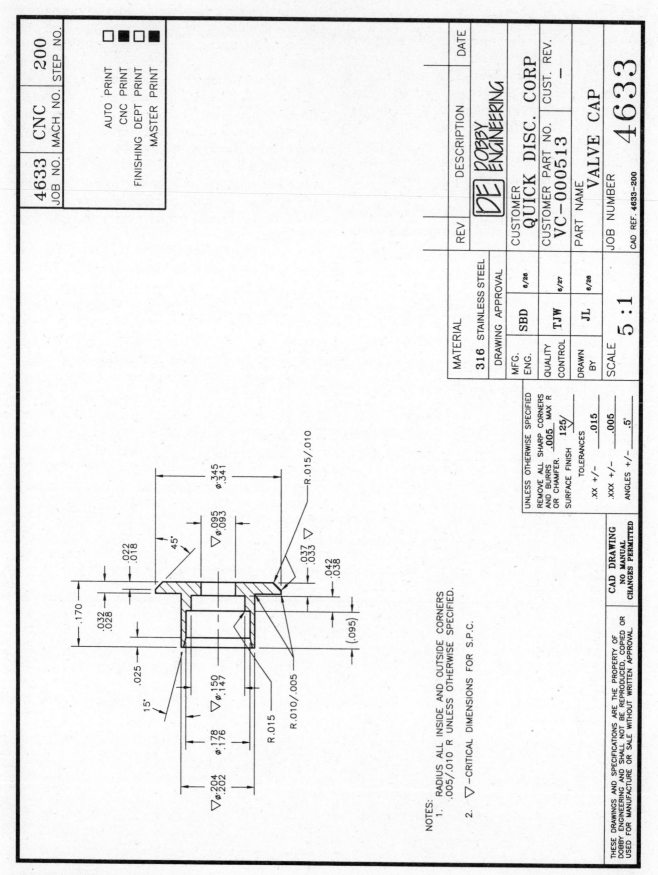

NOTES:
1. RADIUS ALL INSIDE AND OUTSIDE CORNERS
   .005/.010 R UNLESS OTHERWISE SPECIFIED.
2. ▽—CRITICAL DIMENSIONS FOR S.P.C.

	4633	CNC	200
	JOB NO.	MACH NO.	STEP NO.

☐ AUTO PRINT
■ CNC PRINT
☐ FINISHING DEPT PRINT
■ MASTER PRINT

REV	DESCRIPTION	DATE

CUSTOMER
QUICK DISC. CORP

CUSTOMER PART NO. | CUST. REV.
VC-000513 | —

PART NAME
VALVE CAP

JOB NUMBER
4633

CAD REF. 4633-200

MATERIAL		
316 STAINLESS STEEL		

DRAWING APPROVAL		
MFG. ENG.	**SBD**	6/26
QUALITY CONTROL	**TJW**	6/27
DRAWN BY	**JL**	6/26
SCALE	**5:1**	

UNLESS OTHERWISE SPECIFIED
REMOVE ALL SHARP CORNERS
AND BURRS .005 MAX R
OR CHAMFER.
SURFACE FINISH 125√

TOLERANCES
.XX +/— .015
.XXX +/— .005
ANGLES +/— .5°

CAD DRAWING
NO MANUAL
CHANGES PERMITTED

THESE DRAWINGS AND SPECIFICATIONS ARE THE PROPERTY OF
DOBBY ENGINEERING AND SHALL NOT BE REPRODUCED, COPIED OR
USED FOR MANUFACTURE OR SALE WITHOUT WRITTEN APPROVAL.

Activity 9-1 Valve Cap.

Name _____ Date _____ Class _____

Activity 9-1
Valve Cap

Refer to **Activity 9-1**. *Study the drawing and familiarize yourself with the views, dimensions, title block, and notes. Read the questions, refer to the print, and write your answers in the blanks provided.*

1. What is the specified length of the part?.

2. Name the material specified for the part.

3. Including the countersunk section, how deep is the Ø.178 counterbore?

4. What is the maximum size of the largest outside diameter?

5. How deep is the 15° countersunk section?

6. What is the scale of the print?

7. What is the tolerance for unspecified three-place decimals?

8. What is the minimum diameter of the largest hole?

9. What is the dimension of the smallest radius found on the part?

10. What is the maximum diameter of the smallest hole?

11. What is the unspecified tolerance for angles?

12. How many radii are on the part?

13. What is the customer part number?

14. What do the triangles symbolize?

15. What is the tolerance on the largest diameter?

1. _____

2. _____

3. _____

4. _____

5. _____

6. _____

7. _____

8. _____

9. _____

10. _____

11. _____

12. _____

13. _____

14. _____

15. _____

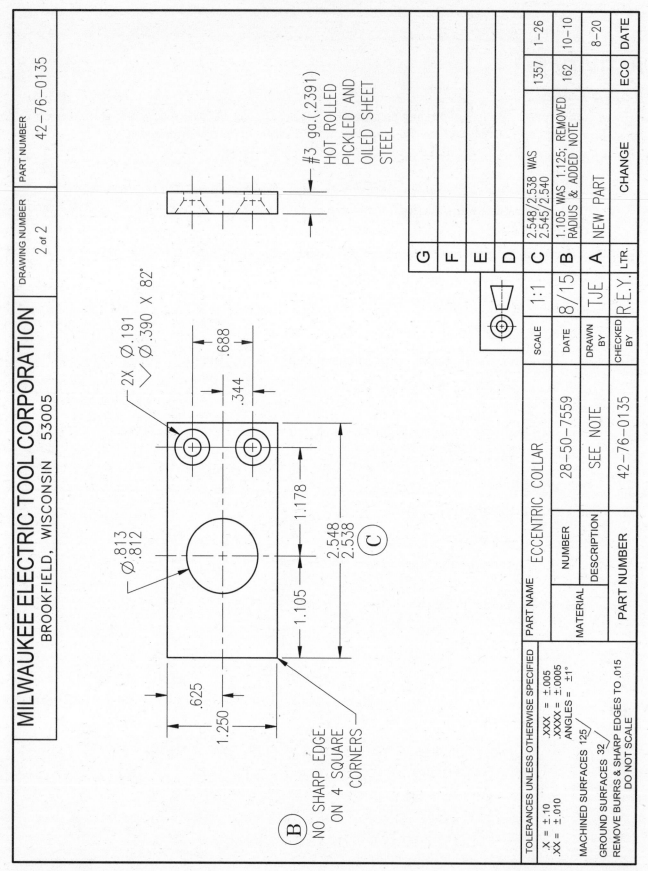

MILWAUKEE ELECTRIC TOOL CORPORATION
BROOKFIELD, WISCONSIN 53005

DRAWING NUMBER	PART NUMBER
2 of 2	42-76-0135

#3 ga.(.2391)
HOT ROLLED
PICKLED AND
OILED SHEET
STEEL

2X Ø.191
Ø.390 X 82°

.688

.344

Ø.813
.812

1.178

2.548
2.538

1.105

C

.625

1.250

B

NO SHARP EDGE
ON 4 SQUARE
CORNERS

G						
F						
E						
D					NEW PART	
C			2.548/2.538 WAS 2.545/2.540	1357		1-26
B			1.105 WAS 1.125; REMOVED RADIUS & ADDED NOTE	162		10-10
A			NEW PART			8-20
LTR.			CHANGE	ECO		DATE

SCALE	1:1
DATE	8/15
DRAWN BY	TJE
CHECKED BY	R.E.Y.

PART NAME	ECCENTRIC COLLAR
MATERIAL	NUMBER 28-50-7559
	DESCRIPTION SEE NOTE
PART NUMBER	42-76-0135

TOLERANCES UNLESS OTHERWISE SPECIFIED
X = ±.10   .XXX = ±.005
.XX = ±.010   .XXXX = ±.0005
              ANGLES = ±1°
MACHINED SURFACES 125/
GROUND SURFACES 32/
REMOVE BURRS & SHARP EDGES TO .015
DO NOT SCALE

Activity 9-2 Eccentric Collar.

Name _____ Date _____ Class _____

Activity 9-2
Eccentric Collar

Refer to **Activity 9-2**. *Study the drawing and familiarize yourself with the views, dimensions, title block, and notes. Read the questions, refer to the print, and write your answers in the blanks provided.*

1. How far are the two small holes from the vertical centerline of the large hole?

    1. _____

2. What is the distance between the centers of the two small holes?

    2. _____

3. What is the maximum length of the part?

    3. _____

4. What is the low limit of the overall length dimension?

    4. _____

5. What is the maximum size of the large hole?

    5. _____

6. How thick is the part's material?

    6. _____

7. What is the minimum height of the part?

    7. _____

8. How far is the vertical centerline of the large hole from the left edge of the part?

    8. _____

9. What was revision C?

    9. _____

10. The drawing is shown in what projection angle?

    10. _____

11. What is the maximum distance the two small holes can be spaced from each other?

    11. _____

12. List two operations able to provide the close tolerance specified for the large hole.

    12. _____

13. What are the maximum and minimum distances from the right edge of the small holes to the right side of the part?

    13. _____

14. What is the distance from the part's horizontal centerline to the horizontal centerline of the upper small hole?

    14. _____

15. What scale is the print?

    15. _____

16. What is the tolerance for three-decimal dimensions?

    16. _____

17. What is the specification for the material of the part?

    17. _____

18. What is the tolerance of the large hole?

    18. _____

19. What is the diameter of the countersink of the two small holes?

    19. _____

20. What is the angle for the countersunk holes?

    20. _____

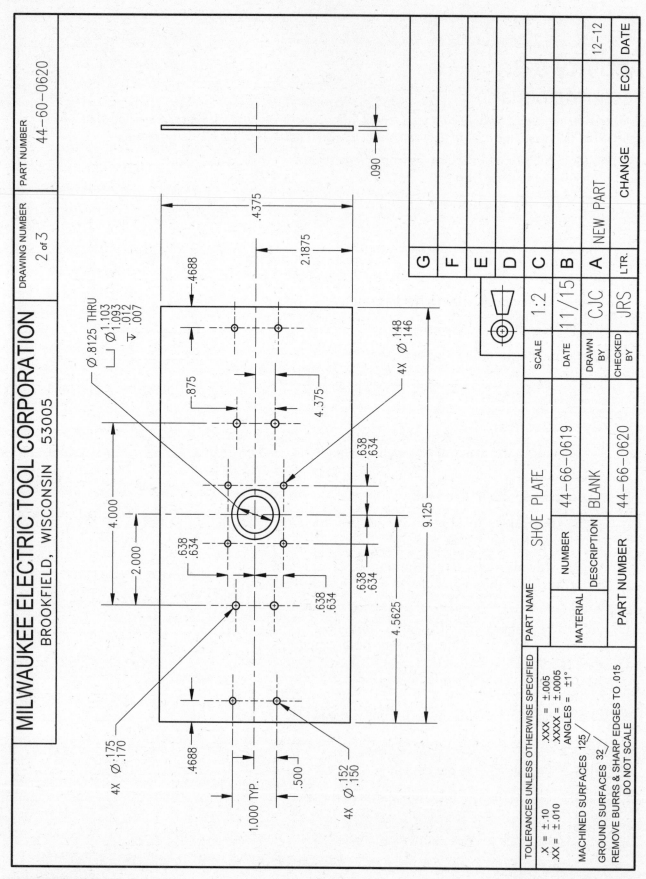

Name _____ Date _____ Class _____

Activity 9-3
Shoe Plate

Refer to **Activity 9-3**. *Study the drawing and familiarize yourself with the views, dimensions, title block, and notes. Read the questions, refer to the print, and write your answers in the blanks provided.*

1. What is the specified length of the part?   1. _____

2. What is the maximum size for the length of the part?   2. _____

3. How thick is the part?   3. _____

4. What diameter is specified for the large hole?   4. _____

5. What is the maximum diameter of the large hole?   5. _____

6. What is the part number?   6. _____

7. How far are the ∅.152/.150 holes from the horizontal centerline of the part?   7. _____

8. How many ∅.152/.150 holes are there?   8. _____

9. How far are the two ∅.152/.150 holes on the left from the two ∅.152/.150 holes on the right from centerpoint to centerpoint?   9. _____

10. How far are the vertical centerlines of the two ∅.175/.170 holes on the right to the vertical centerline of the part?   10. _____

11. How far are the ∅.175/.170 holes from the horizontal centerline of the part?   11. _____

12. How far are the two ∅.152/.150 holes on the left from the two ∅.175/.170 holes on the left measured from their vertical centerlines?   12. _____

13. How far are the two top ∅.148/.146 holes from their centerpoint to the top edge of the part?   13. _____

14. How far are the two ∅.148/.146 holes on the right from their centerpoints to the vertical centerline of the part?   14. _____

15. How far are the two ∅.175/.170 holes on the left from the two ∅.175/.170 holes on the right?   15. _____

16. What is the minimum depth of the counterbore?   16. _____

17. What is the tolerance for the counterbore diameter?   17. _____

18. What is the total number of through holes on the part?   18. _____

19. What is the minimum size for the diameter of the counterbore?   19. _____

20. What are the maximum and minimum distances from the edge of the ∅.152/.150 holes to the sides of the part?   20. _____

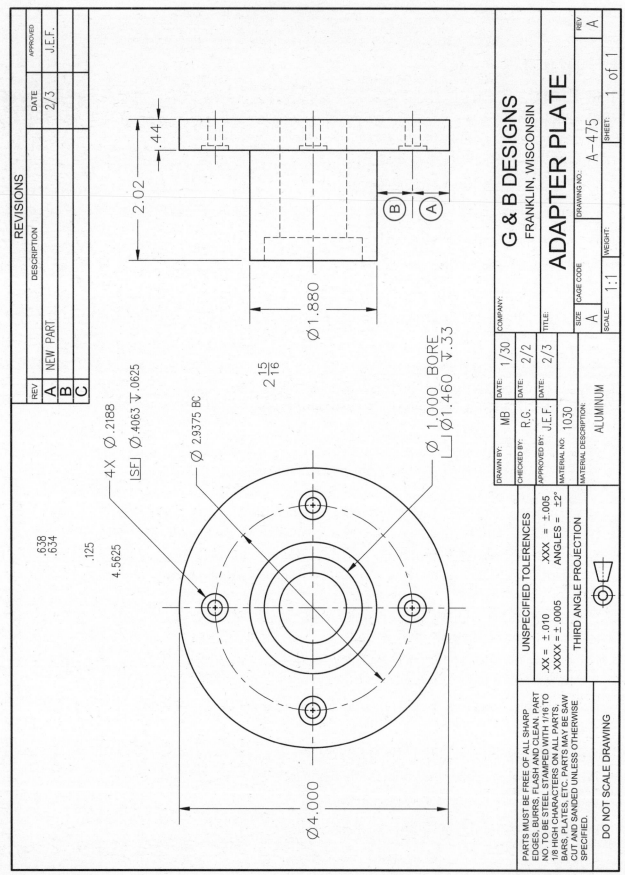

Activity 9-4 Adapter Plate.

Name _____ Date _____ Class _____

Activity 9-4
Adapter Plate

Refer to **Activity 9-4**. *Study the drawing and familiarize yourself with the views, dimensions, title block, and notes. Read the questions, refer to the print, and write your answers in the blanks provided.*

1. What are the maximum and minimum *widths* of the part?   1. _____

2. Name the specified material for the part.   2. _____

3. How deep is the counterbore?   3. _____

4. What is the maximum size for the largest outside diameter?   4. _____

5. What is the minimum length of the Ø1.880 section?   5. _____

6. What is the scale of the print?   6. _____

7. What is the unspecified tolerance for three-place decimals?   7. _____

8. What is the minimum diameter of the bored hole?   8. _____

9. How many holes are spotfaced?   9. _____

10. What is the diameter of the spotfaces?   10. _____

11. How thick is the flange?   11. _____

12. What is the unspecified tolerance for angles?   12. _____

13. Are the spotfaced holes equally spaced?   13. _____

14. What is the maximum diameter of the bored hole?   14. _____

15. On what bolt circle is the Ø.2188 hole located?   15. _____

16. Is distance A equal to distance B?   16. _____

17. How many degrees apart are the spotfaced holes?   17. _____

18. What is the minimum angle between the spotfaced holes?   18. _____

19. If the counterbored hole was machined to a depth of .325, what would be the length of the bored hole?   19. _____

20. What is the wall thickness between the counterbore and the Ø1.880 section?   20. _____

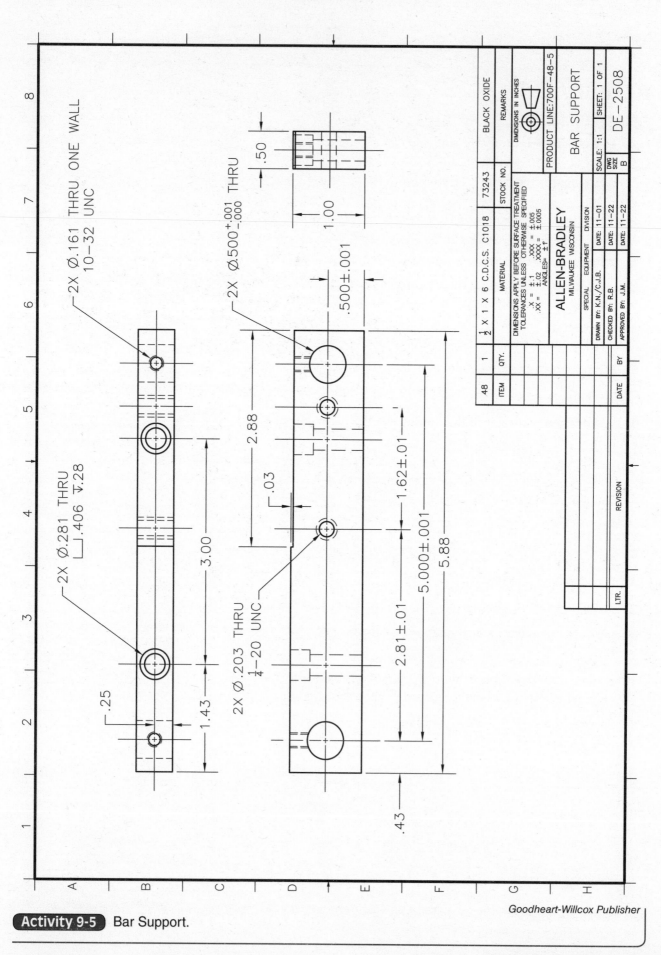

2X Ø.161 THRU ONE WALL
10-32 UNC

2X Ø.500 +.001 -.000 THRU

.500±.001

.50

1.00

2X Ø.281 THRU
⌴.406 ▽.28

2.88

.03

3.00

1.62±.01

2X Ø.203 THRU
¼-20 UNC

5.000±.001

5.88

2.81±.01

1.43

.25

.43

					BLACK OXIDE	
					REMARKS	
			73243		DIMENSIONS IN INCHES	PRODUCT LINE:700F-48-5
			STOCK NO.			BAR SUPPORT
48	1	½ X 1 X 6 C.D.C.S. C1018	MATERIAL			
						SCALE: 1:1	SHEET: 1 OF 1
						DWG SIZE B	DE-2508

DIMENSIONS APPLY BEFORE SURFACE TREATMENT
TOLERANCES UNLESS OTHERWISE SPECIFIED
.X = ±.1   .XXX = ±.005
.XX = ±.02   .XXXX = ±.0005
ANGLES= ±1°

ALLEN-BRADLEY
MILWAUKEE WISCONSIN
SPECIAL EQUIPMENT DIVISION

DRAWN BY: K.N./C.J.B.	DATE: 11-01	
CHECKED BY: R.B.	DATE: 11-22	
APPROVED BY: J.M.	DATE: 11-22	

ITEM QTY.

REVISION

LTR.   DATE   BY

Activity 9-5 Bar Support.

Name _____ Date _____ Class _____

Activity 9-5
Bar Support

Refer to **Activity 9-5**. *Study the drawing and familiarize yourself with the views, dimensions, title block, and notes. Read the questions, refer to the print, and write your answers in the blanks provided.*

1. How many holes are on this part?

    1. _____

2. What is the maximum allowable length of the part?

    2. _____

3. What is the minimum distance allowable between the two ⌀.161 holes?

    3. _____

4. What is the diameter of the counterbore on the ⌀.281 holes?

    4. _____

5. What is the material stock number?

    5. _____

6. How deep is the counterbore on the two ⌀.281 holes?

    6. _____

7. How deep is the 2.88 long step?

    7. _____

8. How is the depth of the two ⌀.161 holes specified?

    8. _____

9. What is the maximum distance allowable between the two ⌀.203 holes?

    9. _____

10. How long is the left ⌀.281 hole if the counterbore is at maximum depth?

    10. _____

11. What tolerance is permitted for four-place decimal values?

    11. _____

12. What is the part number?

    12. _____

13. What is the center-to-center distance between the left ⌀.281 hole and the left ⌀.161 hole?

    13. _____

14. What is the approximate depth of the two ⌀.203 holes?

    14. _____

15. What is the minimum allowable wall thickness between the left ⌀.500 hole and the left end of the part?

    15. _____

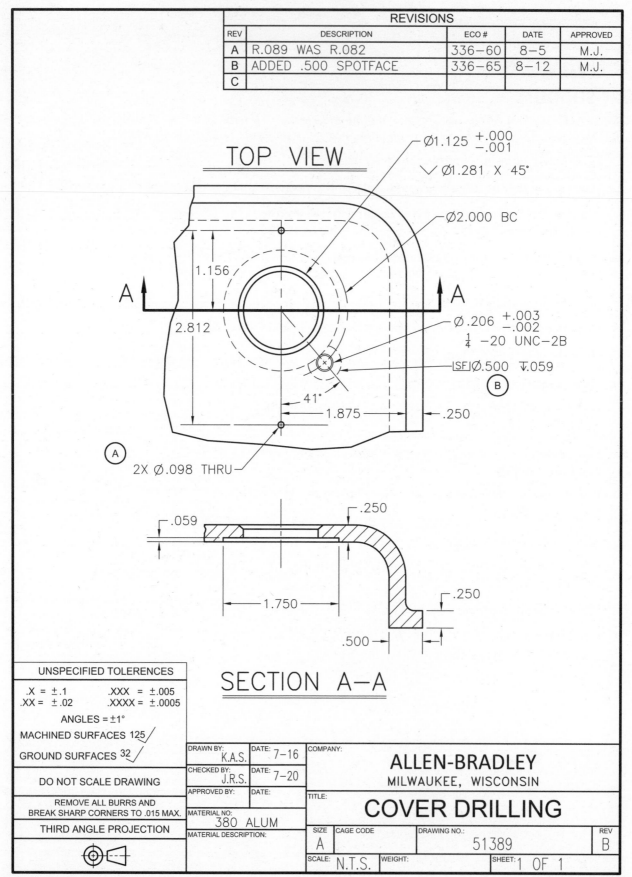

REVISIONS				
REV	DESCRIPTION	ECO #	DATE	APPROVED
A	R.089 WAS R.082	336−60	8−5	M.J.
B	ADDED .500 SPOTFACE	336−65	8−12	M.J.
C				

TOP VIEW

Ø1.125 +.000 −.001

⌵ Ø1.281 X 45°

Ø2.000 BC

1.156

A

2.812

Ø.206 +.003 −.002
¼ −20 UNC−2B

⌴SF⌴Ø.500 ⌵.059

B

41°

1.875

.250

A

2X Ø.098 THRU

.059

.250

1.750

.250

.500

SECTION A−A

UNSPECIFIED TOLERENCES	
.X = ±.1	.XXX = ±.005
.XX = ±.02	.XXXX = ±.0005

ANGLES = ±1°

MACHINED SURFACES 125/

GROUND SURFACES 32/

DO NOT SCALE DRAWING

REMOVE ALL BURRS AND
BREAK SHARP CORNERS TO .015 MAX.

THIRD ANGLE PROJECTION

DRAWN BY: K.A.S.	DATE: 7−16	COMPANY:
CHECKED BY: J.R.S.	DATE: 7−20	**ALLEN-BRADLEY**
APPROVED BY:	DATE:	MILWAUKEE, WISCONSIN
MATERIAL NO: 380 ALUM		TITLE: **COVER DRILLING**

MATERIAL DESCRIPTION:		
SIZE A	CAGE CODE	DRAWING NO.: 51389	REV B
SCALE: N.T.S.	WEIGHT:	SHEET: 1 OF 1

Activity 9-6 Cover Drilling.

Name _____ Date _____ Class _____

Activity 9-6
Cover Drilling

Refer to **Activity 9-6**. *Study the drawing and familiarize yourself with the views, dimensions, title block, and notes. Read the questions, refer to the print, and write your answers in the blanks provided.*

1. Revision B added what feature to the part?

1. _____

2. What is the depth of the ⌀.500 spotface?

2. _____

3. What is the diameter of the bolt circle?

3. _____

4. The part is made of what material?

4. _____

5. What is the angle of the countersink on the ⌀1.125 hole?

5. _____

6. At what angle from the horizontal is the small spotfaced hole located?

6. _____

7. How far apart are the ⌀.098 holes?

7. _____

8. What is the diameter of the spotface on the ⌀1.125 hole?

8. _____

9. What is the unspecified tolerance for angles?

9. _____

10. What is the maximum allowable depth of the spotface on the ⌀1.125 hole?

10. _____

11. Approximately how deep are the ⌀.098 holes?

11. _____

12. What is the total tolerance on the ⌀.206 hole?

12. _____

13. Are the two ⌀.098 holes equidistant from the ⌀1.125 hole?

13. _____

14. What is the tolerance for three-place decimal values?

14. _____

15. What is the diameter of the countersink on the ⌀1.125 hole at its widest point?

15. _____

Notes

Learning Objectives

After studying this unit, you will be able to:

✓ Demonstrate the ability to distinguish different types of angles.

✓ Recognize a bevel and determine its measurement.

✓ Identify a chamfer and interpret its size.

✓ Define the term "taper" and interpret the meaning of TPF and TPI.

✓ Perform calculations using taper data.

Key Terms

angle
bevel

chamfer
taper

vertex

Angular Dimensioning

An *angle* is the amount of rotation or turn measured in degrees (°) between two lines that converge or meet at a point called the *vertex*. The division of a circle into 360° defines the measurement of an angle. Refer to Unit 7—Dimensions and Tolerances for additional information on angular dimensioning.

Applied to machining, an angle is the measurement between planes (surfaces) or locations (such as holes located on a circle) measured in degrees. Two methods are used to dimension angles: the coordinate method and the angular method.

The coordinate method, **Figure 10-1**, uses two linear dimensions to define an angle. The coordinate method is used when a high degree of accuracy is desired. The angular method uses one linear dimension and one angular dimension in degrees to define an angle. The angular dimension can be measured from the centerline or the included angle, **Figure 10-2**. Included angles must be divided by two to obtain the angle from the centerline.

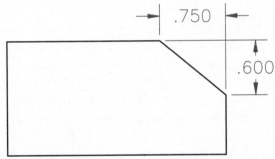

.750

.600

Goodheart-Willcox Publisher

Figure 10-1 The coordinate method of dimensioning angles.

The types of angles applied to parts and their features are bevels, chamfers, and tapers.

Bevel

A *bevel* is the angle one surface makes with another surface when they are not at right angles. It is the slant or inclination of a surface, **Figure 10-3**.

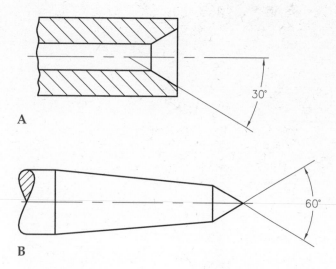

A

B

Goodheart-Willcox Publisher

Figure 10-2 The angular method of dimensioning angles. A—Angle measured from the centerline. B—The included angle.

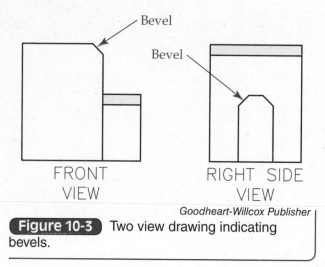

Bevel

Bevel

FRONT VIEW

RIGHT SIDE VIEW

Goodheart-Willcox Publisher

Figure 10-3 Two view drawing indicating bevels.

Bevel is used to denote an angle machined on a part that is not cylindrical. An angle applied to a cylindrical piece or hole is referred to as a chamfer. Sometimes, however, the terms bevel and chamfer are used interchangeably.

Chamfers

A *chamfer* is a beveled edge or angle applied to a hole, shaft, or an edge to remove sharp edges. Chamfers are defined by a linear dimension and an angle or by two linear dimensions. Chamfering aids in the assembly of parts. There are two types of chamfers—interior and exterior.

Interior chamfers are angles applied to edges of interior features such as holes. A chamfer on a hole is dimensioned by stating its diameter and angle. **Figure 10-4** shows two methods for dimensioning interior chamfers of a hole.

Exterior chamfers are angles applied to exterior features such as edges. Chamfers are common on the ends of threaded fasteners. The chamfer aids in starting the fastener in the threaded hole, as well as protecting the start of the thread.

An exterior chamfer, such as a chamfer on a shaft, is dimensioned by stating its length and angle or two linear dimensions. Any chamfer angle less than 45° should have a linear dimension included on the drawing. See **Figure 10-5**.

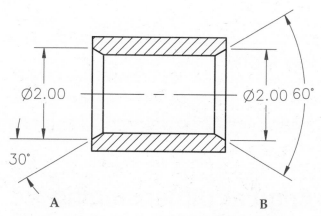

$\varnothing 2.00$

$\varnothing 2.00$ 60°

30°

A

B

Goodheart-Willcox Publisher

Figure 10-4 Two methods for dimensioning internal chamfers. A—Using the angle from the centerline. B—Using the included angle.

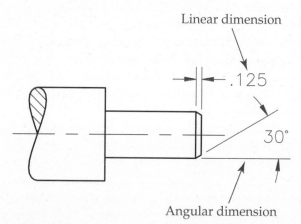

Linear dimension

.125

30°

Angular dimension

Goodheart-Willcox Publisher

Figure 10-5 The chamfer on the shaft dimensioned by its length and angle.

A leader line and note dimension chamfers on 45° perpendicular surfaces. See **Figure 10-6**. Sometimes "chamfer" is included in a note although it does not follow the current ASME standard.

When two perpendicular surfaces intersect and their linear values for the chamfer are different from each other, then both values appear on the drawing. See **Figure 10-7A**. The same dimensioning applies to two non-perpendicular surfaces with a chamfer, as shown in **Figure 10-7B**.

Tapers

A *taper* is a conical surface with a uniformly changing diameter along its length. Tapers are found on cylinders and holes, **Figure 10-8**. The taper is given as the difference in diameters per unit of length, usually as taper per inch (TPI) or taper per foot (TPF).

Example 10-1: Convert Taper per Inch (TPI) to Taper per Foot (TPF)
Convert .125 TPI to TPF.

Solution:
$$TPF = TPI \times 12$$
$$TPF = .125 \times 12$$
$$TPF = 1.500''$$

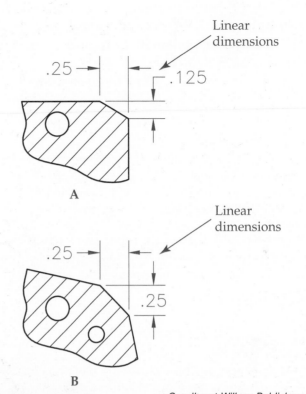

Figure 10-7 External chamfers on different surfaces. A—Dimensioning a chamfer with different measurements on perpendicular surfaces. B—Dimensioning a chamfer on surfaces that are not perpendicular.

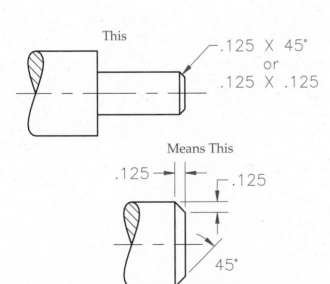

Figure 10-6 Two ways of dimensioning a 45° exterior chamfer.

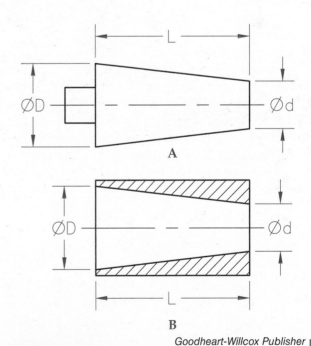

Figure 10-8 Tapers are dimensioned by providing large diameter, small diameter, and length of taper. A—Exterior taper. B—Interior taper.

Example 10-2: Convert Taper per Foot (TPF) to Taper per Inch (TPI)

Convert 2.4 TPF to TPI.

Solution:

TPI = TPF ÷ 12

TPI = 2.4 ÷ 12

TPI = .2″

The formula to calculate taper is the difference between the large diameter (D) and the small diameter (d), divided by the length (L):

Taper (TPI or TPF) = (D – d) / L

Example 10-3: Calculate Taper per Inch (TPI)

See **Figure 10-9**. Calculate the taper per inch (TPI) given large diameter, small diameter, and taper length.

Solution:

TPI = (D – d) / L

TPI = (1.500 – .875) / 4.080

TPI = .625″

Standard machine tapers are used on cutting tools, tool holders, and machine tool spindles. Standard machine tapers include the following:

- Morse—5/8″ taper per foot, Jacobs—.5915″ to .7619″ taper per foot
- Brown and Sharpe—1/2″ taper per foot
- Jarno—.600″ taper per foot
- American National Standard—3 1/2″ taper per foot

Tapers can be dimensioned in several ways, including the following:

- By indicating diameter of both ends, including taper length, **Figure 10-8**.
- By stating large diameter, taper length, and including a note for standard tapers such as No. 5 Morse taper. See **Figure 10-10**.
- By giving diameter at large end, length of taper, and included angle, as shown in **Figure 10-11**.

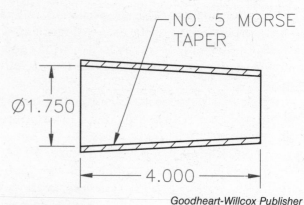

Goodheart-Willcox Publisher

Figure 10-10 The note "No. 5 Morse Taper" represents a standard machine taper of 5/8″ taper per foot (TPF).

- By specifying taper per inch (TPI) or taper per foot (TPF), taper length, and large diameter. See **Figure 10-12**.
- By specifying the taper on diameter per unit of length after the taper symbol, as shown in **Figure 10-13**.

Taper Calculations

Example 10-4: Calculate the Small Diameter

See **Figure 10-14**. Calculate the small diameter (Ød) given the TPI, taper length, and large diameter.

Solution:

TPI = (D – d) / L

TPI × L = D – d

.0625 × 4.625 = 1.875 – d

d = 1.875 – .289

d = 1.586″

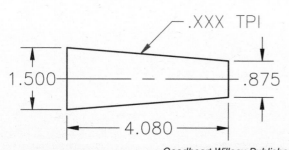

Goodheart-Willcox Publisher

Figure 10-9 Example 10-3: Calculate taper per inch (TPI).

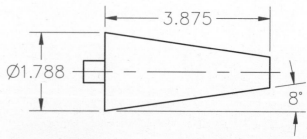

Goodheart-Willcox Publisher

Figure 10-11 Tapers can also be dimensioned by giving large diameter, length of taper, and included angle.

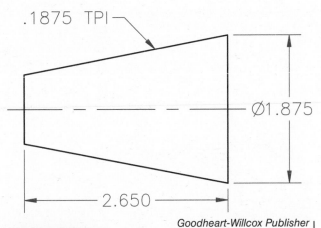

Figure 10-12 Specifying large diameter, length of taper, and taper per inch (TPI) or taper per foot (TPF) is another acceptable way of dimensioning taper.

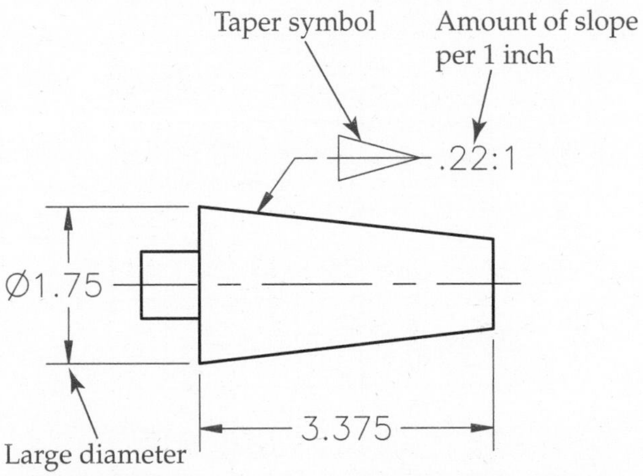

Example 10-5: Calculate the Small Diameter

See **Figure 10-15**. Calculate the small diameter (Ød) given the TPF, taper length, and large diameter.

Solution:
Convert TPF to TPI:
$$TPI = TPF \div 12$$
$$TPI = .600 \div 12$$
$$TPI = .050''$$
Calculate the small diameter:
$$TPI = (D - d) / L$$
$$TPI \times L = D - d$$
$$.050 \times 5.440 = 1.125 - d$$
$$d = 1.125 - .272$$
$$d = .853''$$

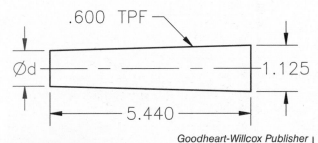

Goodheart-Willcox Publisher

Figure 10-15 Example 10-5: Calculate the small diameter from taper per foot (TPF).

Taper symbol   Amount of slope per 1 inch

.22:1

Ø1.75

Large diameter

3.375

Goodheart-Willcox Publisher

Figure 10-13 Specifying a taper by the amount of slope per unit of length measured from the large diameter.

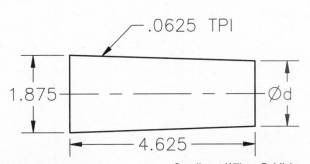

.0625 TPI

1.875

Ød

4.625

Goodheart-Willcox Publisher

Figure 10-14 Example 10-4: Calculate the small diameter.

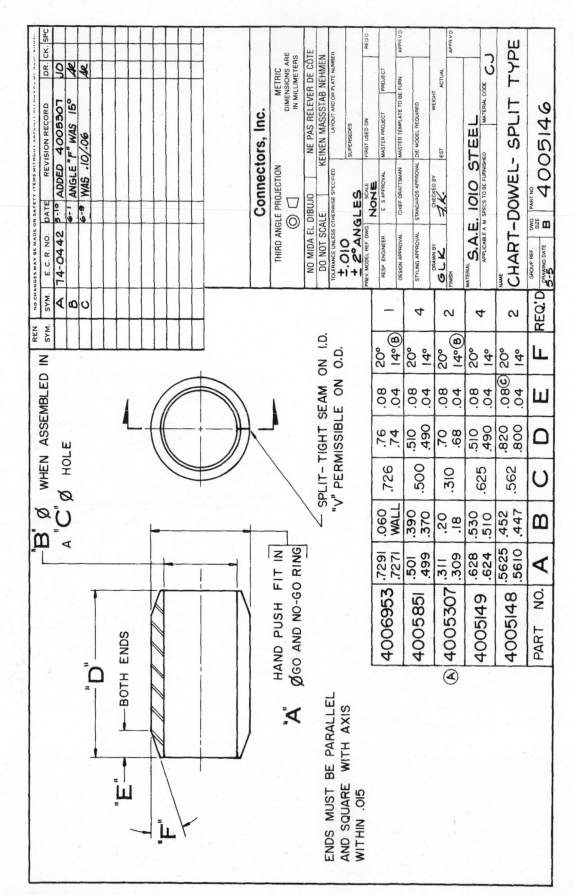

ENDS MUST BE PARALLEL
AND SQUARE WITH AXIS
WITHIN .015

"A" HAND PUSH FIT IN
Ø GO AND NO-GO RING

A "C" Ø HOLE

"B" Ø WHEN ASSEMBLED IN

BOTH ENDS

SPLIT – TIGHT SEAM ON I.D.
"V" PERMISSIBLE ON O.D.

PART NO.	A	B	C	D	E	F	REQ'D
4006953	.7291 / .7271	.060 WALL	.726	.76 / .74	.08 / .04	20° / 14° (B)	1
4005851	.501 / .499	.390 / .370	.500	.510 / .490	.08 / .04	20° / 14°	4
4005307 (A)	.311 / .309	.20 / .18	.310	.70 / .68	.08 / .04	20° / 14° (B)	2
4005149	.628 / .624	.530 / .510	.625	.510 / .490	.08 / .04	20° / 14°	4
4005148	.5625 / .5610	.452 / .447	.562	.820 / .800	.08 (C) / .04	20° / 14°	2

Connectors, Inc.

THIRD ANGLE PROJECTION

METRIC
DIMENSIONS ARE
IN MILLIMETERS

NO MIDA EL DIBUJO
DO NOT SCALE
NE PAS RELEVER DE CÔTE
KEINEN MASSSTAB NEHMEN

TOLERANCE UNLESS OTHERWISE SPECIFIED
±.010
±2° ANGLES

SCALE: NONE

PREV MODEL REF. DWG. | SUPERSEDES | FIRST USED ON | REQ'D

RESP. ENGINEER | E. S. APPROVAL | MASTER PROJECT | PROJECT
DESIGN APPROVAL | CHIEF DRAFTSMAN | MASTER TEMPLATE TO BE FURN. | APPR'V'D
STYLING APPROVAL | STANDARDS APPROVAL | DIE MODEL REQUIRED

DRAWN BY G L K | CHECKED BY J. K. | WEIGHT EST | ACTUAL | APPR'V'D

MATERIAL S.A.E. 1010 STEEL | MATERIAL CODE C J
FINISH | APPLICABLE A.M. SPECS TO BE FURNISHED

NAME CHART–DOWEL– SPLIT TYPE

GROUP REF. | DWG SIZE B | PART NO. 4005146
DRAWING DATE 5-5

REN

| REN SYM. | | | | |

SYM.	E.C.R. NO.	DATE	REVISION RECORD	DR.	CK.	SPC
A	74-0442	5-15	ADDED 4005307	JO		
B		6-1	ANGLE "F" WAS 15°	AL		
C		6-9	WAS .10/.06	AL		

NO CHANGES MAY BE MADE ON SAFETY ITEMS WITHOUT LAYOUT APPROVAL OF PART DWG.

Activity 10-1 Split Type Dowel.

Name _____ Date _____ Class _____

Activity 10-1
Split Type Dowel

Refer to **Activity 10-1**. *Study the drawing and familiarize yourself with the views, dimensions, title block, and notes. Read the questions, refer to the print, and write your answers in the blanks provided.*

1. Which part has the longest length?

1. _____

2. Which part has the smallest tolerance for the outside diameter?

2. _____

3. What diameter will "B" become when assembled in a Ø.500 hole?

3. _____

4. Which part numbers require 4 parts?

4. _____

5. What was revision C?

5. _____

6. What material is used for the split dowels?

6. _____

7. What is the unspecified tolerance for angular dimensions?

7. _____

8. What is the high limit for "F" on Part No. 4005148?

8. _____

9. If the outside diameter of Part No. 4006953 is at the high limit and the wall thickness is .060, what will the high and low limits of diameter "B" become when assembled in a Ø.726 hole?

9. _____

10. How is diameter "A" checked?

10. _____

11. What scale is the print?

11. _____

12. What units are specified on the print?

12. _____

13. What is the maximum tolerance for diameter A among the various parts?

13. _____

14. Why is one-half of the front view sectioned?

14. _____

15. What is the tolerance on "F," the chamfer angle?

15. _____

16. How square must the ends of the dowel be with the axis?

16. _____

17. If the chamfer lengths of Part No. 4006953 are at the high limit and the total dowel length at low limit, how long will the bearing surface of the dowel be?

17. _____

18. If the chamfer lengths of Part No. 4005851 are at the low limit and the total dowel length at low limit, how long will the bearing surface of the dowel be?

18. _____

19. What is the maximum allowable length for Part No. 4005149?

19. _____

20. Which part number was added last to the print?

20. _____

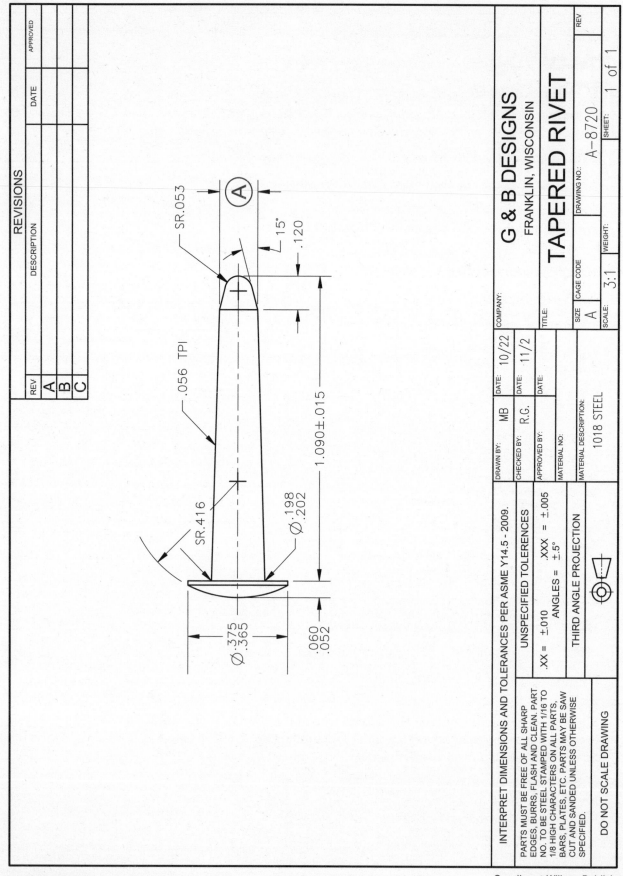

Activity 10-2 Tapered Rivet.

Name _____ Date _____ Class _____

Activity 10-2
Tapered Rivet

Refer to **Activity 10-2**. *Study the drawing and familiarize yourself with the views, dimensions, title block, and notes. Read the questions, refer to the print, and write your answers in the blanks provided.*

1. What is the maximum overall length of the rivet?   1. _____

2. What is the diameter of the rivet's head?   2. _____

3. What is the spherical radius (SR) of the tip of the rivet?   3. _____

4. What taper per inch is specified?   4. _____

5. What tolerance is specified for three-place decimal dimensions?   5. _____

6. If the large diameter of the shank is .202, what is the diameter at A?   6. _____

7. What is the spherical radius (SR) of the head of the rivet?   7. _____

8. What is the tolerance of the large diameter of the tapered shank?   8. _____

9. What is the scale of the print?   9. _____

10. What type of tolerance is found on the 1.090 dimension?   10. _____

11. Including the spherical radius, how long is the 15° taper?   11. _____

12. How is the part number to be indicated on the part?   12. _____

13. What kind of material is used to make the part?   13. _____

14. What is the high limit on the 15° taper?   14. _____

15. How long is the rivet head?   15. _____

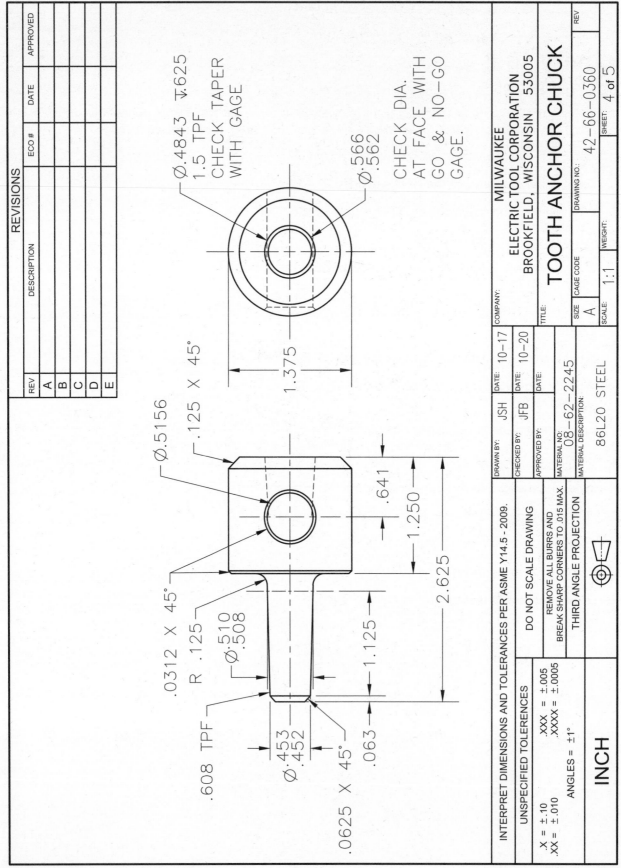

REVISIONS

REV	DESCRIPTION	ECO #	DATE	APPROVED
A				
B				
C				
D				
E				

∅.4843 ⌴.625
1.5 TPF
CHECK TAPER
WITH GAGE

∅.566
.562

CHECK DIA.
AT FACE WITH
GO & NO-GO
GAGE.

.125 X 45°

1.375

∅.5156

.0312 X 45°

R .125

∅.510
.508

.641

1.250

2.625

1.125

.608 TPF

∅.453
.452

.063

.0625 X 45°

INTERPRET DIMENSIONS AND TOLERANCES PER ASME Y14.5 - 2009.

UNSPECIFIED TOLERANCES

.X = ±.10
.XX = ±.010

.XXX = ±.005
.XXXX = ±.0005

ANGLES = ±1°

INCH

DO NOT SCALE DRAWING

REMOVE ALL BURRS AND
BREAK SHARP CORNERS TO .015 MAX.

THIRD ANGLE PROJECTION

COMPANY:

MILWAUKEE
ELECTRIC TOOL CORPORATION
BROOKFIELD, WISCONSIN 53005

TITLE:

TOOTH ANCHOR CHUCK

DRAWN BY: JSH	DATE: 10-17
CHECKED BY: JFB	DATE: 10-20
APPROVED BY:	DATE:

MATERIAL NO: 08-62-2245

MATERIAL DESCRIPTION:

86L20 STEEL

| SIZE A | CAGE CODE | DRAWING NO.: 42-66-0360 | REV |
| SCALE: 1:1 | WEIGHT: | SHEET: 4 of 5 |

Activity 10-3 Tooth Anchor Chuck.

Name _____ Date _____ Class _____

Activity 10-3
Tooth Anchor Chuck

Refer to **Activity 10-3**. *Study the drawing and familiarize yourself with the views, dimensions, title block, and notes. Read the questions, refer to the print, and write your answers in the blanks provided.*

1. What size chamfer is found on the end of the tapered shank?

2. What is the diameter of the internal taper at the face of the chuck?

3. What is the diameter of the large end of the tapered shank?

4. What is the taper per inch of the tapered shank?

5. What is the low limit for the total length of the part?

6. How far is the ∅.5156 hole from the right end of the part?

7. What is the diameter of the small end of the tapered shank?

8. How deep must you drill the ∅.4843 hole?

9. How is the inside diameter at the face of the chuck checked?

10. What kind of material is used for the part?

11. What is the total length of the shank?

12. What tolerance is applied to the 1.125 dimension?

13. What is the maximum diameter for the ∅.5156 hole?

14. What is the taper per inch of the internal hole?

15. What scale is the print?

16. What size chamfer is put on the ∅.5156 diameter hole?

17. What size radius is found on the print?

18. What is the material number of the part?

19. What size chamfer is found on the right end of the part?

20. What is the minimum diameter of the chuck body?

1. _____

2. _____

3. _____

4. _____

5. _____

6. _____

7. _____

8. _____

9. _____

10. _____

11. _____

12. _____

13. _____

14. _____

15. _____

16. _____

17. _____

18. _____

19. _____

20. _____

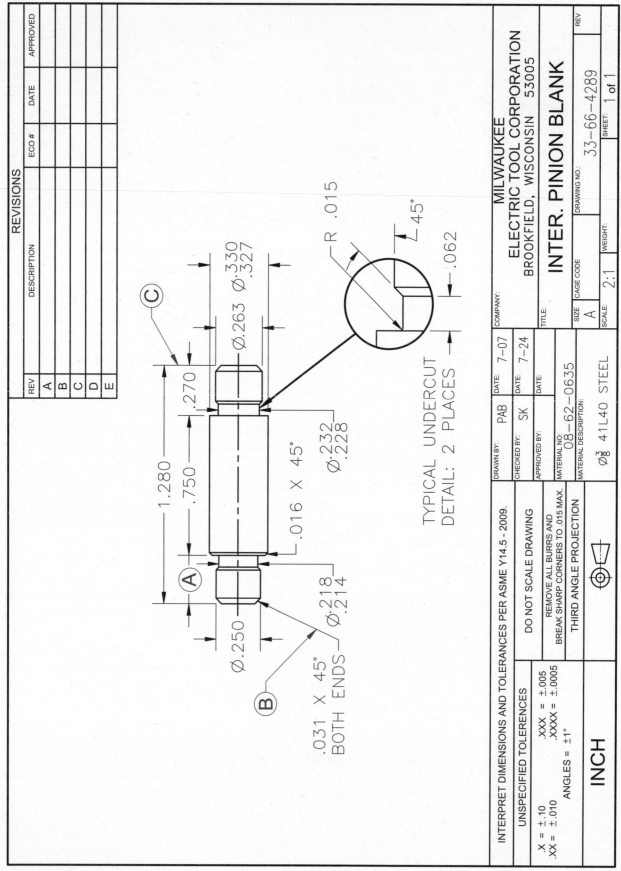

TYPICAL UNDERCUT
DETAIL: 2 PLACES

R .015

∟45°

.062

C

Ø.330
Ø.327

Ø.263

.270

Ø.232
Ø.228

.016 X 45°

1.280

.750

A

Ø.218
Ø.214

Ø.250

.031 X 45°
BOTH ENDS

B

REVISIONS				
REV	DESCRIPTION	ECO #	DATE	APPROVED
A				
B				
C				
D				
E				

INTERPRET DIMENSIONS AND TOLERANCES PER ASME Y14.5 - 2009.

UNSPECIFIED TOLERANCES	DO NOT SCALE DRAWING
X = ±.10 .XXX = ±.005	REMOVE ALL BURRS AND
.XX = ±.010 .XXXX = ±.0005	BREAK SHARP CORNERS TO .015 MAX.
ANGLES = ±1°	THIRD ANGLE PROJECTION

INCH

DRAWN BY: PAB	DATE: 7-07	COMPANY:
CHECKED BY: SK	DATE: 7-24	
APPROVED BY:	DATE:	

MILWAUKEE
ELECTRIC TOOL CORPORATION
BROOKFIELD, WISCONSIN   53005

TITLE:

INTER. PINION BLANK

MATERIAL NO: 08-62-0635	SIZE A	CAGE CODE	DRAWING NO: 33-66-4289	REV
MATERIAL DESCRIPTION: Ø³⁄₈ 41L40 STEEL		SCALE: 2:1	WEIGHT:	SHEET: 1 of 1

Activity 10-4 Inter. Pinion Blank.

Name _____ Date _____ Class _____

Activity 10-4
Inter. Pinion Blank

Refer to **Activity 10-4**. *Study the drawing and familiarize yourself with the views, dimensions, title block, and notes. Read the questions, refer to the print, and write your answers in the blanks provided.*

1. How long is the middle section minus the chamfer?   1. _____

2. What is the dimension of the chamfer on the middle   2. _____
    section?

3. How wide are the two grooves in the part?   3. _____

4. What is the maximum allowable length of the part?   4. _____

5. What scale is the print?   5. _____

6. Determine dimension A.   6. _____

7. What are the low limits for the groove diameters?   7. _____

8. What type of line is B?   8. _____

9. Give the material description for the part.   9. _____

10. What type of line is C?   10. _____

11. What is the part number?   11. _____

12. What is the tolerance for three-place decimal dimensions?   12. _____

13. What is the dimension of the chamfers on the ends of   13. _____
    the part?

14. List the radii found on the part.   14. _____

15. What is the maximum allowable diameter of the part?   15. _____

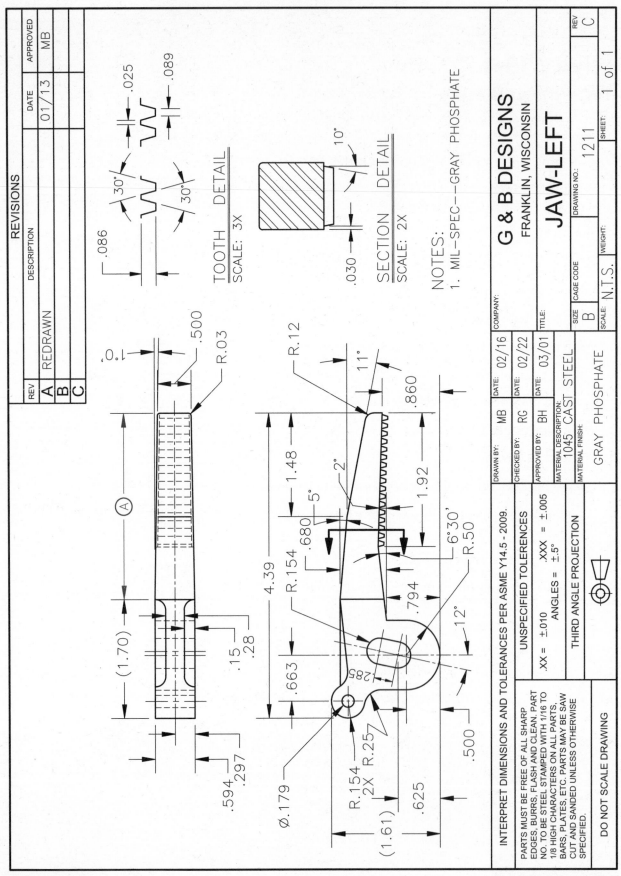

TOOTH DETAIL
SCALE: 3X

SECTION DETAIL
SCALE: 2X

NOTES:
1. MIL-SPEC—GRAY PHOSPHATE

REVISIONS			
REV	DESCRIPTION	DATE	APPROVED
A	REDRAWN	01/13	MB
B			
C			

DRAWN BY:	MB	DATE: 02/16
CHECKED BY:	RG	DATE: 02/22
APPROVED BY:	BH	DATE: 03/01

MATERIAL DESCRIPTION:
1045 CAST STEEL
MATERIAL FINISH:
GRAY PHOSPHATE

COMPANY:

G & B DESIGNS
FRANKLIN, WISCONSIN

TITLE:
JAW-LEFT

SIZE	CAGE CODE	DRAWING NO:	REV
B		1211	C

SCALE: N.T.S. | WEIGHT: | SHEET: 1 of 1

INTERPRET DIMENSIONS AND TOLERANCES PER ASME Y14.5 - 2009.

UNSPECIFIED TOLERENCES	
.XX = ±.010	.XXX = ±.005
ANGLES = ±.5°	

THIRD ANGLE PROJECTION

PARTS MUST BE FREE OF ALL SHARP EDGES, BURRS, FLASH AND CLEAN. PART NO. TO BE STEEL STAMPED WITH 1/16 TO 1/8 HIGH CHARACTERS ON ALL PARTS, BARS, PLATES, ETC. PARTS MAY BE SAW CUT AND SANDED UNLESS OTHERWISE SPECIFIED.

DO NOT SCALE DRAWING

Activity 10-5 Jaw-Left

Name _____ Date _____ Class _____

Activity 10-5
Jaw-Left

Refer to **Activity 10-5**. *Study the drawing and familiarize yourself with the views, dimensions, title block, and notes. Read the questions, refer to the print, and write your answers in the blanks provided.*

1. List all of the angles shown on the print.

   1. _____

2. What material is the part to be made from?

   2. _____

3. What is the total length of the slot?

   3. _____

4. What finish is to be applied to the part?

   4. _____

5. What two dimensions are used to locate the position of the slot?

   5. _____

6. What does (1.70) mean?

   6. _____

7. What is the length of the 11° angled surface?

   7. _____

8. What is the maximum allowable total length of the part?

   8. _____

9. What is the approximate distance for A?

   9. _____

10. What is the centerline angle between each tooth?

    10. _____

11. What is the length of the 6° 30′ angled surface?

    11. _____

12. What is the approximate height of the part?

    12. _____

13. What is the length of the tip of each tooth?

    13. _____

14. What is the maximum allowable width of the part?

    14. _____

15. What dimensions locate the start of the R.50 radius?

    15. _____

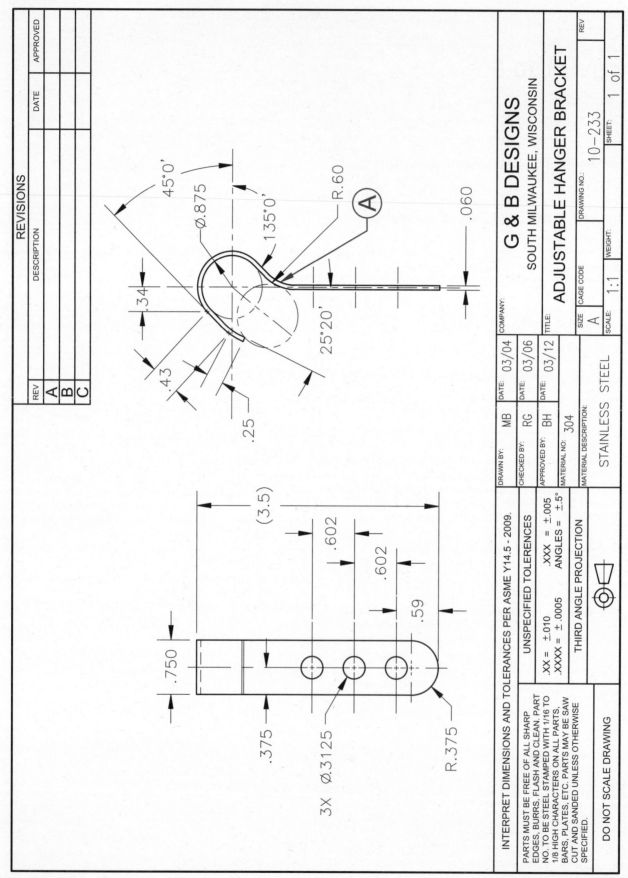

Activity 10-6 Adjustable Hanger Bracket

Name _____ Date _____ Class _____

Activity 10-6
Adjustable Hanger Bracket

*Refer to **Activity 10-6**. Study the drawing and familiarize yourself with the views, dimensions, title block, and notes. Read the questions, refer to the print, and write your answers in the blanks provided.*

1. What dimensions locate the bottommost hole on the part?

    1. _____

2. What is the inside radius of the hook?

    2. _____

3. What angle is the tip of the hook?

    3. _____

4. What dimension locates the start of the 45° angle?

    4. _____

5. What would be the value of the outside radius A?

    5. _____

6. What is the maximum diameter of the three holes?

    6. _____

7. What kind of material is used for the part?

    7. _____

8. List all of the radii on the part.

    8. _____

9. Convert 25°20′ into decimal degrees.

    9. _____

10. The center-to-center distance from the top hole to the bottom hole?

    10. _____

11. List all of the diameters.

    11. _____

12. What is the maximum allowable thickness of the material?

    12. _____

13. What is the total allowance for the height of the part?

    13. _____

14. What dimension locates the start of the 25°20′ angle?

    14. _____

15. What standard is referenced for the drawing?

    15. _____

Notes

Learning Objectives

After studying this unit, you will be able to:

✓ Describe similarities and differences between several 60° thread forms.
✓ Identify the different types of thread representations.
✓ Identify common thread series.
✓ Interpret the meaning of thread specifications on a print.

Key Terms

detailed representation
external thread
internal thread
lead
major diameter
minor diameter

nominal pipe size
pipe thread
pitch
pitch diameter
schematic representation
screw thread

simplified representation
tap
thread
thread class
thread form
thread series

A *thread* or *screw thread* is a "V" shaped groove machined on a shaft or in a hole that follows a helical (spiral) path. Threads are a common component of machined parts and they appear on many machine shop prints. Threads are used in several situations, including the following:

■ Fastening or joining of two or more parts.
■ Conveying power, as commonly found on machine tools in the form of a lead screw.
■ Providing motion or travel, as in the case of measuring tools.

This unit describes common thread terms and definitions, representations, forms, series, classes, specifications, tapped holes, ISO metric threads, and pipe threads.

Thread Terms and Definitions

Threads are external or internal in form. *External threads* are threads on the outside of a shaft, such as bolts and screws. See **Figure 11-1**. *Internal threads* are threads on the inside of a hole, such as hexagonal nuts and tapped holes. See **Figure 11-2**.

The *major diameter* is the largest diameter of the thread. The *minor diameter* is the smallest diameter of the thread. The *pitch diameter* is the diameter of an imaginary cylinder that would pass through the thread at a point halfway between the top (crest) and the bottom (root) of the thread.

The *pitch* is the distance from one thread point to the next thread point, measured parallel to the thread axis. The thread depth is the difference between the crest and root of the

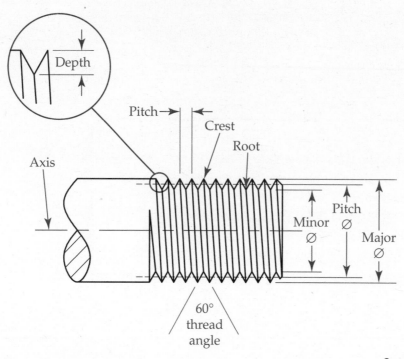

Goodheart-Willcox Publisher

Figure 11-1 Parts of an external thread.

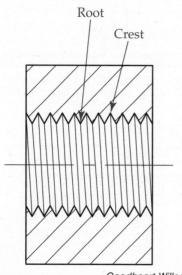

Goodheart-Willcox Publisher

Figure 11-2 Parts of an internal thread.

thread. The *lead* is the distance a nut will travel during one complete rotation of the screw thread. The lead equals the pitch on single thread screws. Multiple screw threads increase the lead.

Thread Representations

A drawing that represents screw threads is a thread representation. The ASME Y14.6-2001, *Screw Thread Representation* standard recommends the use of the simplified, schematic, and detailed methods of representation.

Simplified Representation

The *simplified representation* method uses hidden lines to represent straight and tapered 60° threads. Hidden lines replace detailed line work. The hidden lines are drawn parallel to the center axis to represent the minor diameter of the thread, as shown in **Figure 11-3A**.

Schematic Representation

A *schematic representation* uses alternating short and long lines drawn perpendicular to the center axis to represent the major and minor diameters of the threads. The added line work in a schematic representation communicates the form more effectively than a simplified drawing. See **Figure 11-3B**.

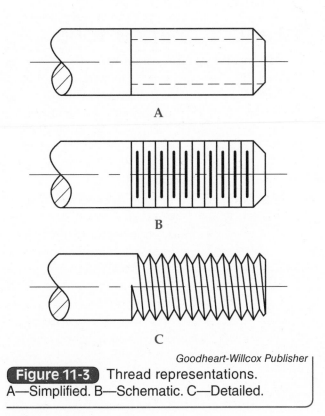

Goodheart-Willcox Publisher

Figure 11-3 Thread representations.
A—Simplified. B—Schematic. C—Detailed.

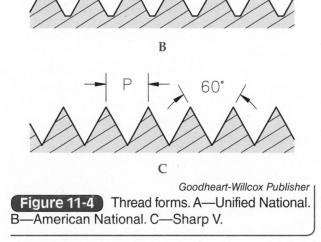

Goodheart-Willcox Publisher

Figure 11-4 Thread forms. A—Unified National.
B—American National. C—Sharp V.

Detailed Representation

A *detailed representation* is a thread drawing that shows an approximate true representation of a thread in a detailed, but simplified form. Detailed representations are used when additional information is needed, such as for new forms or modified threads. See **Figure 11-3C**.

Thread Forms

Thread form is the shape and characteristics that define a thread. Two of the most common thread forms used in the Unites States are the Unified National (UN) and the American National (N) threads. These threads are similar in form, looking like a "V" and having an included angle of 60° between the sides of the thread. The only differences between the two forms are the shape of the crests and roots of their threads. See **Figure 11-4A** and **4B**.

Another thread form similar to the Unified National and American National threads, but not as widely used, is the Sharp V thread, **Figure 11-4C**. The sharp crests and roots of the Sharp V thread are used in applications that require a tight fit and seal.

Thread Series

The *thread series* defines the type of thread based on the number of threads per inch for external and internal threads. Threads are available in various series that match specific applications. These series include coarse threads, fine threads, extra-fine threads, and constant pitch threads. The following series are approved Unified National thread standards.

Coarse Threads

Coarse threads, specified as Unified National Coarse (UNC), are used for general applications and automatic assembly. Nuts, bolts, and screws that have coarse threads are resistant to stripping.

Fine Threads

Fine threads, specified as Unified National Fine (UNF), have more threads per square inch than coarse threads, which give them more holding power. Situations calling for high strength and vibration resistance require fine threads. A fine thread is also suited for uses that require a short length of engagement or a small lead angle.

Extra-Fine Threads

Extra-fine threads, specified as Unified National Extra-Fine (UNEF), are used for special applications, such as thin wall tubing, nuts, or couplings. They are also suited for uses that require very short lengths of engagement.

Constant-Pitch Threads

Constant-pitch threads, specified as Unified National Constant-Pitch (UN), are used when the other thread series will not work, such as in high-pressure applications. UN threads are specified by a number representing the number of threads per inch (4, 6, 8, 12, 16, 20, 28, or 32 threads per inch). The 8-, 12-, and 16-UN thread series act as larger versions of the coarse, fine, and extra-fine series, respectively.

Thread Class

The *thread class* defines the fit or tolerance between an external thread and the mated internal thread. Fits come in three designations. Class 1 indicates a loose fit for easy assembly, Class 2 fits are the most commonly used, and Class 3 designates close tolerance, tight fits. The numeral designation is followed by a letter, either "A" for external threads or "B" for internal threads.

Thread Specifications

Specifications for standard threads are included in a thread note attached to a thread representation, as shown in **Figure 11-5**. A standard thread note contains the following information:

1. **Thread size.** Nominal diameter of the thread (major diameter) expressed as a fraction, decimal, or whole number followed by a dash (–).
2. **Number of threads per inch.** Unless designated, all screw threads are single threads. To identify multiple-start threads, the "threads per inch" note is replaced with the pitch (P) and lead (L) in inches, and the number of threads in parentheses. Double threads are noted "2 STARTS" and triple threads are noted "3 STARTS." See **Figure 11-6**.
3. **Thread form and series.** Example: UNC for Unified National Coarse series.
4. **Thread class.** Class 1, 2, or 3 fit designation and the letter "A" for external threads or "B" for internal threads.
5. **Direction of threads.** Unless designated, threads are right-handed. Left-hand threads are designated by the letters "LH."
6. **Thread depth.** A specification for the depth of internal threads or length of external threads is optional. No specification or a "THRU" note indicates the threads go through the part.

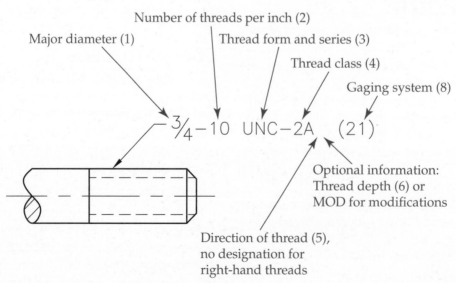

Figure 11-5   Basic thread specifications.

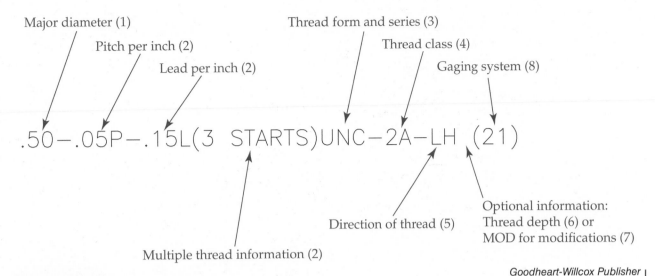

Goodheart-Willcox Publisher

Figure 11-6 Multiple start thread specifications.

The following are additional specifications for newer standards:

7. **Modifications.** The optional "MOD" specification placed at the end of the note indicates the threads are modified in form. The specified information for the modifications is listed on additional lines or in a separate drawing note.

8. **Gaging system information.** Information included for inspection purposes. Placed in parentheses or included in a separate drawing note.

Tapped Holes

A tapped hole is a hole with internal threads cut into its surface with a thread-cutting tool called a *tap*. The size of the hole must meet the specifications for the size of the tap. A tap drill is a drill that cuts the required size hole for a specified tap. A drawing with a note specifies a threaded hole on a drawing, as shown in **Figure 11-7**. The depth of the hole is specified in the top view note or it is dimensioned on a front, side, or section view.

ISO Metric Threads

The International Organization for Standardization (ISO) established metric screw thread standards in 1949. ISO metric threads are specified in millimeters and are similar to the UN thread forms except for their diameter and pitch measurements.

Metric Thread Series

The following are the three series of metric threads:

- ISO metric coarse pitch
- ISO metric fine pitch
- ISO metric constant pitch

ISO metric coarse pitch threads are common on fasteners. ISO metric fine pitch threads are chiefly used on precision measuring tools and instruments. The ISO metric constant pitch series are used on machine parts and spark plugs.

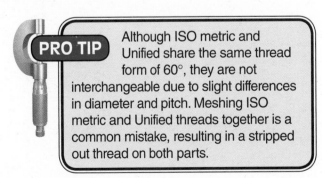

PRO TIP Although ISO metric and Unified share the same thread form of 60°, they are not interchangeable due to slight differences in diameter and pitch. Meshing ISO metric and Unified threads together is a common mistake, resulting in a stripped out thread on both parts.

Metric Thread Specifications

Specifications for ISO metric threads are included in a thread note, as shown in **Figure 11-8**, and are similar to standard threads. An ISO

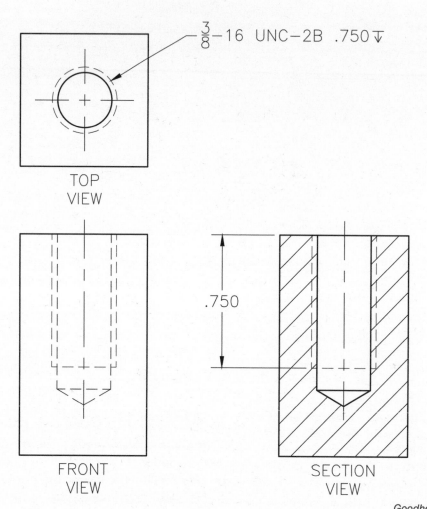

$\frac{3}{8}$–16 UNC–2B .750 ▽

TOP
VIEW

.750

FRONT
VIEW

SECTION
VIEW

Goodheart-Willcox Publisher

Figure 11-7   Simplified representation of a threaded hole.

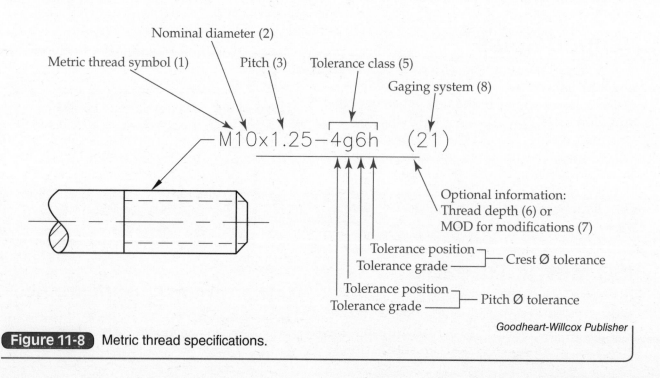

Nominal diameter (2)

Metric thread symbol (1)   Pitch (3)   Tolerance class (5)

Gaging system (8)

M10x1.25—4g6h   (21)

Optional information:
Thread depth (6) or
MOD for modifications (7)

Tolerance position ──┐
Tolerance grade ──── └── Crest Ø tolerance

Tolerance position ──┐
Tolerance grade ──── └── Pitch Ø tolerance

Goodheart-Willcox Publisher

Figure 11-8   Metric thread specifications.

metric screw thread note contains the following information:

1. **Metric thread symbol.** Metric thread specifications always begin with the capital letter "M."
2. **Thread size.** Nominal diameter of the thread (major diameter) expressed in millimeters, followed by an "×".
3. **Pitch.** Instead of specifying the number of threads per length, the pitch is specified in millimeters, followed by a dash (–). Unless designated, all screw threads are single thread. To identify multiple-start threads, the lead (L) and pitch (P) are given in millimeters, followed by the number of threads in parentheses. Double threads are noted as "2 STARTS" and triple threads are noted as "3 STARTS." See **Figure 11-9**.
4. **Direction of threads.** Unless designated, threads are right-handed. Left-hand threads are designated by the letters "LH."
5. **Tolerance class.** To specify the tolerance class, the grade and position of the pitch diameter is listed first and the grade and position of the crest diameter is listed second. External tolerance grades range from 3 to 9 (fine to coarse), and use the letter "g" or "h" to signify the position. Internal tolerance grades range from 4 to 8 (fine to coarse), and use the letter "G" or "H" to signify the position. The "g" and "G" positions indicate a small

allowance. The "h" and "H" positions indicate no allowance. If the two diameters have the same tolerance class, only one is given. Examples: External tolerance class of 4g6h. Internal tolerance class of 5H.

6. **Thread depth.** A specification for the depth of internal threads or length of external threads is optional. No specification or a "THRU" note indicates the threads go through the part.

The following are additional specifications for newer standards:

7. **Modifications.** The optional "MOD" specification placed at the end of the note indicates the threads are modified in form. The specified information for the modifications is listed on additional lines or in a separate drawing note.
8. **Gaging system information.** Information included for inspection purposes. Placed in parentheses or included in a separate drawing note.

Pipe Threads

Pipe threads are specifically used for joining pipes and fittings. There is a variety of pipe threads used for different purposes. The two common types approved by the American National Standard are tapered and straight pipe threads.

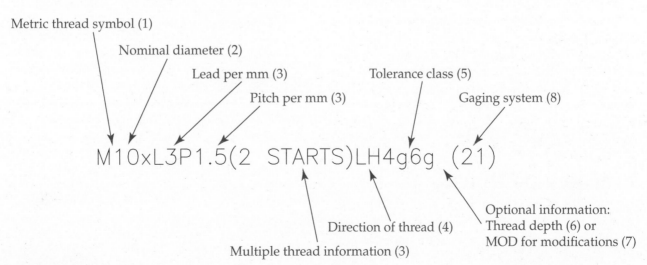

Metric thread symbol (1)

Nominal diameter (2)

Lead per mm (3)

Pitch per mm (3)

Tolerance class (5)

Gaging system (8)

M10xL3P1.5(2 STARTS)LH4g6g (21)

Direction of thread (4)

Multiple thread information (3)

Optional information: Thread depth (6) or MOD for modifications (7)

Goodheart-Willcox Publisher

Figure 11-9   Metric multiple start thread specifications.

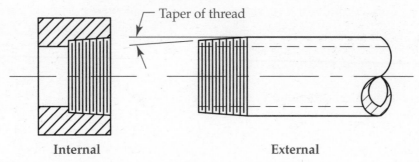

Goodheart-Willcox Publisher

Figure 11-10 Tapered pipe thread representations.

Tapered Pipe Threads

Tapered pipe threads differ from straight machine threads because they have a taper along their threaded length. See **Figure 11-10**. The American National Standard for joining tapered pipe threads and fittings is the National Pipe Taper (NPT). NPT conforms to the ANSI/ASME B1.20.1 standard, which specifies a 0.75 inch per foot taper, 60° included angle thread form, and flat crests and roots. The specified taper yields an angle of 1°47′ measured from the centerline of the threads.

NPT threads are specified for normal use, including situations where pressure is involved. The taper ensures easy starting and forms a tight seal with its mated fitting. Pipe thread sealant tape or sealant compound is required when joining parts to provide a leak free fit. NPT threads are used in plumbing, heating, ventilation, air-conditioning, and hydraulic applications.

The National Pipe Taper Fuel (NPTF) variant, also called the Dryseal American Standard pipe thread, produces a leak free seal without the use of sealing compounds. The thread is similar in design to the NPT thread, but the roots of the thread have wider flats than the crests. This causes the crests to deform during assembly, creating a seal. Dryseal pipes are used for fuel applications.

Straight Pipe Threads

Straight pipe threads have a 60° thread parallel to the centerline axis of the thread. Common American National Standard straight pipe thread designations include the National Pipe Straight Couplings (NPSC), for use in applications with low internal pressures, and the

National Pipe Straight Mechanical (NPSM), for use in mechanical assemblies with no internal pressures.

Pipe Thread Specifications

Tapered and straight pipe threads are specified by their nominal pipe size. The *nominal pipe size* refers to a set of standard pipe sizes related to the inside diameter of the pipe. The nominal sizes range from 1/16″ to 24″. A typical pipe thread note specifies its nominal size, followed by the number of threads per inch, and ending with its thread series and form. See **Figure 11-11**.

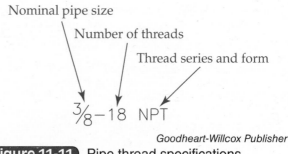

Goodheart-Willcox Publisher

Figure 11-11 Pipe thread specifications.

Notes

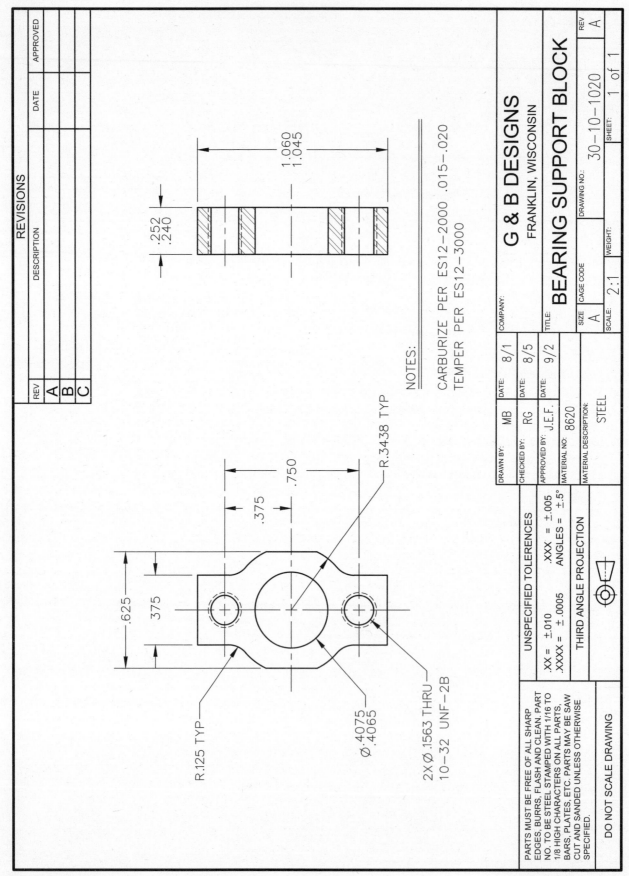

NOTES:

CARBURIZE PER ES12-2000 .015-.020
TEMPER PER ES12-3000

R.3438 TYP

R.125 TYP

.750

.375

.625

375

Ø.4075
.4065

2X Ø.1563 THRU
10-32 UNF-2B

1.060
1.045

.252
.240

REVISIONS			
REV	DESCRIPTION	DATE	APPROVED
A			
B			
C			

COMPANY:

G & B DESIGNS
FRANKLIN, WISCONSIN

TITLE:

BEARING SUPPORT BLOCK

DRAWN BY: MB	DATE: 8/1	
CHECKED BY: RG	DATE: 8/5	
APPROVED BY: J.E.F.	DATE: 9/2	
MATERIAL NO: 8620		
MATERIAL DESCRIPTION: STEEL		

SIZE A	CAGE CODE	DRAWING NO: 30-10-1020	REV A
	SCALE: 2:1	WEIGHT:	SHEET: 1 of 1

UNSPECIFIED TOLERENCES

.XX = ±.010   .XXX = ±.005
.XXXX = ±.0005   ANGLES = ±.5°

THIRD ANGLE PROJECTION

PARTS MUST BE FREE OF ALL SHARP
EDGES, BURRS, FLASH AND CLEAN. PART
NO. TO BE STEEL STAMPED WITH 1/16 TO
1/8 HIGH CHARACTERS ON ALL PARTS,
BARS, PLATES, ETC. PARTS MAY BE SAW
CUT AND SANDED UNLESS OTHERWISE
SPECIFIED.

DO NOT SCALE DRAWING

Activity 11-1 Bearing Support Block.

Name _____ Date _____ Class _____

Activity 11-1
Bearing Support Block

Refer to **Activity 11-1**. *Study the drawing and familiarize yourself with the views, dimensions, title block, and notes. Read the questions, refer to the print, and write your answers in the blanks provided.*

1. What is the diameter of the large hole?

2. What is the tolerance for the reamed hole?

3. What is the distance between the two tapped holes?

4. What are the specifications for the two threaded holes?

5. How far is the center point of the bottom-threaded hole to the horizontal centerline of the large hole?

6. What is the maximum height of the part?

7. What is the maximum width allowed on the part?

8. What is the minimum width allowed on the part?

9. What scale is the print?

10. What material is used for the part?

11. What is the minimum thickness of the part?

12. What is the maximum thickness allowed for the part?

13. What is the unspecified tolerance for three-place decimal dimensions?

14. What fit is called for on the threaded holes?

15. To what depth is the part carburized?

16. Have there been any print revisions?

17. List all of the radii on the print.

18. List all of the fillets on the print.

19. What is the thread series for the small holes?

20. What heat treatment is performed according to ES 12-3000?

1. _____

2. _____

3. _____

4. _____

5. _____

6. _____

7. _____

8. _____

9. _____

10. _____

11. _____

12. _____

13. _____

14. _____

15. _____

16. _____

17. _____

18. _____

19. _____

20. _____

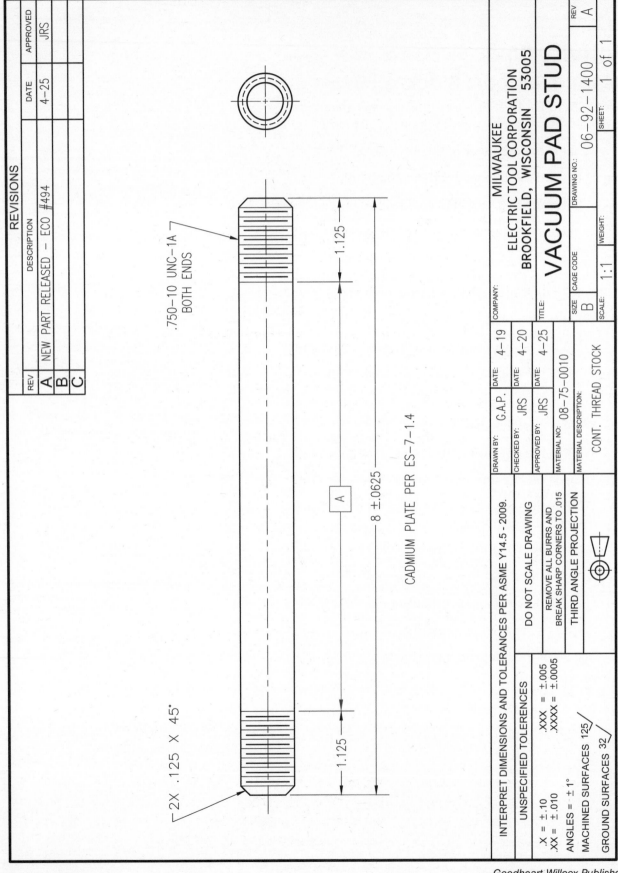

REVISIONS

REV	DESCRIPTION	DATE	APPROVED
A	NEW PART RELEASED – ECO #494	4-25	JRS
B			
C			

.750-10 UNC-1A
BOTH ENDS

1.125

8 ±.0625

A

CADMIUM PLATE PER ES-7-1.4

2X .125 X 45°

1.125

INTERPRET DIMENSIONS AND TOLERANCES PER ASME Y14.5 - 2009.

UNSPECIFIED TOLERENCES

.X = ±.10   .XXX = ±.005
.XX = ±.010   .XXXX = ±.0005
ANGLES = ±1°
MACHINED SURFACES 125
GROUND SURFACES 32

DO NOT SCALE DRAWING

REMOVE ALL BURRS AND
BREAK SHARP CORNERS TO .015

THIRD ANGLE PROJECTION

DRAWN BY: G.A.P.	DATE: 4-19
CHECKED BY: JRS	DATE: 4-20
APPROVED BY: JRS	DATE: 4-25

MATERIAL NO: 08-75-0010

MATERIAL DESCRIPTION:
CONT. THREAD STOCK

COMPANY:

MILWAUKEE
ELECTRIC TOOL CORPORATION
BROOKFIELD, WISCONSIN 53005

TITLE:
VACUUM PAD STUD

| SIZE B | CAGE CODE | DRAWING NO: 06-92-1400 | REV A |

SCALE: 1:1   WEIGHT: SHEET: 1 of 1

Activity 11-2   Vacuum Pad Stud.

Name _____ Date _____ Class _____

Activity 11-2
Vacuum Pad Stud

Refer to **Activity 11-2.** *Study the drawing and familiarize yourself with the views, dimensions, title block, and notes. Read the questions, refer to the print, and write your answers in the blanks provided.*

1. What is the scale of the print?

2. What two views are shown on the print?

3. What size are the two chamfers on the part?

4. What does "1A" mean when referring to threads?

5. What does UNC signify?

6. How long are the threads on each end of the part?

7. What kind of thread representation is shown on the print?

8. How many threads per inch is the thread?

9. What is the material number?

10. Approximately how many threads are specified for one end of the stud?

11. What is the high limit on the overall length?

12. What diameter is the stud?

13. What operation is performed according to ES-7-1.4?

14. What is the material description?

15. What tolerance is given on the length of the stud?

16. What is the value of dimension A?

17. What class fit is the thread?

18. What print change was made?

19. What is the part number of the part?

20. What is the major diameter of the thread?

1. _____

2. _____

3. _____

4. _____

5. _____

6. _____

7. _____

8. _____

9. _____

10. _____

11. _____

12. _____

13. _____

14. _____

15. _____

16. _____

17. _____

18. _____

19. _____

20. _____

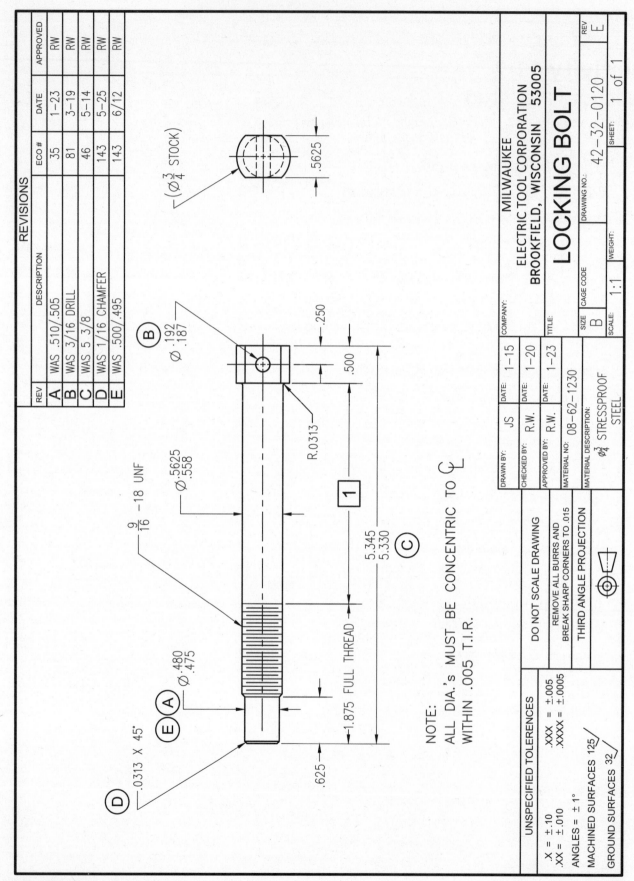

Activity 11-3 Locking Bolt.

Name _____ Date _____ Class _____

Activity 11-3
Locking Bolt

Refer to **Activity 11-3**. *Study the drawing and familiarize yourself with the views, dimensions, title block, and notes. Read the questions, refer to the print, and write your answers in the blanks provided.*

1. What size hole is drilled in the part?

2. What is the largest outside diameter?

3. How much tolerance is given on the overall length of the part?

4. What print change was made at B?

5. What size radius is specified on the print?

6. How long is the thread?

7. What size is the thread diameter?

8. What does UNF mean?

9. What did ECO 46 specify?

10. What material is the part?

11. What is the low limit on the smallest outside diameter?

12. What is the value of dimension 1?

13. What are the minimum and maximum allowable diameters for the main body?

14. What is the width across the flats of the locking bolt head?

15. How far is the drilled hole from center point to the right end of the part?

16. What is the scale of the print?

17. What was the original size of the smallest outside diameter?

18. How concentric must the diameters be to the centerline of the part?

19. What is the low limit on the length of the bolt head dimension?

20. What is the tolerance for the .625 length dimension?

1. _____

2. _____

3. _____

4. _____

5. _____

6. _____

7. _____

8. _____

9. _____

10. _____

11. _____

12. _____

13. _____

14. _____

15. _____

16. _____

17. _____

18. _____

19. _____

20. _____

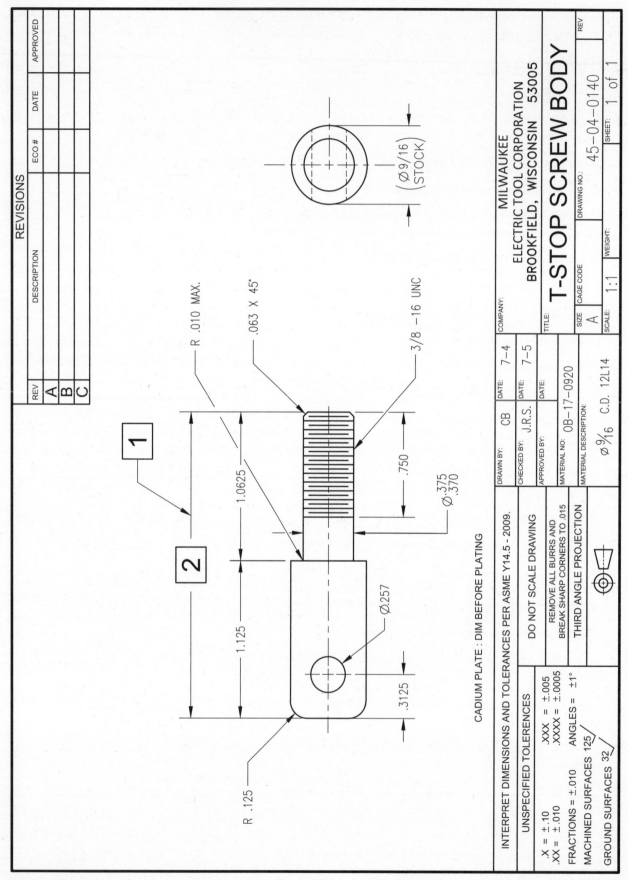

Activity 11-4 T-Stop Screw Body.

Name _____ Date _____ Class _____

Activity 11-4
T-Stop Screw Body

Refer to **Activity 11-4**. *Study the drawing and familiarize yourself with the views, dimensions, title block, and notes. Read the questions, refer to the print, and write your answers in the blanks provided.*

1. What material number is specified for the part?
2. What size is the large diameter of the part?
3. What two views are shown on the print?
4. What kind of line is line 1?
5. What size fillet is specified on the print?
6. What is the maximum allowable size for the hole?
7. Does the hole go through the part?
8. Including the chamfer, how long is the thread?
9. How far is the center point of the hole from start of the Ø.375/.370 section?
10. What material is the part?
11. How many threads per inch are specified for the screw?
12. How long is the Ø.375 section?
13. What is the value of dimension 2?
14. What thread series is specified?
15. What is the distance from the left edge of the hole to the left end of the part?
16. What size radius is specified on the end of the part?
17. What finish material is specified for the part?
18. What size thread is called for on the print?
19. What is the maximum length allowed on the Ø.5625 section?
20. What size chamfer is specified on the print?

1. _____
2. _____
3. _____
4. _____
5. _____
6. _____
7. _____
8. _____
9. _____
10. _____
11. _____
12. _____
13. _____
14. _____
15. _____
16. _____
17. _____
18. _____
19. _____
20. _____

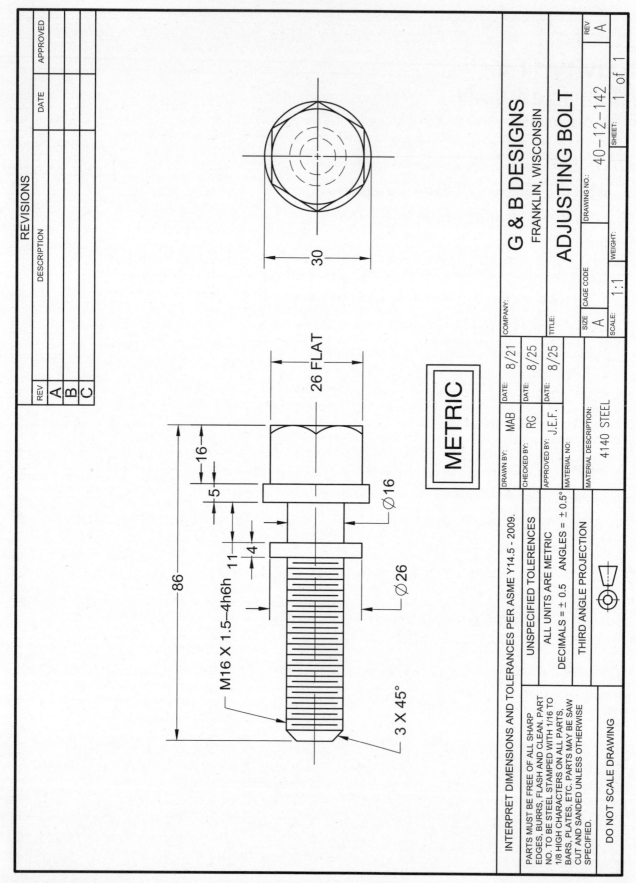

METRIC

G & B DESIGNS
FRANKLIN, WISCONSIN

ADJUSTING BOLT

DRAWING NO.: 40-12-142

REV A

SHEET: 1 of 1

SIZE A   CAGE CODE   WEIGHT:

SCALE: 1:1

COMPANY:

TITLE:

DRAWN BY: MAB   DATE: 8/21
CHECKED BY: RG   DATE: 8/25
APPROVED BY: J.E.F.   DATE: 8/25
MATERIAL NO.:
MATERIAL DESCRIPTION: 4140 STEEL

INTERPRET DIMENSIONS AND TOLERANCES PER ASME Y14.5 - 2009.

UNSPECIFIED TOLERENCES
ALL UNITS ARE METRIC
DECIMALS = ± 0.5   ANGLES = ± 0.5°

THIRD ANGLE PROJECTION

PARTS MUST BE FREE OF ALL SHARP
EDGES, BURRS, FLASH AND CLEAN. PART
NO. TO BE STEEL STAMPED WITH 1/16 TO
1/8 HIGH CHARACTERS ON ALL PARTS,
BARS, PLATES, ETC. PARTS MAY BE SAW
CUT AND SANDED UNLESS OTHERWISE
SPECIFIED.

DO NOT SCALE DRAWING

30

26 FLAT

M16 X 1.5–4h6h

16
5
4
11
86

Ø16
Ø26

3 X 45°

Activity 11-5   Adjusting Bolt.

Name _____ Date _____ Class _____

Activity 11-5
Adjusting Bolt

Refer to **Activity 11-5**. *Study the drawing and familiarize yourself with the views, dimensions, title block, and notes. Read the questions, refer to the print, and write your answers in the blanks provided.*

1. What is the high limit on the largest diameter?

    1. _____

2. What is the low limit on the total length of the part?

    2. _____

3. What is the major diameter of the threads?

    3. _____

4. What tolerance is specified for angles on a metric print?

    4. _____

5. Including the chamfer, what is the length of the threaded portion of the bolt?

    5. _____

6. What is the pitch of the thread?

    6. _____

7. What signifies that the thread is metric?

    7. _____

8. What is the width of the groove on the bolt?

    8. _____

9. What is the part number?

    9. _____

10. What is the flat to flat width of the bolt head?

    10. _____

11. What is the diameter of the groove?

    11. _____

12. What is the direction of the threads?

    12. _____

13. The part is made from what material?

    13. _____

14. What is the diameter of the thread?

    14. _____

15. What is lowest allowable dimension for the flats?

    15. _____

16. What tolerance is specified for the width of the groove?

    16. _____

17. Approximately how many threads are specified for the bolt?

    17. _____

18. What is the distance from the top of the bolt head to the bottom of the ∅26 mm collar?

    18. _____

19. What thread chamfer is specified?

    19. _____

20. What tolerance class is specified for the pitch diameter?

    20. _____

Notes

Learning Objectives

After studying this unit, you will be able to:

✓ Determine the difference between a neck and a groove.
✓ Determine the sizes of necks and grooves.
✓ Identify common types of slots.
✓ Identify type and size of a keyway and keyseat.
✓ Identify a flat.
✓ Determine the difference between a boss and a pad.

Key Terms

boss	key	pad
dovetail slot	keyseat	slot
flat	keyway	T-slot
groove	neck	V-groove

This unit groups together various topics that are important to interpreting elementary level industrial prints. The first six topics share commonality and can be grouped under two headings: the fastening group and the assembly group. Each of the six topics (necks, grooves, slots, keyways, keyseats, and flats) is produced by a machining operation. All six are used in the fastening and assembly of manufactured parts. The remaining two topics, bosses and pads, deal with material allowance found on parts.

Necks

Necks (sometimes called "grooves") are recesses in the outside diameters of parts to allow mating objects to fit flush to each other. See **Figure 12-1**. Necks are frequently used when it is necessary to have a threaded diameter assemble flush to a shoulder. See **Figure 12-2**.

Usually, a radius is found between stepped diameters of a part. However, when that radius

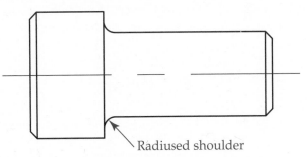

Radiused shoulder

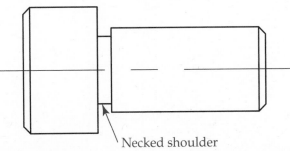

Necked shoulder

Goodheart-Willcox Publisher

Figure 12-1 Drawings show radiused and necked shoulders on machined parts.

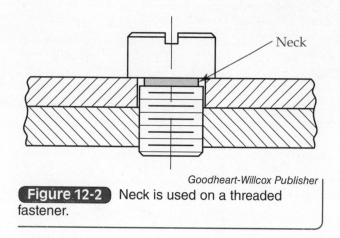

Goodheart-Willcox Publisher

Figure 12-2 Neck is used on a threaded fastener.

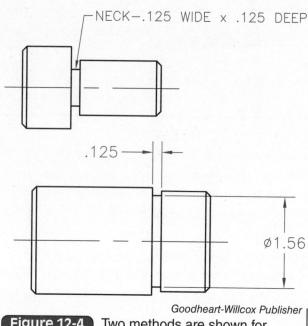

NECK—.125 WIDE x .125 DEEP

.125

Ø1.56

Goodheart-Willcox Publisher

Figure 12-4 Two methods are shown for dimensioning necks.

interferes with the proper assembly of two objects, a necking operation is performed, **Figure 12-3**.

Necks can be dimensioned by stating the width and depth of the recess, using a note, or by stating the width and diameter of the recess, **Figure 12-4**. When the size of the neck is not important, no dimension is given.

Grooves

Annular (ring-like) *grooves* can be found on both inside and outside diameters of parts, **Figure 12-5**. Grooves are frequently used for mounting fasteners, such as snap rings and retaining rings. They are also used for installing seals, such as O-rings, and for passageways for lubricating oils.

Annular grooves may be dimensioned by stating the diameter and width or by providing the depth and width. See **Figure 12-6**.

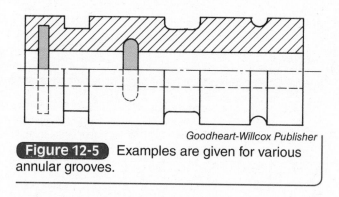

Goodheart-Willcox Publisher

Figure 12-5 Examples are given for various annular grooves.

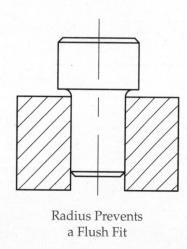

Radius Prevents
a Flush Fit

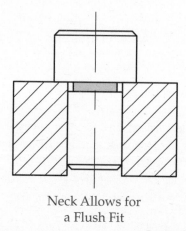

Neck Allows for
a Flush Fit

Goodheart-Willcox Publisher

Figure 12-3 Using a neck for a flush fit.

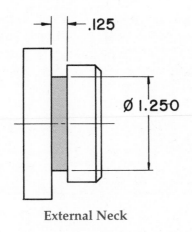

External Neck

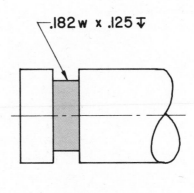

External Groove

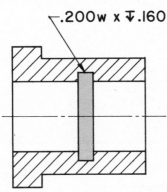

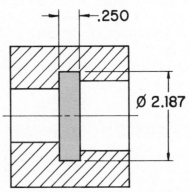

Recess or Internal Groove

Goodheart-Willcox Publisher

Figure 12-6 Drawings present methods of dimensioning external and internal grooves.

The *V-groove* is a familiar type of groove found on pulleys used with V-belts. V-grooves are produced in many forms, which vary in angles, widths, and depth. See **Figure 12-7**.

Some shafts have specially designed grooves as shown in **Figure 12-8**. This requires the use of special ground (machined) tooling.

An internal groove is sometimes called a "recess" or "undercut." Some external grooves on cylindrical parts may be called a "neck" when they occur at a shoulder.

Slots

Two principal types of *slots* used on machines and other parts include the T-slot (tee slot) and the dovetail slot, **Figure 12-9**. *T-slots* are used on machine tables for the purpose of fastening down devices such as vises, clamps, straps, and fixtures. T-bolts are used with the table to fasten down the holding devices and workpieces.

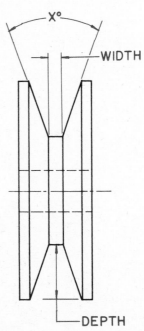

Goodheart-Willcox Publisher

Figure 12-7 Dimensioning requirements are given for a V-groove on a pulley.

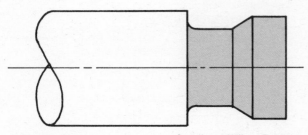

Goodheart-Willcox Publisher

Figure 22-8 Drawing shows an example of a special groove on a shaft end.

Dovetail slots are used on machine tools as slides, creating an interlocking assembly between two machine parts to provide a reciprocating movement.

Common applications of dovetails include cross-slide and compound rest of a lathe, table and saddle of a milling machine, and tool-slide on an adjustable boring head.

Keyways and Keyseats

A *keyway* is an internal groove machined in a hole along its length which provides a slot for inserting a key.

A *keyseat* is an external groove cut into a shaft along its length to provide a seat for a key.

There are many types of keys used in keyways and keyseats. Common keys include the flat, square, and Woodruff types. See **Figures 12-10**, **12-11**, and **12-12**.

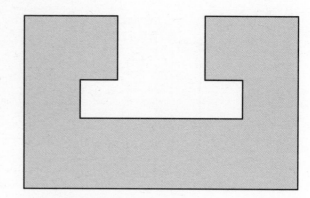

T-Slot

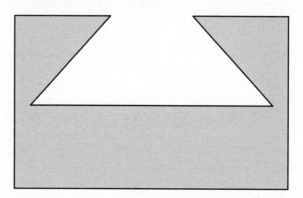

Dovetail

Goodheart-Willcox Publisher

Figure 12-9 Examples are given for a T-slot and a dovetail slot, used on machine tools.

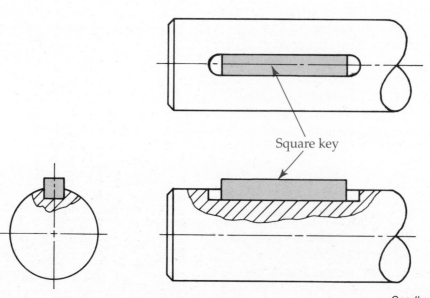

Square key

Goodheart-Willcox Publisher

Figure 12-10 Three views show how a square key fits into a keyseat in a shaft.

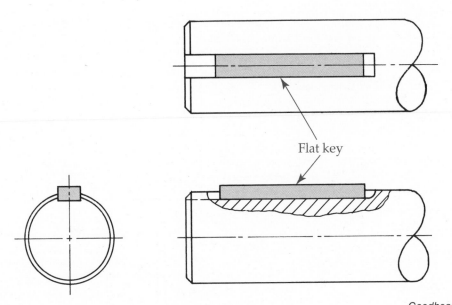

Flat key

Goodheart-Willcox Publisher

Figure 12-11 Three views illustrate fit of a flat key in a keyseat.

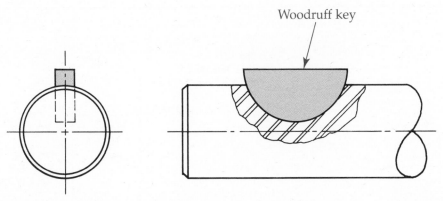

Woodruff key

Goodheart-Willcox Publisher

Figure 12-12 Two views picture a Woodruff key in a keyseat in a shaft.

A *key* is used to secure a part to a shaft. Parts secured in this manner include gears, pulleys, collars (spacers), cranks, handles, handwheels, and cutting tools. A key provides a positive force and prevents the part from slipping on the shaft.

Various tolerances are used for keys, keyways, and keyseats. These specifications can be found in reference texts such as the *Machinery's Handbook*.

Keyways and keyseats for flat and square keys are dimensioned with leader lines. The width dimension is given first, followed by the depth. The length is given by using a direct dimension on the print. See **Figure 12-13**.

Woodruff keyseats, which are semicircular in shape, are dimensioned with leader lines. The

number of the Woodruff keyseat and its location along the shaft are given. See **Figure 12-14**. These keyseats are machined with special cutters called Woodruff cutters. For positive identification, the number of the Woodruff cutter corresponds identically with the number of the Woodruff key.

The number of the Woodruff key identifies its size. The last two digits of the number give the diameter of the key in eighths of an inch. The digits preceding the last two digits give the width of the key in 1/32nds of an inch.

For example, the numbers in a #807 key mean the key is 8/32″ × 7/8″ or 1/4″ wide × 7/8″ diameter.

For precision fits such as required for interchangeable assemblies, keyway and keyseat

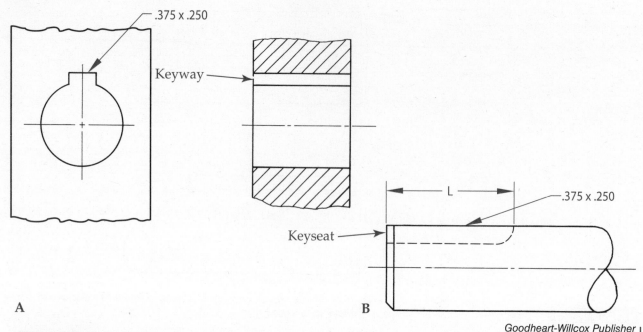

Figure 12-13 A—Dimensioning requirements for a keyway (flat key). B—Dimensioning requirements for a keyseat (flat key).

Goodheart-Willcox Publisher

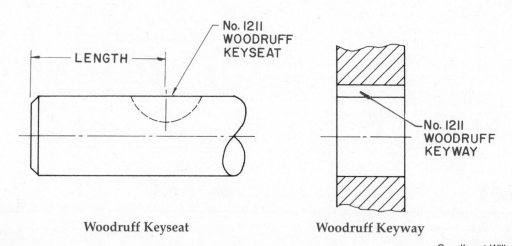

Figure 12-14 Woodruff key number (1211) identifies its size: 12/32 × 11/8 or 3/8 × 1 3/8 (see text for explanation).

Goodheart-Willcox Publisher

dimensions are given with limit dimensions, **Figure 12-15**, to provide proper fits.

Flats

A *flat* is a depression found on a shaft or shank that provides a seat for a setscrew. The setscrew is used with another part to hold that object in place on the shaft or shank. Many straight tool shanks have flats on them, **Figure 12-16**, to secure the tool to a tool holder.

Bosses and Pads

A *boss* is a cylindrical (round) raised surface found on a casting or forging which provides additional material to the part. A raised surface with a shape other than round is called a *pad*.

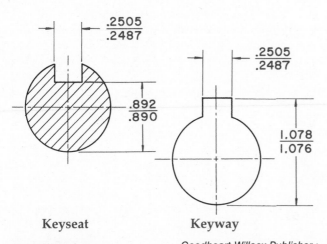

Keyseat **Keyway**

Goodheart-Willcox Publisher

Figure 12-15 Precision fits of keys, keyseats, and keyways require limit dimensioning on drawing.

Flats

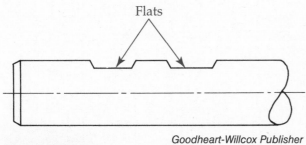

Goodheart-Willcox Publisher

Figure 12-16 Drawing depicts flats on a tool shank.

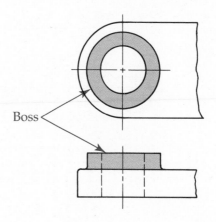

Boss

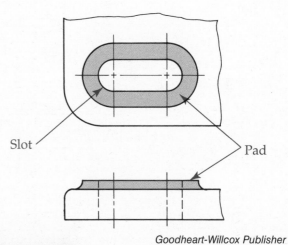

Slot

Pad

Goodheart-Willcox Publisher

Figure 12-17 Examples show a boss and a pad in two views.

Both bosses and pads are machined to provide smooth surfaces for mating parts. Bosses usually have holes machined in or through them, while pads are commonly found both with and without a slot milled through them. See **Figure 12-17**.

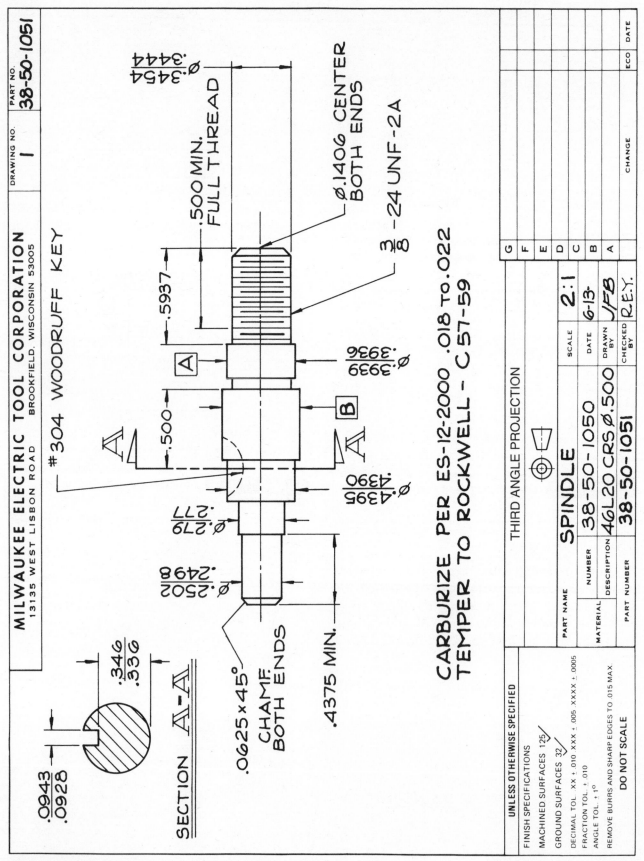

MILWAUKEE ELECTRIC TOOL CORPORATION
13135 WEST LISBON ROAD   BROOKFIELD, WISCONSIN 53005

DRAWING NO. 1

PART NO. 38-50-1051

#304 WOODRUFF KEY

Ø.3454/.3444

.500 MIN. FULL THREAD

.5937

Ø.1406 CENTER BOTH ENDS

3/8 - 24 UNF-2A

A

Ø.3939/.3936

.500

B

Ø.4395/.4390

Ø.279/.277

Ø.2502/.2498

.0625×45° CHAMF. BOTH ENDS

.4375 MIN.

CARBURIZE PER ES-12-2000 .018 to .022
TEMPER TO ROCKWELL - C 57-59

.346/.336

.0943/.0928

SECTION A-A

			CHANGE	ECO	DATE
G					
F					
E					
D					
C					
B					
A					

THIRD ANGLE PROJECTION

SCALE **2:1**

PART NAME **SPINDLE**

	NUMBER	38-50-1050	DATE 6-13-	DRAWN BY *JFB*
MATERIAL	DESCRIPTION	46L20 CRS Ø.500		CHECKED BY *R.E.Y.*

PART NUMBER 38-50-1051

UNLESS OTHERWISE SPECIFIED

FINISH SPECIFICATIONS
MACHINED SURFACES 125/
GROUND SURFACES 32/
DECIMAL TOL. XX ± .010 XXX ± .005 XXXX ± .0005
FRACTION TOL. ± .010
ANGLE TOL. ± 1°
REMOVE BURRS AND SHARP EDGES TO .015 MAX.
DO NOT SCALE

Goodheart-Willcox Publisher

Activity 12-1 Spindle.

Name _____ Date _____ Class _____

Activity 12-1
Spindle

Refer to **Activity 12-1**. *Study the drawing and familiarize yourself with the views, dimensions, title block, and notes. Read the questions, refer to the print, and write your answers in the blanks provided.*

1. What size is the thread?

2. What tolerance is given for the .250 diameter?

3. How wide is the keyseat?

4. Which diameter has the smallest tolerance?

5. What is dimension A?

6. What size is the Woodruff key?

7. What size is the outside diameter of the thread?

8. What scale is the print?

9. What material is the part?

10. What is the dimension of the chamfer on the thread?

11. What is dimension B?

12. To what hardness is the part tempered?

13. Does this part feature a neck?

14. What dimension shows the location of the keyseat?

15. What is the minimum length of the thread?

16. To what depth is the part hardened?

17. What does ES mean?

18. State what finish is required for ground surfaces.

19. What thread series is specified on the print?

20. What does 2A mean on the print?

1. _____

2. _____

3. _____

4. _____

5. _____

6. _____

7. _____

8. _____

9. _____

10. _____

11. _____

12. _____

13. _____

14. _____

15. _____

16. _____

17. _____

18. _____

19. _____

20. _____

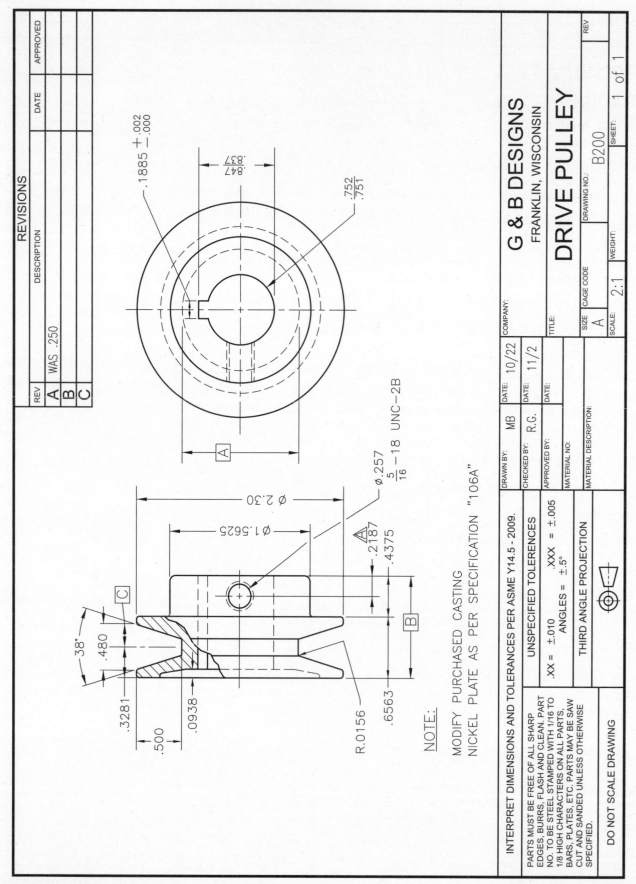

NOTE:

MODIFY PURCHASED CASTING
NICKEL PLATE AS PER SPECIFICATION "106A"

REV	DESCRIPTION	DATE	APPROVED
A	WAS .250		
B			
C			

REVISIONS

INTERPRET DIMENSIONS AND TOLERANCES PER ASME Y14.5 - 2009.

UNSPECIFIED TOLERENCES		
.XX = ±.010	.XXX = ±.005	
ANGLES = ±.5°		

THIRD ANGLE PROJECTION

PARTS MUST BE FREE OF ALL SHARP EDGES, BURRS, FLASH AND CLEAN. PART NO. TO BE STEEL STAMPED WITH 1/16 TO 1/8 HIGH CHARACTERS ON ALL PARTS, BARS, PLATES, ETC. PARTS MAY BE SAW CUT AND SANDED UNLESS OTHERWISE SPECIFIED.

DO NOT SCALE DRAWING

COMPANY:

G & B DESIGNS
FRANKLIN, WISCONSIN

TITLE:

DRIVE PULLEY

DRAWN BY: MB	DATE: 10/22
CHECKED BY: R.G.	DATE: 11/2
APPROVED BY:	DATE:
MATERIAL NO:	
MATERIAL DESCRIPTION:	

| SIZE A | CAGE CODE | DRAWING NO.: B200 | REV |
| SCALE: 2:1 | WEIGHT: | SHEET: 1 of 1 |

Name _____ Date _____ Class _____

Activity 12-2
Drive Pulley

Refer to **Activity 12-2**. *Study the drawing and familiarize yourself with the views, dimensions, title block, and notes. Read the questions, refer to the print, and write your answers in the blanks provided.*

1. How wide is the drive pulley groove at the outside?   1. _____

2. How wide is the hub of the pulley?   2. _____

3. How wide is the pulley keyway?   3. _____

4. How deep is the pulley groove?   4. _____

5. What is the maximum outside diameter of the pulley?   5. _____

6. What is dimension A?   6. _____

7. How long is the threaded hole?   7. _____

8. What is dimension B?   8. _____

9. What is the approximate length of the keyway?   9. _____

10. What kind of material is used for the pulley?   10. _____

11. What angle is the groove?   11. _____

12. What diameter is the hub of the pulley?   12. _____

13. What is the low limit of the hole with the keyway?   13. _____

14. What does the large hidden circle represent?   14. _____

15. How far is the tapped hole from the end of the hub?   15. _____

16. What was revision A?   16. _____

17. What finish is required on the pulley?   17. _____

18. What type of tolerance is specified for the keyway?   18. _____

19. What is the maximum dimension from the center of the pulley to the bottom of the keyway?   19. _____

20. What is dimension C?   20. _____

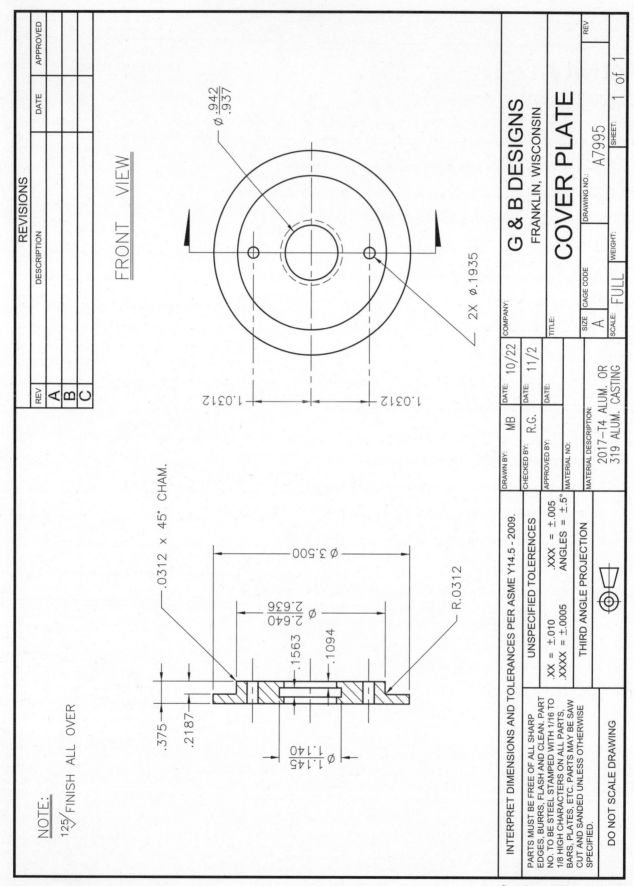

FRONT VIEW

Ø .942
  .937

2X Ø.1935

1.0312

1.0312

NOTE:
125/ FINISH ALL OVER

.0312 × 45° CHAM.

R.0312

Ø 3.500

Ø 2.640
  2.636

.1563

.1094

.375

.2187

Ø 1.145
  1.140

INTERPRET DIMENSIONS AND TOLERANCES PER ASME Y14.5 - 2009.

PARTS MUST BE FREE OF ALL SHARP EDGES, BURRS, FLASH AND CLEAN. PART NO. TO BE STEEL STAMPED WITH 1/16 TO 1/8 HIGH CHARACTERS ON ALL PARTS, BARS, PLATES, ETC. PARTS MAY BE SAW CUT AND SANDED UNLESS OTHERWISE SPECIFIED.

DO NOT SCALE DRAWING

UNSPECIFIED TOLERANCES

.XX = ±.010   .XXX = ±.005
.XXXX = ±.0005   ANGLES = ±.5°

THIRD ANGLE PROJECTION

DRAWN BY:	MB	DATE:	10/22	COMPANY:
CHECKED BY:	R.G.	DATE:	11/2	
APPROVED BY:		DATE:		
MATERIAL NO:				
MATERIAL DESCRIPTION:				

2017–T4 ALUM. OR
319 ALUM. CASTING

G & B DESIGNS
FRANKLIN, WISCONSIN

TITLE:
COVER PLATE

SIZE	CAGE CODE	DRAWING NO.:	A7995	REV
A				
SCALE: FULL	WEIGHT:	SHEET: 1 of 1		

Activity 12-3 Cover Plate.

Name _____ Date _____ Class _____

Activity 12-3
Cover Plate

Refer to **Activity 12-3**. *Study the drawing and familiarize yourself with the views, dimensions, title block, and notes. Read the questions, refer to the print, and write your answers in the blanks provided.*

1. What is the dimension of the chamfer?   1. _____

2. What does the dashed line circle in the front view represent?   2. _____

3. How many holes are in the part?   3. _____

4. How long is the 3.500 diameter?   4. _____

5. What is the scale of this print?   5. _____

6. Determine the high limit on the largest diameter.   6. _____

7. How far apart are the two #10 holes from each other?   7. _____

8. What finish is required on the part?   8. _____

9. What is the maximum diameter allowed on the groove?   9. _____

10. How wide is the groove?   10. _____

11. What is the overall length (thickness) of the part?   11. _____

12. How many surfaces are shown in the front view?   12. _____

13. What does the number 319 refer to on the print?   13. _____

14. Is the groove in the exact center (middle) of the part?   14. _____

15. What size is the large hole on the print?   15. _____

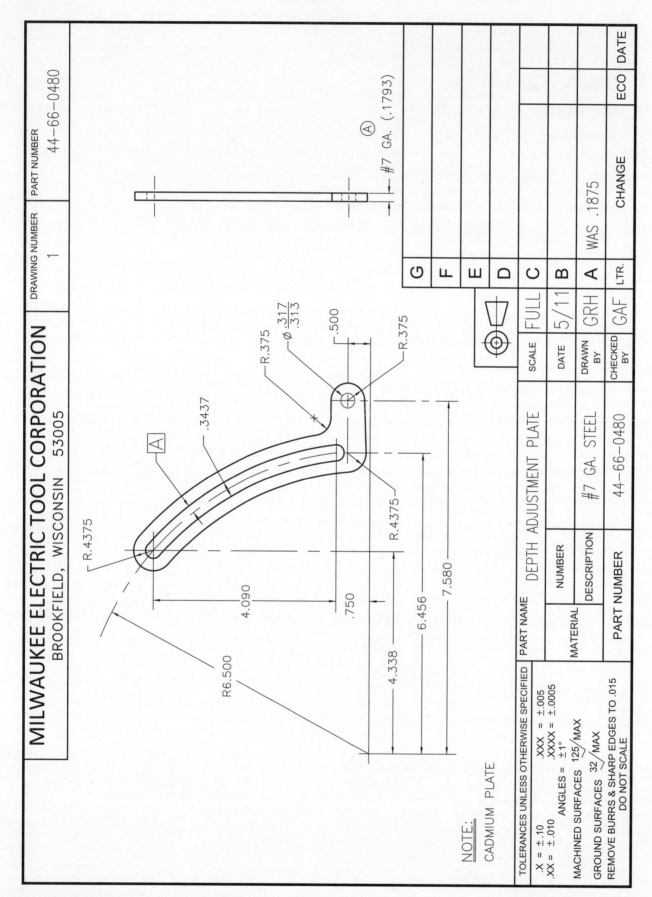

MILWAUKEE ELECTRIC TOOL CORPORATION
BROOKFIELD, WISCONSIN   53005

DRAWING NUMBER   1

PART NUMBER   44-66-0480

#7 GA. (.1793)   Ⓐ

Ø .317 / .313

R.375

.500

R.375

.3437

Ⓐ

R.4375

R.4375

R.375

R6.500

4.090

.750

4.338

6.456

7.580

NOTE:

CADMIUM PLATE

TOLERANCES UNLESS OTHERWISE SPECIFIED

.X = ±.10   .XXX = ±.005
.XX = ±.010   .XXXX = ±.0005
        ANGLES = ±1°

MACHINED SURFACES 125/MAX
GROUND SURFACES 32/MAX
REMOVE BURRS & SHARP EDGES TO .015
DO NOT SCALE

PART NAME	DEPTH ADJUSTMENT PLATE			SCALE	FULL
MATERIAL	NUMBER			DATE	5/11
	DESCRIPTION	#7 GA. STEEL		DRAWN BY	GRH
PART NUMBER	44-66-0480			CHECKED BY	GAF

G				
F				
E				
D				
C				
B				
A		WAS .1875		
LTR.		CHANGE	ECO	DATE

Goodheart-Willcox Publisher

Name _____ Date _____ Class _____

Activity 12-4
Depth Adjustment Plate

Refer to **Activity 12-4**. *Study the drawing and familiarize yourself with the views, dimensions, title block, and notes. Read the questions, refer to the print, and write your answers in the blanks provided.*

1. List all radii found on the print.

2. What gage steel is used to make this part?

3. Give the two dimensions describing the location of the hole.

4. How wide is the slot?

5. On what radius is the slot located?

6. What is the actual height of the part?

7. What type of finish does the part receive?

8. What thickness was the part originally?

9. What is the part number?

10. What is dimension A?

11. What is the width of the part?

12. What is the tolerance of the hole?

13. What is the scale of the print?

14. How far is the horizontal centerline of the hole from the upper horizontal centerline of the slot?

15. How far apart are the horizontal centerlines of the slot?

16. What is the overall length of the part?

17. At what distance is the horizontal centerline of the hole from the lower horizontal centerline of the slot?

18. How far apart are the vertical centerlines of the slot?

19. How many centerlines are shown on the print?

20. What is the low limit for the 7.580 dimension?

1. _____

2. _____

3. _____

4. _____

5. _____

6. _____

7. _____

8. _____

9. _____

10. _____

11. _____

12. _____

13. _____

14. _____

15. _____

16. _____

17. _____

18. _____

19. _____

20. _____

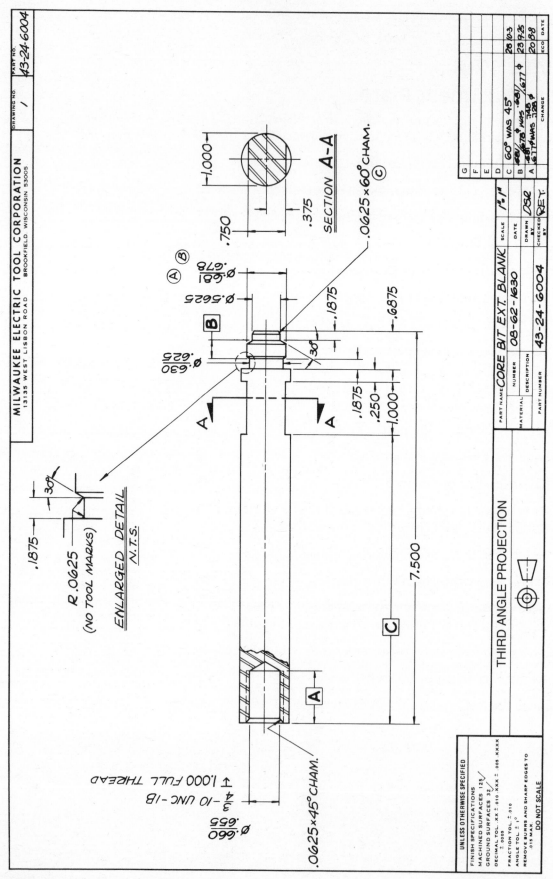

MILWAUKEE ELECTRIC TOOL CORPORATION
13135 WEST LISBON ROAD   BROOKFIELD, WISCONSIN 53005

DRAWING NO   1
PART NO.   43-24-6004

SECTION A-A

.0625×60° CHAM.

ENLARGED DETAIL
N.T.S.

R.0625
(NO TOOL MARKS)

SECTION A-A

.0625×45° CHAM.

ø .660 / .655
3/4 -10 UNC-1B
↑ 1.000 FULL THREAD

THIRD ANGLE PROJECTION

UNLESS OTHERWISE SPECIFIED			
FINISH SPECIFICATIONS			
MACHINED SURFACES 125/			
GROUND SURFACES 32/			
DECIMAL TOL. .XX ± .010 .XXX ± .005 .XXXX ± .0005			
FRACTION TOL. ± .010			
ANGLE TOL. ± ½°			
REMOVE BURRS AND SHARP EDGES TO .015 MAX.			
DO NOT SCALE			

PART NAME CORE BIT EXT. BLANK
NUMBER 08-62-1630
MATERIAL
DESCRIPTION
PART NUMBER 43-24-6004
SCALE 1" = 1"
DATE
DRAWN BY DSL
CHECKED BY
REV.

				CHANGE	ECO	DATE
G						
F						
E						
D						
C				60° WAS 45°	28	10-3
B				.681 / .678 ø WAS .681 / .677 ø	23	9-25
				.748 / .745		
A				.6719 WAS .7128 ø	20	88

Goodheart-Willcox Publisher

Name _____ Date _____ Class _____

Activity 12-5
Core Bit Ext. Blank

Refer to **Activity 12-5**. *Study the drawing and familiarize yourself with the views, dimensions, title block, and notes. Read the questions, refer to the print, and write your answers in the blanks provided.*

1. How long are the two flats shown on the print?   1. _____

2. What is the maximum allowable total length of the part?   2. _____

3. What size is the groove diameter?   3. _____

4. What size chamfer is used on the thread?   4. _____

5. How wide is the groove?   5. _____

6. Is there a fit designated for the thread? If yes, what is it?   6. _____

7. What is dimension A?   7. _____

8. What is the low limit for the smallest diameter shown on the print?   8. _____

9. What note refers to the external groove on the print?   9. _____

10. Determine dimension B.   10. _____

11. What is the vertical distance between the two flats?   11. _____

12. What does N.T.S. mean?   12. _____

13. What is the outside diameter of the part?   13. _____

14. What is revision C?   14. _____

15. Does this part feature any internal grooves?   15. _____

16. What is the number of the material used for the part?   16. _____

17. What engineering change order initiated Revision B?   17. _____

18. What kind of shoulder is found between the .5625 diameter and the .681/.678 diameter?   18. _____

19. Determine dimension C.   19. _____

20. How many centerlines are found on the print?   20. _____

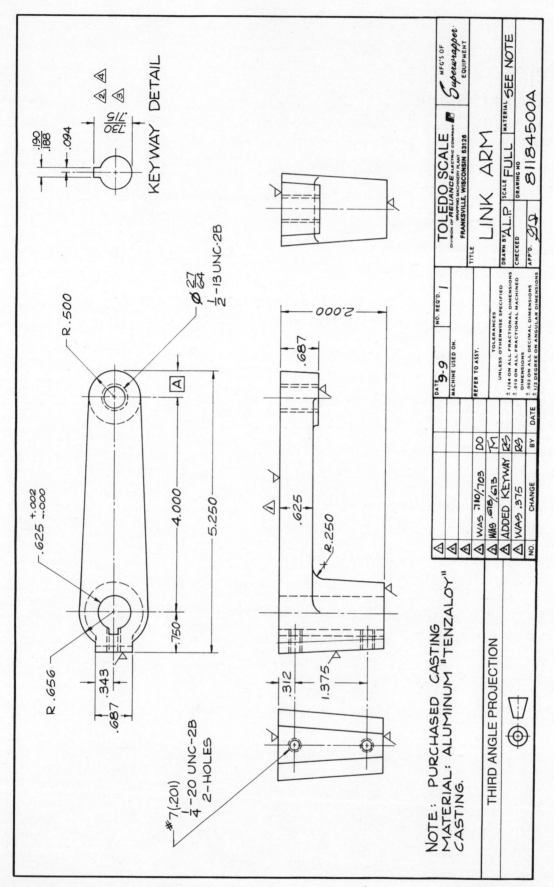

Activity 12-6 Link Arm.

Name _____ Date _____ Class _____

Activity 12-6
Link Arm

Refer to **Activity 12-6**. *Study the drawing and familiarize yourself with the views, dimensions, title block, and notes. Read the questions, refer to the print, and write your answers in the blanks provided.*

1. What is the minimum total height of the part?

1. _____

2. What is the total length of the part?

2. _____

3. What is revision 1?

3. _____

4. How long (deep) is the 1/2-13 UNC thread?

4. _____

5. What diameter is the boss on the part?

5. _____

6. How much clearance is there between the sides of the keyway and a 3/16 key?

6. _____

7. How far apart are the 1/4 tapped holes?

7. _____

8. Was the keyway depth increased or decreased from the last revision?

8. _____

9. How far is the hole with the keyway from the 1/2-13 UNC threaded hole?

9. _____

10. Is the right side view necessary?

10. _____

11. How much does the part taper along its length? Give answer in TPI.

11. _____

12. Determine dimension A.

12. _____

13. How much was the thickness of the link arm base increased?

13. _____

14. What kind of material is "TENZALOY"?

14. _____

15. What is the maximum possible depth of the keyway?

15. _____

16. How many surfaces are finished?

16. _____

17. What is the lower limit of the hole with the keyway?

17. _____

18. What size taps are used on the part?

18. _____

19. How far does the top of the larger boss extend above the top of the base?

19. _____

20. Where is Toledo Scale located?

20. _____

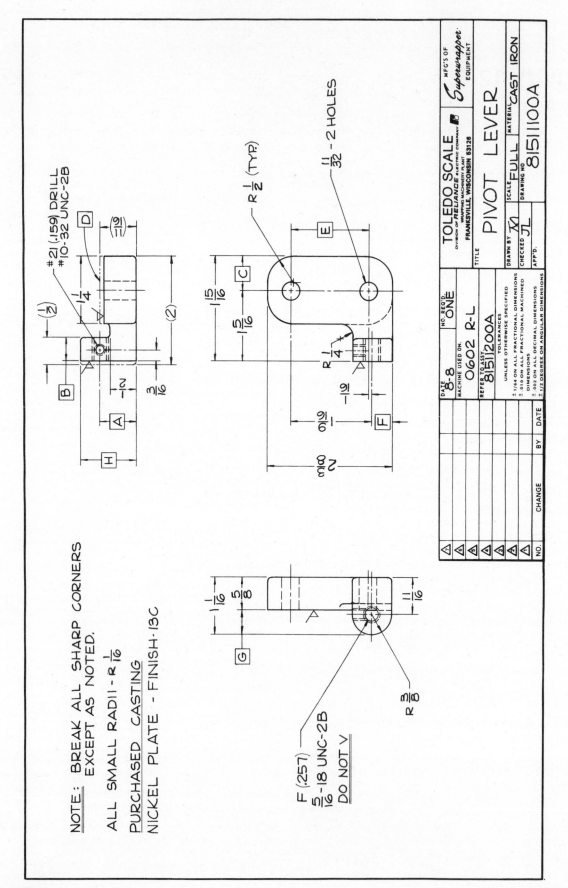

NOTE: BREAK ALL SHARP CORNERS
EXCEPT AS NOTED.

ALL SMALL RADII - R $\frac{1}{16}$

PURCHASED CASTING

NICKEL PLATE - FINISH·13C

F (.257)
$\frac{5}{16}$-18 UNC-2B
DO NOT ∨

R $\frac{3}{8}$

#21 (.159) DRILL
#10-32 UNC-2B

R $\frac{1}{2}$ (TYP.)

$\frac{11}{32}$ - 2 HOLES

TOLEDO SCALE
DIVISION OF *RELIANCE* ELECTRIC COMPANY
WRAPPING MACHINERY PLANT
FRANKSVILLE, WISCONSIN 53126

MFG'S OF
Superwrapper
EQUIPMENT

TITLE
PIVOT LEVER

SCALE **FULL**	MATERIAL **CAST IRON**	
DRAWN BY Ⅺ	DRAWING NO	**81511OOA**
CHECKED ⅬⅬ		
APP'D.		

DATE **8-8**	NO. REQ'D **ONE**
MACHINE USED ON. **0602 R-L**	
REFER TO ASSY. **81511200A**	

TOLERANCES
UNLESS OTHERWISE SPECIFIED
± 1/64 ON ALL FRACTIONAL DIMENSIONS
± .010 ON ALL FRACTIONAL MACHINED
DIMENSIONS
± .002 ON ALL DECIMAL DIMENSIONS
± 1/2 DEGREE ON ANGULAR DIMENSIONS

NO.	CHANGE	BY	DATE

Activity 12-7 Pivot Leader.

Name _____ Date _____ Class _____

Activity 12-7
Pivot Leader

Refer to **Activity 12-7**. *Study the drawing and familiarize yourself with the views, dimensions, title block, and notes. Read the questions, refer to the print, and write your answers in the blanks provided.*

1. Determine dimension A.

2. How many parts are required?

3. What size are the small radii on the print?

4. Determine dimension B.

5. How far are the two drilled holes from the centerline of the small tapped hole?

6. What is dimension C?

7. What finish is required on the casting?

8. What kind of line is D?

9. What do the reference dimensions represent?

10. Determine dimension E.

11. What size is the fillet on the casting?

12. Determine dimension F.

13. What does the "B" signify on the thread designation 5/16—18 UNC—2B?

14. What machine is this part used on?

15. What views are shown on this print?

16. Determine dimension G.

17. What is the width of the large machined surface?

18. What tolerance is used on fractional dimensions?

19. Determine dimension H.

20. Which view shows the shape of the small machined surface?

1. _____

2. _____

3. _____

4. _____

5. _____

6. _____

7. _____

8. _____

9. _____

10. _____

11. _____

12. _____

13. _____

14. _____

15. _____

16. _____

17. _____

18. _____

19. _____

20. _____

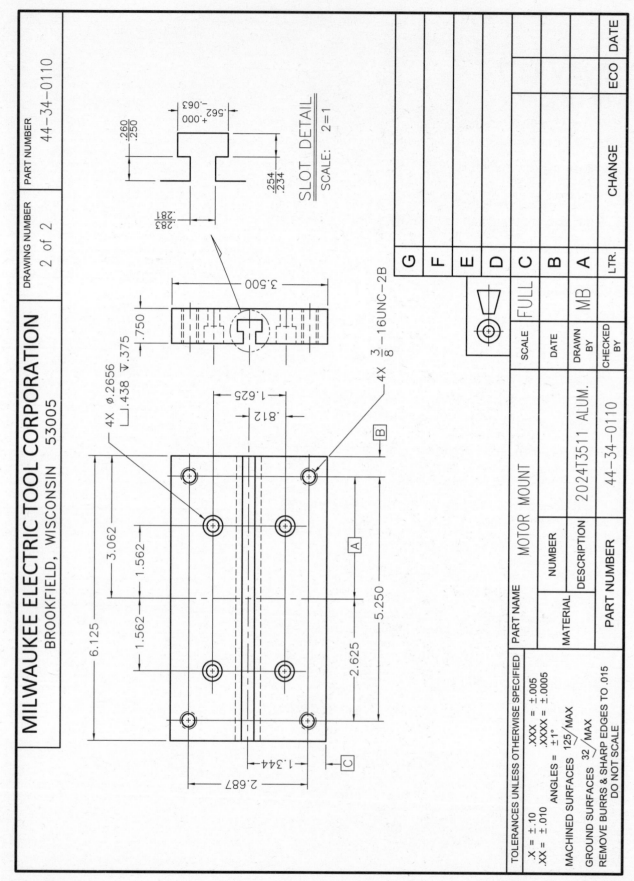

MILWAUKEE ELECTRIC TOOL CORPORATION
BROOKFIELD, WISCONSIN   53005

DRAWING NUMBER   2 of 2

PART NUMBER   44-34-0110

SLOT DETAIL   SCALE: 2=1

.260
.250

.562 +.000 −.063

.254
.234

.283
.281

3.500

.750

4X ⌀.2656
⌴.438 ▽.375

1.625

.812

3.062

1.562

1.562

6.125

2.687

1.344

2.625

5.250

4X 3/8 −16UNC−2B

A

B

C

TOLERANCES UNLESS OTHERWISE SPECIFIED
.X = ±.10
.XX = ±.010
.XXX = ±.005
.XXXX = ±.0005
ANGLES = ±1°
MACHINED SURFACES 125/MAX
GROUND SURFACES 32/MAX
REMOVE BURRS & SHARP EDGES TO .015
DO NOT SCALE

PART NAME   MOTOR MOUNT

MATERIAL   NUMBER
            DESCRIPTION   2024T3511 ALUM.

PART NUMBER   44-34-0110

SCALE FULL

DATE

DRAWN BY   MB

CHECKED BY

G
F
E
D
C
B
A
LTR.   CHANGE   ECO   DATE

Name _____ Date _____ Class _____

Activity 12-8
Motor Mount

Refer to **Activity 12-8**. *Study the drawing and familiarize yourself with the views, dimensions, title block, and notes. Read the questions, refer to the print, and write your answers in the blanks provided.*

1. How wide is the opening of the T-slot?   1. _____

2. What is the total length of the part?   2. _____

3. How many holes are tapped?   3. _____

4. Determine dimension A.   4. _____

5. How thick is the part?   5. _____

6. How far apart horizontally are the two tapped holes on the left from the two counterbore holes on the left?   6. _____

7. What is the minimum total depth of the T-slot?   7. _____

8. What is the maximum total depth of the T-slot?   8. _____

9. Determine dimension B.   9. _____

10. What size are the counterbores?   10. _____

11. How far apart are the two counterbore holes on the left from the two counterbore holes on the right?   11. _____

12. What is the upper limit dimension for the width at the bottom of the T-slot?   12. _____

13. How far are the 3/8-16 tapped holes from the horizontal centerline of the part?   13. _____

14. Determine dimension C.   14. _____

15. How deep are the counterbores?   15. _____

16. What is the distance between the two upper counterbore holes and the two lower tapped holes?   16. _____

17. What is the lower limit dimension for the width at the bottom of the T-slot?   17. _____

18. What is the upper limit dimension on the height of the part?   18. _____

19. What size are the tapped holes?   19. _____

20. What is the maximum thickness of the part below the bottom of the T-slot?   20. _____

Notes

Learning Objectives

After studying this unit, you will be able to:

✓ Explain the terms sectional view, cutting-plane line, and section lines.

✓ Identify various section lines.

✓ Complete various sectional views.

✓ Identify sectional views such as full, offset, half, revolved, removed, and broken-out.

Key Terms

break line

broken-out section

cutting plane

cutting-plane line

full section

half section

offset section

removed section

revolved section

section line

sectional view

A *sectional view* of an object in a drawing is created by the imaginary cutting away of its front portion to reveal its interior. The exposed (cut) surface, then, is emphasized by the use of section lines (cross-hatching). See **Figure 13-1**.

Working with Sectional Views

The use of sectional views is a graphic method of exposing the interior details of a part. It is an effective way of showing inside features that would be complicated or confusing if described entirely by hidden lines. A sectional view may serve as one of the principal views—front, top, or side—on a print or it can be used as an additional view.

A sectional view is developed by first passing an imaginary *cutting plane* through the part. Then, the section of the part nearest the "reader" is removed, thereby revealing a direct and clear view of the interior shape. See **Figure 13-1**.

Cutting-Plane Line

The location of the imaginary cutting plane is indicated by a heavy line called a *cutting-plane line*, **Figure 13-2**. The ends of the cutting-plane line are bent at 90° with arrowheads on the ends that point in the direction of viewing sight. The side of the part toward the arrows is the side that will be sectioned. The position of the cutting-plane line on the part will determine the type of sectioning (full, offset, half, revolved, removed, or broken-out).

When two or more sectional views are shown on a print, letters are placed at each end of the cutting-plane line. These letters match the letters shown directly below the sectional view identified with that cutting-plane line. See **Figure 13-2**.

Several types of cutting-plane lines are used on industrial prints. Examples are shown in **Figure 13-3**.

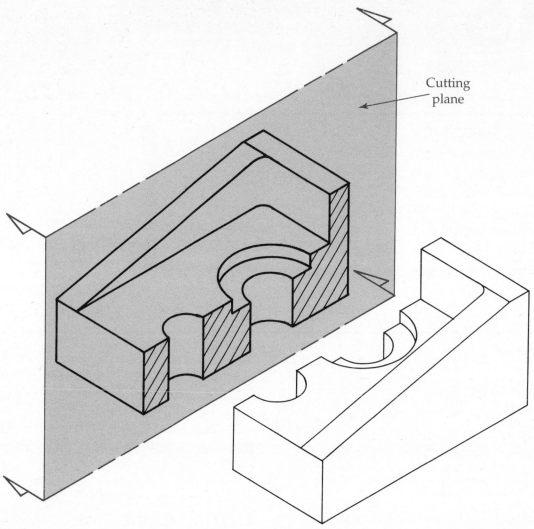

Goodheart-Willcox Publisher

Figure 13-1 Sectioned view shows how imaginary cutting plane cuts away front portion of part to reveal interior details.

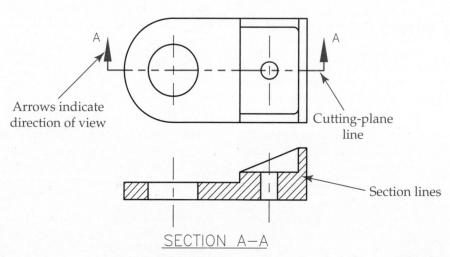

Goodheart-Willcox Publisher

Figure 13-2 Location of cutting plane in top view is indicated by a heavy cutting-plane line marked A-A. Sectioned view below is identified as Section A-A.

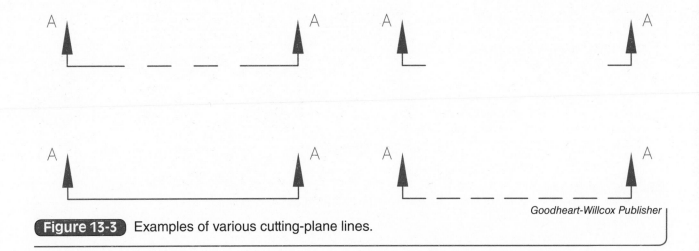

Goodheart-Willcox Publisher

Figure 13-3 Examples of various cutting-plane lines.

Section Lines

Section lines (cross-hatch lines) are used to identify and emphasize the surfaces that have been cut and exposed by a cutting-plane line.

Section lines are thin, parallel, slanted lines, usually drawn at an angle of 45°, **Figure 13-4**. However, if the section lines become parallel with part of the section outline, then they must be drawn at some other angle (usually 30° or 60°). Both poor practice and preferred method are shown in **Figure 13-5**.

When it becomes necessary to show different kinds of materials (on assembly drawings, for example), various patterns of section lines are used. Section lines are used on all types of materials, nonmetallic as well as metallic. See **Figure 13-6**.

In general use, however, the section lines for cast iron are applied to working (detail) drawings of a separate part.

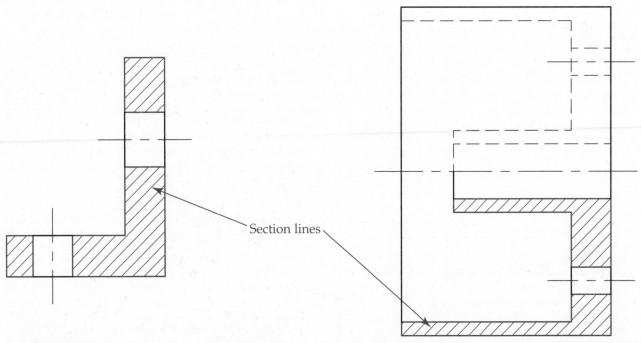

Section lines

Goodheart-Willcox Publisher

Figure 13-4 Section lines emphasize surfaces that have been cut. Usually, they are drawn at a 45° angle.

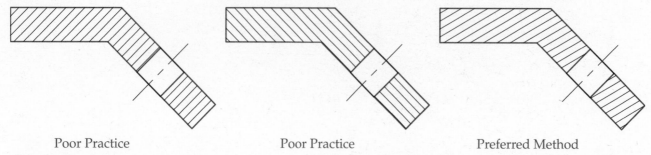

Poor Practice   Poor Practice   Preferred Method

Goodheart-Willcox Publisher

Figure 13-5 Left and center views illustrate poor practice because section lines are drawn parallel with the part object lines. Right view shows preferred method of section lining.

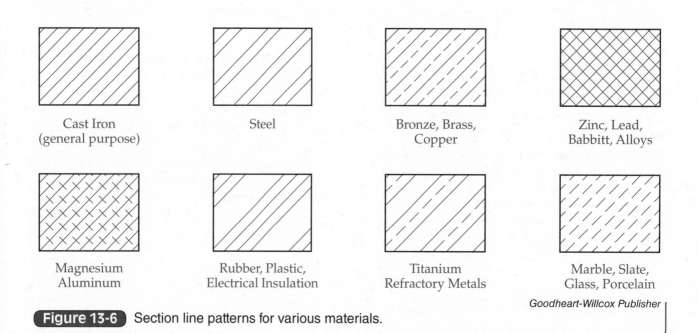

Cast Iron
(general purpose)

Steel

Bronze, Brass,
Copper

Zinc, Lead,
Babbitt, Alloys

Magnesium
Aluminum

Rubber, Plastic,
Electrical Insulation

Titanium
Refractory Metals

Marble, Slate,
Glass, Porcelain

Goodheart-Willcox Publisher

Figure 13-6 Section line patterns for various materials.

Full Sections

A *full section* is created by passing the cutting-plane line through the entire object in a straight line, as shown in **Figure 13-7**. Lines which were hidden are now exposed and shown as visible lines.

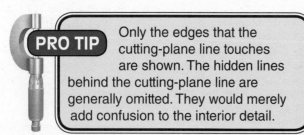

PRO TIP Only the edges that the cutting-plane line touches are shown. The hidden lines behind the cutting-plane line are generally omitted. They would merely add confusion to the interior detail.

When reading a sectional view, first determine the direction of the cutting-plane line on the corresponding view. Then transfer the details behind the cutting-plane line to the sectional view. The transferring of points on the cutting-plane line to the sectional view locates the interior detail. Note that the outer edges of the part are always defined by visible lines. Hidden lines are generally not permissible.

With practice, this technique will be most helpful in visualizing the other types of sectional views.

Offset Sections

An *offset section* has the characteristics of a full section, but the cutting-plane line is slightly diverted to include nearby details.

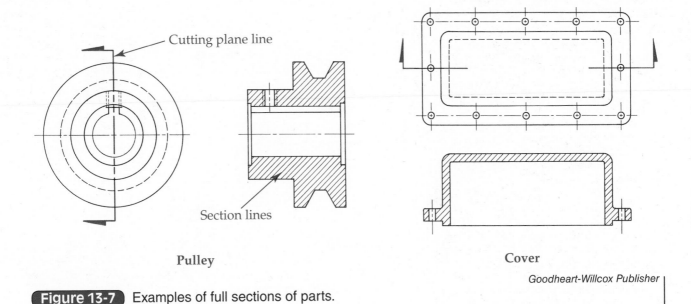

Pulley

Cover

Goodheart-Willcox Publisher

Figure 13-7   Examples of full sections of parts.

The same method for transferring interior details used for full sections can be applied to an offset section. The path of the cutting-plane line in an offset section should be reasonably close to the details. If the details are not reasonably close, a second cutting-plane line should be used instead of an offset section.

In **Figure 13-8**, the cutting-plane line passes through two holes and a slot. The actual section has the appearance of a full section. The bend lines of the cutting-plane line are not shown in the sectional view.

Half Sections

A *half section* of a symmetrical (both halves the same) object shows the interior and exterior features in the same view. See **Figure 13-9**. One half of the view will show the sectioning of the interior of the object, while the remaining half will detail the exterior. Again, section views are drawn with visible lines. Hidden lines may tend to add more confusion to the section lines.

The omitted hidden lines are standard practice if the part is not complicated. However, hidden lines may be of value when used with forged parts or complicated cast parts for dimensioning purposes.

The cutting-plane line in **Figure 13-9** is drawn at a 90° angle toward the right side of the top view. Follow the transfer lines from the top view into the front view. The right side of

Offset cutting-plane line

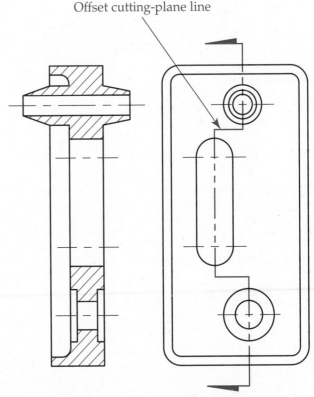

Goodheart-Willcox Publisher

Figure 13-8   An offset section uses bends in the cutting-plane line to include nearby details. The sectional view has the appearance of a full section.

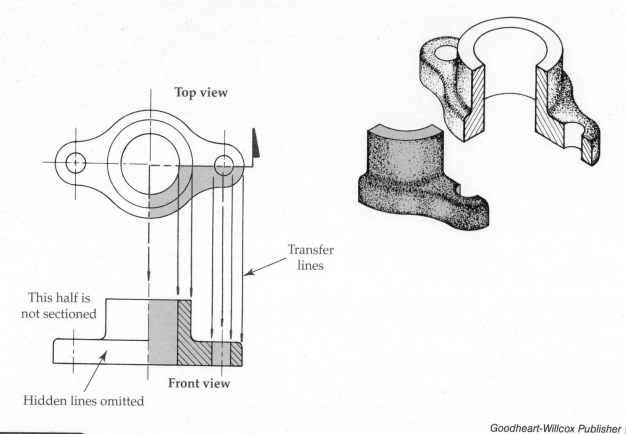

Goodheart-Willcox Publisher

Figure 13-9 Top view of housing is projected downward to show half section of front view. Isometric view of half section is pictured at right.

the front view has been sectioned by the use of the cutting-plane line and sectioning lines. The main centerline in the front view will remain unchanged as the center of the part in **Figure 13-9**.

The half section can easily be seen in the front view. As in all half sections, the sectioned portion will only be one half of that view.

Revolved Section

A *revolved section* is a sectional drawing that represents a single portion of the part. Revolved sections are cross sections of the shape not shown in the conventional views. See **Figure 13-10**.

A centerline is drawn through the portion that will be revolved and sectioned. The centerline acts as an imaginary viewing plane (cutting plane) that revolves 90° to the original view on the centerline. The section is revolved directly on the surface of the original view and drawn into true size and shape. Keep in mind, the drafter or designer must provide the true size and shape of the contours. The lines from the view may

overlap with the revolved section. These should be omitted.

Revolved sections are commonly applied to arms, ribs, spokes, bars, and irregularly contoured parts.

Removed Section

A *removed section* is similar to the revolved section, but "removed" from its position within a view to a new location elsewhere on the print. The relocating of this removed section allows the drafter to enlarge that particular section for clarity. Also, by enlarging the removed section, dimensioning may be added for further clarification.

To maintain clarity when more than one removed section is shown, each section should be marked separately. The removed sections are marked in alphabetical order, starting with Section A-A (then Section B-B, Section C-C, etc.) with each of the corresponding letters on the cutting-plane line ends. See **Figure 13-11**. The use

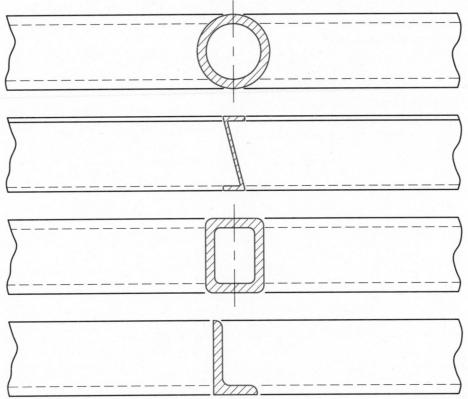

Goodheart-Willcox Publisher

Figure 13-10 Revolved sections represent a cross section of a portion of part.

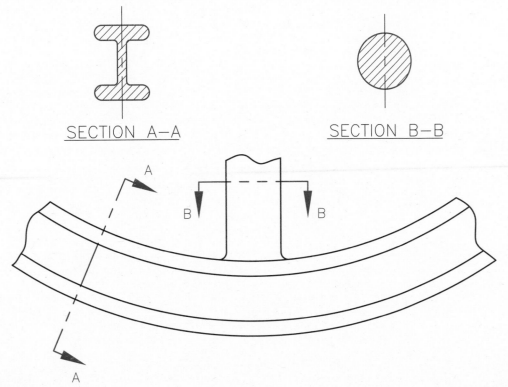

SECTION A–A

SECTION B–B

Goodheart-Willcox Publisher

Figure 13-11 Removed sections A-A and B-B are sectional views relocated on print away from part.

of a cutting-plane line marked with letters at both ends is the main difference between a removed section and a revolved section.

Broken-Out Section (Partial)

The *broken-out section* is a special type of a sectioning application, **Figure 13-12**. Often, a small, single-detail portion of the part's interior needs more clarification. The broken-out section enables the print reader to concentrate directly on that detailed portion of the part. By concentrating only on that portion, the detail would be more visible to the print reader. Otherwise, the detail may be lost if a full section or half section were applied.

A broken-out section can be located on a print by the use of a wavy line called a *break line*. See **Figure 13-12**.

In this case, the break line acts as an imaginary cutting-plane line. The large and heavy cutting-plane line with arrows would not be placed on the drawing. The area within the break line to the part outline is section lined according to material type.

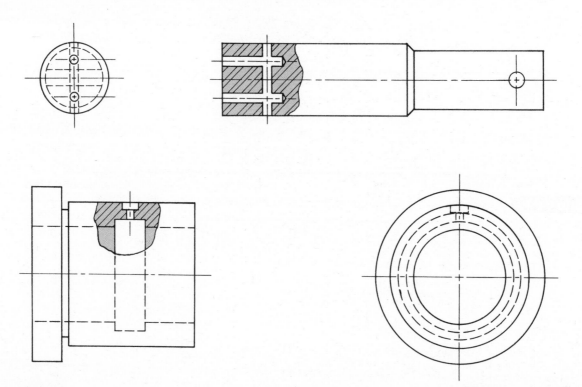

Figure 13-12 Broken-out sections are sectioned areas of parts' interiors outlined by wavy break lines.

Name _____ Date _____ Class _____

Sectioning Problem 13-1

Draw the front view as a full section. Use a straightedge or rule to complete the sketch.

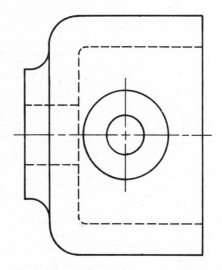

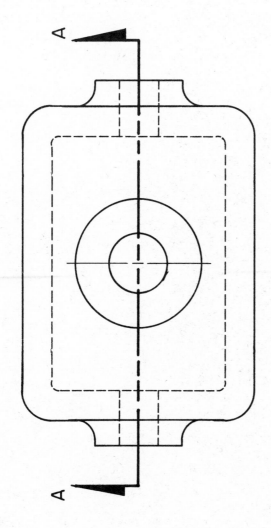

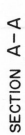

SECTION A–A

Name _____ Date _____ Class _____

Sectioning Problem 13-2

Complete the left-side view as an offset section. Use a straightedge or rule to complete the sketch.

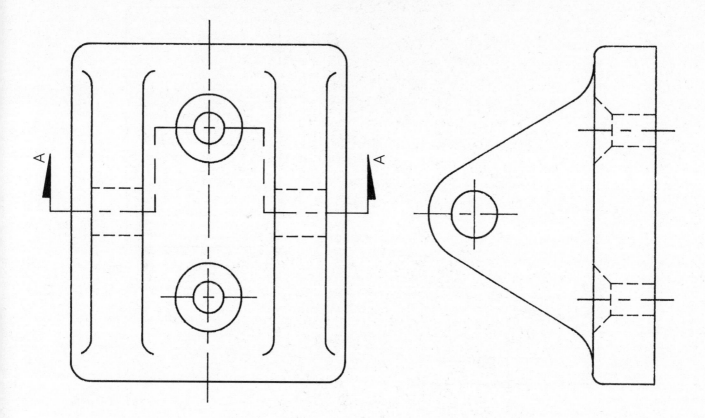

SECTION A—A

Name _____ Date _____ Class _____

Sectioning Problem 13-3

Complete this view as a half section. Use a straightedge or rule to complete the sketch.

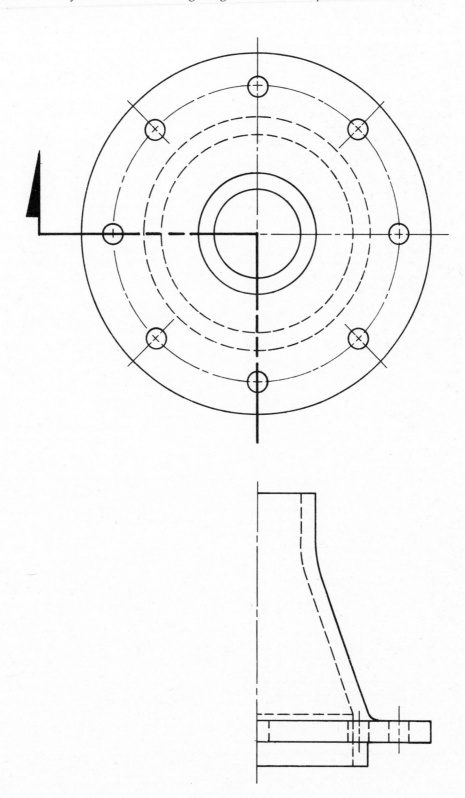

Name _____ Date _____ Class _____

Sectioning Problem 13-4

Make a half section of front view. Use a straightedge or rule to complete the sketch.

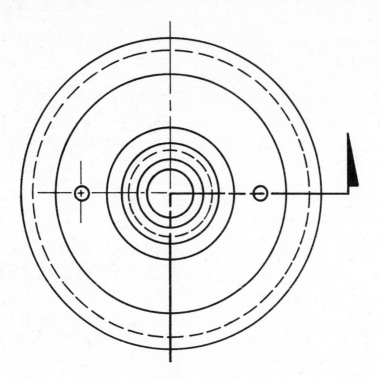

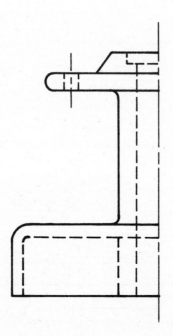

Name _____   Date _____   Class _____

Sectioning Problem 13-5

Draw revolved section on centerlines provided. Use a straightedge or rule to complete the sketch.

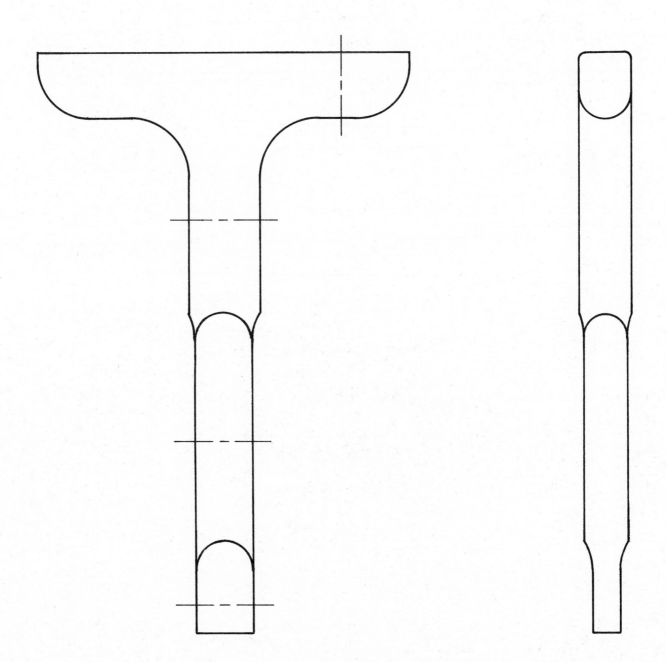

Name _____ Date _____ Class _____

Sectioning Problem 13-6

Draw removed sections on centerlines provided. Use rounded edges where necessary. Use a straightedge or rule to complete the sketch.

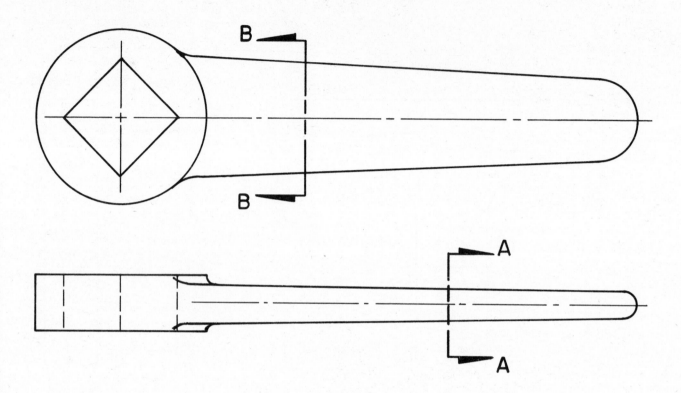

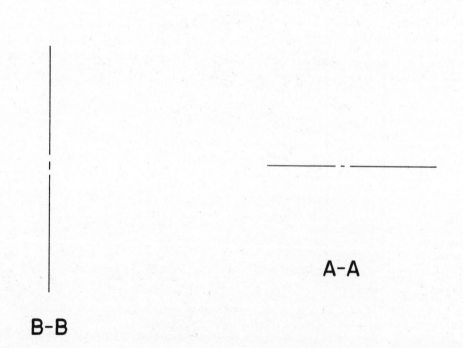

B-B

A-A

Name _____ Date _____ Class _____

Sectioning Problem 13-7

Complete the left-side view as a broken-out section. Use a straightedge or rule to complete the sketch.

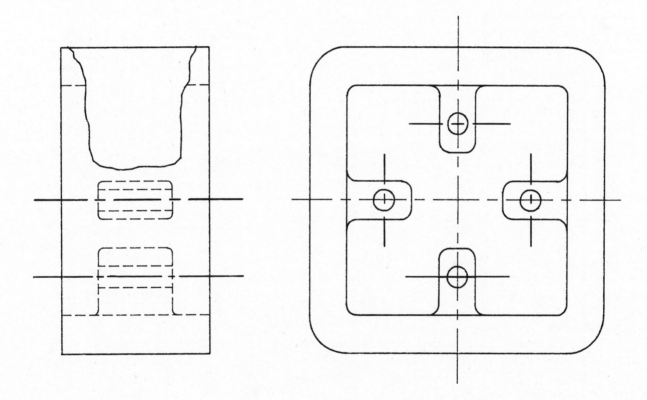

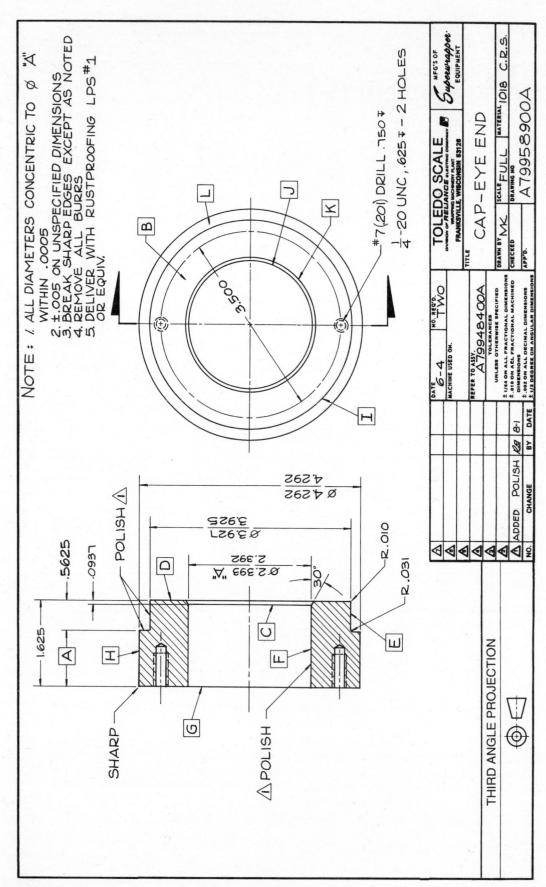

NOTE: 1. ALL DIAMETERS CONCENTRIC TO ⌀ "A"
WITHIN .0005
2. ±.005 ON UNSPECIFIED DIMENSIONS
3. BREAK SHARP EDGES EXCEPT AS NOTED
4. REMOVE ALL BURRS
5. DELIVER WITH RUSTPROOFING LPS #1
OR EQUIV.

3.500

#7 (201) DRILL .750 ▼
¼-20 UNC, .625 ▼ – 2 HOLES

4.292
4.292

⌀ 4.292

3.925
3.921

⌀ 3.921

2.392
2.393

⌀ 2.393 "A"

30°

R.010

R.031

.5625

.0937

1.625

POLISH

SHARP

POLISH

POLISH

THIRD ANGLE PROJECTION

TOLEDO SCALE		
DIVISION OF **RELIANCE** ELECTRIC COMPANY		
WEIGHING MACHINERY PLANT		
FRANKSVILLE, WISCONSIN 53126		

TITLE
CAP-EYE END

DRAWN BY MK	SCALE FULL	MATERIAL 1018 C.R.S.
CHECKED		DRAWING NO
APP'D.		A79958900A

MFG'S OF
Superwrapper
EQUIPMENT

DATE 6-4	NO. REQ'D. TWO
	MACHINE USED ON.
REFER TO ASSY. A799484OOA	

TOLERANCES
UNLESS OTHERWISE SPECIFIED
±1/64 ON ALL FRACTIONAL DIMENSIONS
±.010 ON AEL FRACTIONAL MACHINED
DIMENSIONS
±.001 ON ALL DECIMAL DIMENSIONS
±1/2 DEGREE ON ANGULAR DIMENSIONS

NO.	CHANGE	BY	DATE
⚠	ADDED POLISH 🖊 8-1		
⚠			
⚠			
⚠			
⚠			
⚠			
⚠			

Activity 13-1 Cap-Eye End.

Name _____ Date _____ Class _____

Activity 13-1
Cap-Eye End

Refer to **Activity 13-1**. *Study the drawing and familiarize yourself with the views, dimensions, title block, and notes. Read the questions, refer to the print, and write your answers in the blanks provided.*

1. What kind of sectional view is shown?   1. _____

2. How many surfaces are polished?   2. _____

3. How long is the 3.927 diameter surface?   3. _____

4. What size tap drill is used?   4. _____

5. How deep should the holes be tapped?   5. _____

6. What assembly print is this part used on?   6. _____

7. What kind of material is used for the part?   7. _____

8. What does note 4 state?   8. _____

9. What specification is given for the rear outer edge of the part?   9. _____

10. What size is the bore on the part?   10. _____

11. What is the low limit on the large outside diameter of the part?   11. _____

12. Which diameters must be concentric to diameter "A"? 12. _____

13. What size chamfer is called for on the part?   13. _____

14. What is dimension A?   14. _____

15. What surface or line in the section view is the same as I in the front view?   15. _____

16. Line D in the section view is what surface in the front view?   16. _____

17. How many parts are required?   17. _____

18. Line F in the section view is what line in the front view?   18. _____

19. Is surface G shown in the front view?   19. _____

20. What tolerance is used on unspecified dimensions?   20. _____

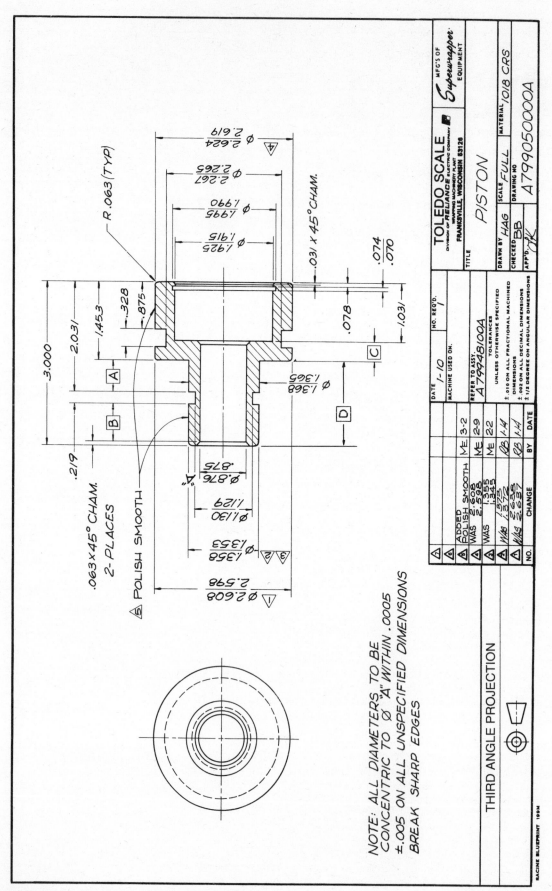

NOTE: ALL DIAMETERS TO BE
CONCENTRIC TO Ø "A" WITHIN .0005
±.005 ON ALL UNSPECIFIED DIMENSIONS
BREAK SHARP EDGES

THIRD ANGLE PROJECTION

RACINE BLUEPRINT 1934

NO.	CHANGE	BY	DATE
	ADDED POLISH SMOOTH	ME	3-2
	WAS 2.608 2.598	ME	2-9
	WAS 1.355 .345	ME	2-2
	WAS 1.373 1.363	RB	1-4
	WAS 2.638 2.637	RB	1-4

DATE	1-10	NO. REQ'D
MACHINE USED ON.		1

REFER TO ASSY.
A79948100A

TOLERANCES
UNLESS OTHERWISE SPECIFIED
±.010 ON ALL FRACTIONAL MACHINED
DIMENSIONS
±.005 ON ALL DECIMAL DIMENSIONS
±.002 ON ALL DECIMAL DIMENSIONS
±1/2 DEGREE ON ANGULAR DIMENSIONS

TOLEDO SCALE
DIVISION OF RELIANCE ELECTRIC COMPANY
WRAPPING MACHINERY PLANT
FRANKSVILLE, WISCONSIN 53126

MFG'S OF *Superwrapper* EQUIPMENT

TITLE
PISTON

DRAWN BY HAG	SCALE *FULL*	MATERIAL *1018 CRS*
CHECKED BB	DRAWING NO	
APP'D JK	A799050000A	

Activity 13-2 Piston.

Name _____ Date _____ Class _____

Activity 13-2
Piston

Refer to **Activity 13-2**. *Study the drawing and familiarize yourself with the views, dimensions, title block, and notes. Read the questions, refer to the print, and write your answers in the blanks provided.*

1. How wide is the internal groove in the part?   1. _____

2. List all the outside groove diameters.   2. _____

3. What diameter is the internal groove?   3. _____

4. How long is the 1.358/1.353 diameter?   4. _____

5. How far is the .219 wide groove from the left end of the part?   5. _____

6. Including the chamfers, how long is the .876/.875 hole?   6. _____

7. Determine dimension A.   7. _____

8. What is the mean dimension on the counterbore diameter?   8. _____

9. What are the specifications for chamfers?   9. _____

10. How many pieces are required?   10. _____

11. How far is the internal groove located from the right end of the part?   11. _____

12. How many chamfers are required on the part?   12. _____

13. Determine dimension B.   13. _____

14. What tolerance is required on unspecified dimensions?   14. _____

15. How deep is the counterbore?   15. _____

16. What was revision 2?   16. _____

17. Determine dimension C.   17. _____

18. What is the radius for rounds?   18. _____

19. What maximum possible wall thickness between the bottom of the .219 wide groove and the .876/.875 diameter?   19. _____

20. Determine distance D.   20. _____

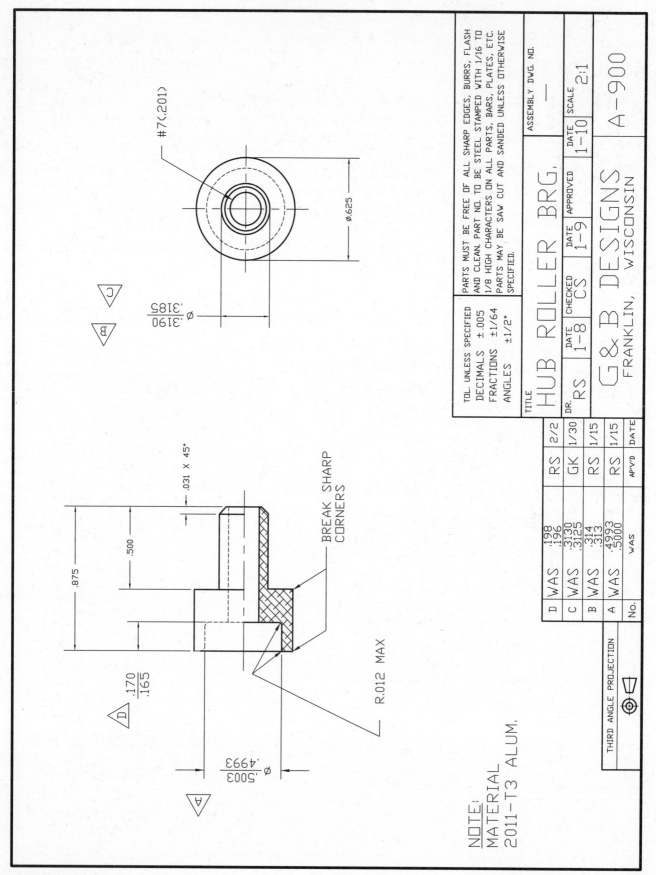

#7(.201)

Ø.625

Ø .3190 / .3185

C
B

.031 X 45°

.500

.875

.170 / .165

BREAK SHARP CORNERS

R.012 MAX

Ø .5003 / .4993

D

A

NOTE:
MATERIAL
2011-T3 ALUM.

D	WAS	.198 / .196	RS	2/2
C	WAS	.3130 / .3125	GK	1/30
B	WAS	.314 / .313	RS	1/15
A	WAS	.4993 / .5000	RS	1/15
No.		WAS	APV'D	DATE

TOL. UNLESS SPECIFIED
DECIMALS   ±.005
FRACTIONS   ±1/64
ANGLES   ±1/2°

PARTS MUST BE FREE OF ALL SHARP EDGES, BURRS, FLASH AND CLEAN. PART NO. TO BE STEEL STAMPED WITH 1/16 TO 1/8 HIGH CHARACTERS ON ALL PARTS, BARS, PLATES, ETC. PARTS MAY BE SAW CUT AND SANDED UNLESS OTHERWISE SPECIFIED.

ASSEMBLY DWG. NO.
—

TITLE
HUB ROLLER BRG.

DR. RS	DATE 1-8	CHECKED CS	DATE 1-9	APPROVED	DATE 1-10	SCALE 2:1

G&B DESIGNS
FRANKLIN, WISCONSIN

A-900

THIRD ANGLE PROJECTION

Activity 13-3   Hub Roller Bearing.

Name _____ Date _____ Class _____

Activity 13-3
Hub Roller Bearing

Refer to **Activity 13-3**. *Study the drawing and familiarize yourself with the views, dimensions, title block, and notes. Read the questions, refer to the print, and write your answers in the blanks provided.*

1. How many corners are broken?

1. _____

2. How deep is the counterbore?

2. _____

3. What was revision B?

3. _____

4. What scale is the print?

4. _____

5. What does the scale used on this print actually mean?

5. _____

6. What tolerance is used on unspecified decimal dimensions?

6. _____

7. What size is the drilled hole?

7. _____

8. What kind of sectional view is shown?

8. _____

9. How long is the .625 diameter?

9. _____

10. How wide is the 45° chamfer?

10. _____

11. What size radii are specified on the print?

11. _____

12. What type of material is 2011-T3?

12. _____

13. What is the print number?

13. _____

14. What is the minimum overall length allowed on the part?

14. _____

15. What was revision D?

15. _____

16. How much tolerance is given on the small outside diameter?

16. _____

17. What is the high limit on the counterbore diameter?

17. _____

18. What do the hidden lines in the right-side view represent?

18. _____

19. What shape is the part?

19. _____

20. What chamfer is called for on the print?

20. _____

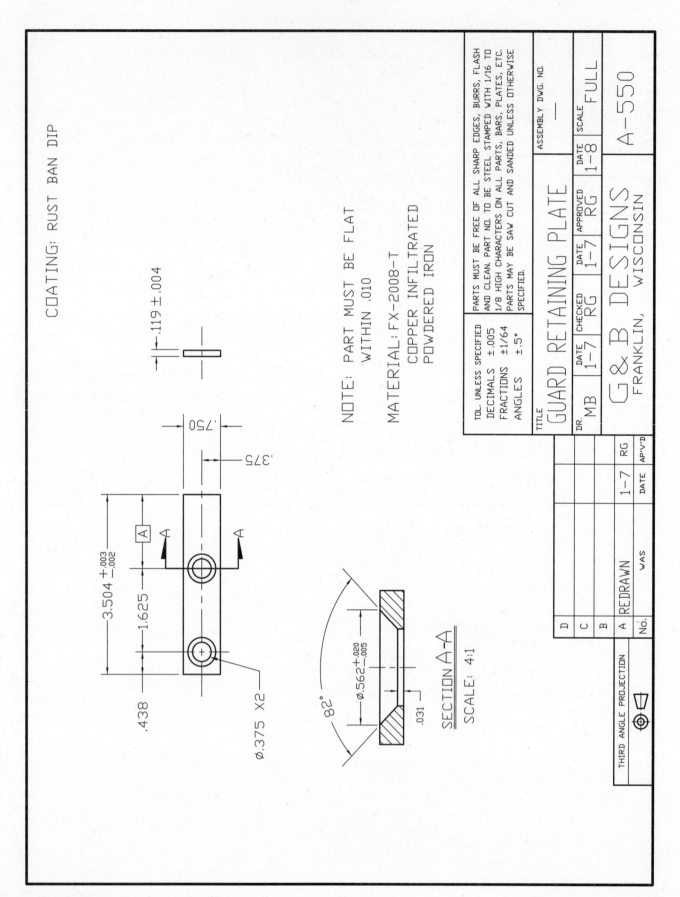

COATING: RUST BAN DIP

.119 ±.004

NOTE: PART MUST BE FLAT
WITHIN .010

MATERIAL: FX-2008-T
COPPER INFILTRATED
POWDERED IRON

.750

.375

3.504 +.003 −.002

1.625

A

Ø.375 X2

.438

82°

Ø.562 +.020 −.005

.031

SECTION A-A
SCALE: 4:1

TOL. UNLESS SPECIFIED	PARTS MUST BE FREE OF ALL SHARP EDGES, BURRS, FLASH
DECIMALS ±.005	AND CLEAN. PART NO. TO BE STEEL STAMPED WITH 1/16 TO
FRACTIONS ±1/64	1/8 HIGH CHARACTERS (IN ALL PARTS, BARS, PLATES, ETC.
ANGLES ±.5°	PARTS MAY BE SAW CUT AND SANDED UNLESS OTHERWISE SPECIFIED.

ASSEMBLY DWG. NO.
—

TITLE
GUARD RETAINING PLATE

DR. MB	DATE 1-7	CHECKED RG	DATE 1-7	APPROVED RG	DATE 1-8	SCALE FULL

G&B DESIGNS
FRANKLIN, WISCONSIN

A-550

D				
C				
B				
A	REDRAWN		1-7	RG
No.	WAS		DATE	AP'V'D

THIRD ANGLE PROJECTION

Name _____ Date _____ Class _____

Activity 13-4
Guard Retaining Plate

Refer to **Activity 13-4**. *Study the drawing and familiarize yourself with the views, dimensions, title block, and notes. Read the questions, refer to the print, and write your answers in the blanks provided.*

1. What is the maximum length of the part?

2. What is the distance between the two countersunk holes?

3. How thick is the part?

4. How deep are the countersunk holes?

5. What is the drawing number?

6. What scale is the print?

7. What is the high limit on the countersink angles?

8. What is the height of the part?

9. Calculate dimension A.

10. Are any unilateral decimal tolerances shown?

11. What kind of sectional view is shown?

12. Why is the sectional view so large in comparison to the other views?

13. What kind of material is used for the part?

14. What surface treatment is required?

15. What is the maximum distance allowed between the two countersunk holes?

16. Were any revisions made to the drawing?

17. What size are the two drilled holes?

18. If the part is made to a thickness of .116, is the part over, under, or within allowable tolerance?

19. What size is the large diameter of the countersink?

20. How flat must the part be?

1. _____

2. _____

3. _____

4. _____

5. _____

6. _____

7. _____

8. _____

9. _____

10. _____

11. _____

12. _____

13. _____

14. _____

15. _____

16. _____

17. _____

18. _____

19. _____

20. _____

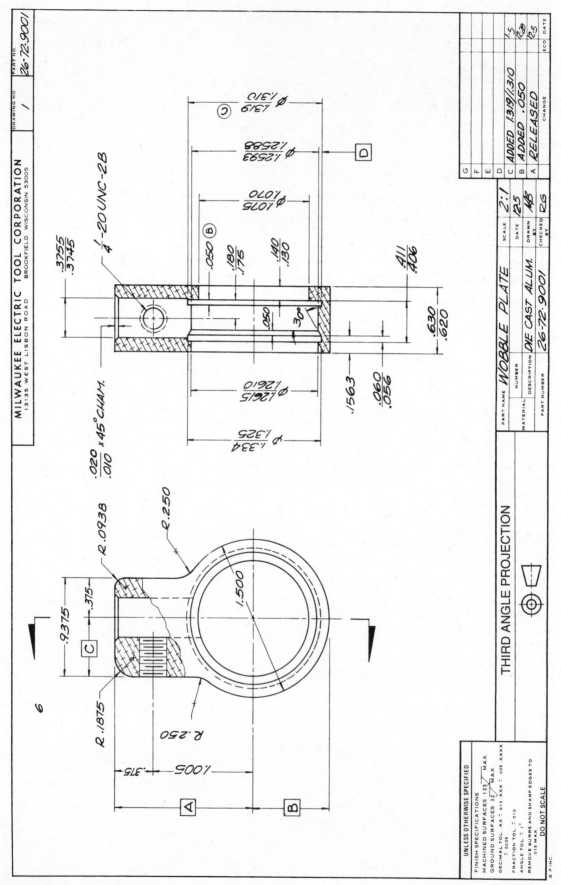

MILWAUKEE ELECTRIC TOOL CORPORATION
13135 WEST LISBON ROAD   BROOKFIELD WISCONSIN 53005

DRAWING NO 1
PART NO 26-72-9001

	CHANGE			ECO DATE
G				
F				
E				
D				
C	ADDED .139/.310			1-5
B	ADDED .050			12-8
A	RELEASED			12-5

PART NAME **WOBBLE PLATE**   SCALE **2:1**
NUMBER   DATE **12-5**
MATERIAL DESCRIPTION *DIE CAST ALUM.*   DRAWN BY **MB**
PART NUMBER **26-72-9001**   CHECKED BY **RG**

THIRD ANGLE PROJECTION

UNLESS OTHERWISE SPECIFIED
FINISH SPECIFICATIONS
MACHINED SURFACES 125 / MAX.
GROUND SURFACES 32 / MAX.
DECIMAL TOL. XX ± .010 XXX ± .005 XXXX ± ____
        ± .0005
FRACTION TOL. ± .010
ANGLE TOL. ± 1°
REMOVE BURRS AND SHARP EDGES TO
.015 MAX
DO NOT SCALE
B P INC.

.3755/.3745
¼-20UNC-2B
.020/.010 x 45° CHAM.

Ø 1.319/1.310 ©
Ø 1.2593/1.2588 Ⓓ
Ø 1.075/1.070 Ⓑ
.180/.175
.140/.130
.050
.050
30°
.411/.406
.630/.620
.563
.060/.056
Ø 1.2615/1.2610
Ø 1.334/1.325

R .250
R .0938
.9375
.315
Ⓒ
R .1875
R .250
1.500
.375
1.005
Ⓐ
Ⓑ
6

Activity 13-5 Wobble Plate.

Name _____ Date _____ Class _____

Activity 13-5
Wobble Plate

Refer to **Activity 13-5**. *Study the drawing and familiarize yourself with the views, dimensions, title block, and notes. Read the questions, refer to the print, and write your answers in the blanks provided.*

1. How deep is the 1.2615/1.2610 hole?   1. _____

2. List the diameters of all grooves.   2. _____

3. How long is the 1.075/1.070 diameter bore?   3. _____

4. Determine distance A.   4. _____

5. What kind of material is used for the part?   5. _____

6. What kind of sectional views are used on this print?   6. _____

7. Which decimal diameter on the print allows the most tolerance?   7. _____

8. How many threaded holes are asked for on the print?   8. _____

9. What is the thread size of the tapped hole?   9. _____

10. What is the maximum allowable size (high limit) for the 1.005 dimension?   10. _____

11. List the fillets shown on the print.   11. _____

12. Determine dimension B.   12. _____

13. How wide is the 30° angle (chamfer)?   13. _____

14. How long is the 1.2593/1.2588 diameter bore?   14. _____

15. What scale is the print?   15. _____

16. What class fit is the tapped hole?   16. _____

17. Determine dimension C.   17. _____

18. How much wall thickness is there between the outside diameter of the part and the diameter of the narrow groove?   18. _____

19. Determine distance D.   19. _____

20. What finish is required on the machined surfaces?   20. _____

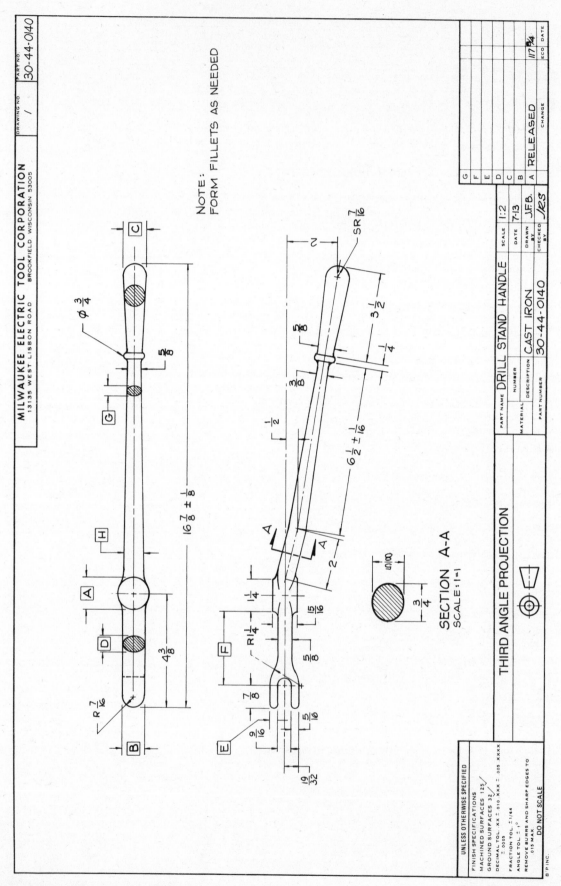

NOTE:
FORM FILLETS AS NEEDED

SECTION A-A
SCALE:1=1

THIRD ANGLE PROJECTION

MILWAUKEE ELECTRIC TOOL CORPORATION
13135 WEST LISBON ROAD   BROOKFIELD WISCONSIN 53005

DRAWING NO /   PART NO 30-44-0140

PART NAME DRILL STAND HANDLE   SCALE 1:2
NUMBER   DATE 7-13
MATERIAL DESCRIPTION CAST IRON   DRAWN BY J.F.B.
PART NUMBER 30-44-0140   CHECKED BY

A RELEASED
CHANGE

UNLESS OTHERWISE SPECIFIED
FINISH SPECIFICATIONS
MACHINED SURFACES 125
GROUND SURFACES 32
DECIMAL TOL .XX ± .010 .XXX ± .005 .XXXX ± .0005
FRACTION TOL ± 1/44
ANGLE TOL ± 1°
REMOVE BURRS AND SHARP EDGES TO .015 MAX
DO NOT SCALE
B P INC.

Activity 13-6 Drill Stand Handle.

Name _____ Date _____ Class _____

Activity 13-6
Drill Stand Handle

*Refer to **Activity 13-6**. Study the drawing and familiarize yourself with the views, dimensions, title block, and notes. Read the questions, refer to the print, and write your answers in the blanks provided.*

1. What is the maximum overall length of the part?   1. _____

2. How wide is the slot on the fork end of the handle?   2. _____

3. What kind of sectional views are shown on this print?   3. _____

4. What is dimension A?   4. _____

5. What scale is the print?   5. _____

6. How deep is the slot?   6. _____

7. What is dimension B?   7. _____

8. How long or wide is the 3/4 diameter shown in the   8. _____
   top view?

9. What type of material is used to make the drill stand handle? 9. _____

10. What is dimension C?   10. _____

11. How long is the short angular bend on the handle?   11. _____

12. What is dimension D?   12. _____

13. What is the radius at the end of the handle?   13. _____

14. Is Section A-A the same scale as the print?   14. _____

15. What is dimension E?   15. _____

16. What is the total drop on the handle from the main   16. _____
    centerline?

17. What is dimension F?   17. _____

18. What is dimension G?   18. _____

19. What is dimension H?   19. _____

20. What tolerance is used on fractional dimensions?   20. _____

Notes

Key Terms

auxiliary plane
auxiliary view

partial auxiliary view
primary auxiliary view

secondary auxiliary view

Auxiliary Views

If the true size and shape of parts with angled surfaces cannot be represented by using basic views, the surfaces or features must be drawn in an auxiliary view. *Auxiliary views* are drawing views that are not parallel to any of the six principal viewing planes and are used to show the true size and shape of an inclined surface or feature.

The block in **Figure 14-1** features a surface labeled A-B-C-D. This surface is inclined and not parallel to the other surfaces. When the surface is shown in the top view or the right side view, it is projected as a foreshortened representation. An auxiliary view is required to view the surface's true size and shape.

Auxiliary projection is the technique used to create auxiliary views. The auxiliary view introduces a new viewing direction called the auxiliary plane, **Figure 14-2**. An *auxiliary plane* is a viewing plane that is parallel to an inclined surface. Perpendicular projectors that extend from the inclined surface transfer the surface's shape and features onto the auxiliary plane.

The auxiliary plane in **Figure 14-2** is "attached" to the frontal plane and swung into alignment with the other planes from the frontal plane. An auxiliary view projected from a principal plane is known as a *primary auxiliary view*. On actual prints, auxiliary views are not labeled, just as principal views are not labeled.

In addition to a foreshortened rectangular surface, **Figure 14-3** features two holes that appear as ellipses, not circles, in the top view. The ellipses are graphic representations of foreshortened circular holes. In the auxiliary view, both the surface and holes appear as true size and shape. Note that the width of the object is true size in both views. Visualizing the details of an auxiliary view requires practice, patience, and the same projection skills as visualizing principal views.

Other Types of Auxiliary Views

Partial auxiliary views are auxiliary views that do not show the entire object. Partial auxiliary views are useful for when only a section

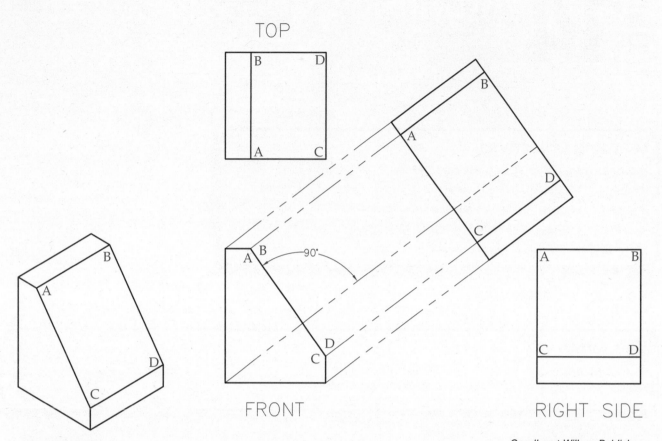

Figure 14-1 To show the true size and shape of the inclined surface, an auxiliary view is needed.

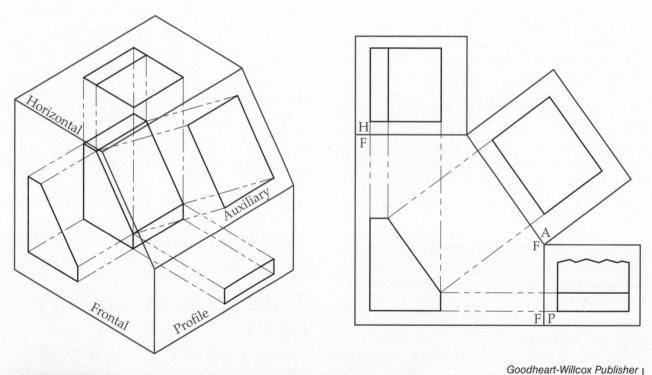

Figure 14-2 The auxiliary plane is parallel to the inclined surface.

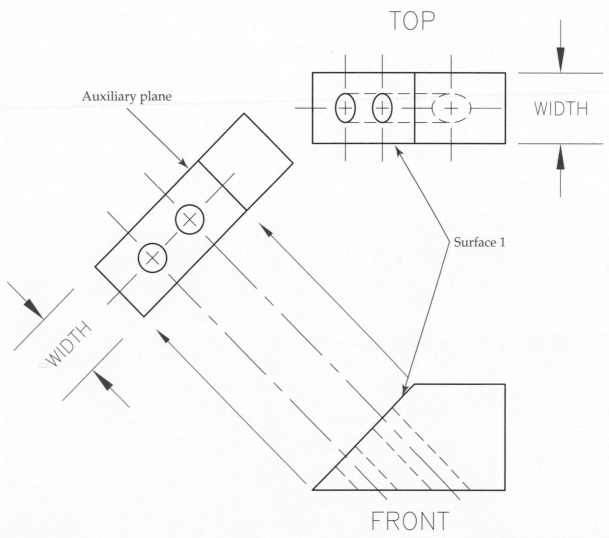

TOP

WIDTH

Auxiliary plane

Surface 1

WIDTH

FRONT

Figure 14-3 Foreshortened holes appear as ellipses on the top view. The auxiliary view shows their true size and shape. Note that some hidden lines have been removed for clarity.

of a larger part is inclined to the principal views. See **Figure 14-4**. Partial auxiliary views use short break lines, much like partial sectional views, to break out a section of the part to project into a new view. No cutting-plane or viewing-plane lines are required.

Auxiliary views can also be moved out of projection, but the auxiliary view must remain oriented as if it was still in projection. A viewing-plane line indicates the orientation of the auxiliary view. See **Figure 14-4**.

An oblique surface is not parallel to any of the principal planes. To view an oblique surface as true size and shape, a *secondary auxiliary view* must be projected from a primary auxiliary view. See **Figure 14-5**.

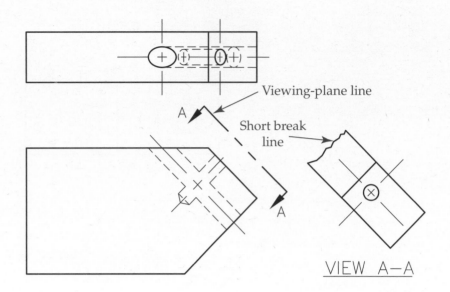

Viewing-plane line

Short break line

VIEW A—A

Goodheart-Willcox Publisher

Figure 14-4 This auxiliary view is a partial auxiliary view that has also been placed out of projection.

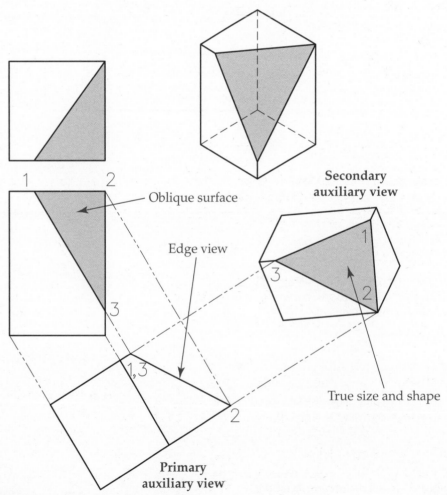

Oblique surface

Edge view

Secondary auxiliary view

True size and shape

Primary auxiliary view

Goodheart-Willcox Publisher

Figure 14-5 A secondary auxiliary view.

Name _____ Date _____ Class _____

Drawing Problem 14-1

Draw the primary auxiliary views. Some lines have been removed for clarity. Use a straightedge or rule to complete the sketch.

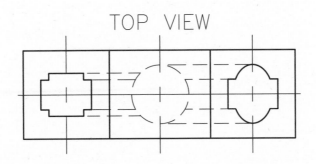

TOP VIEW

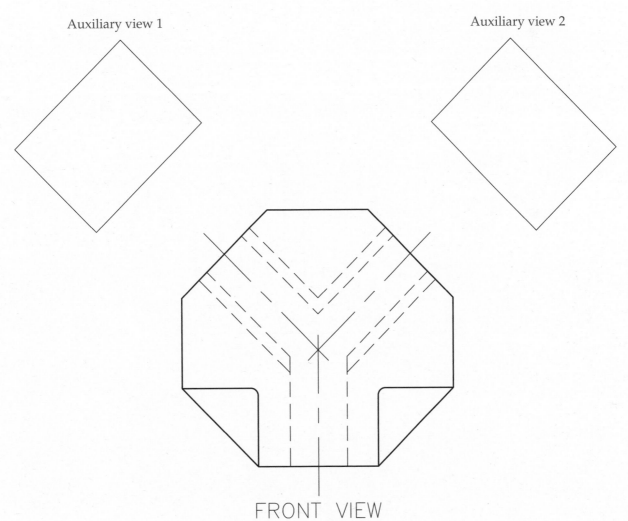

Auxiliary view 1

Auxiliary view 2

FRONT VIEW

Name _____ Date _____ Class _____

Drawing Problem 14-2

Draw the partial auxiliary views. Some lines have been removed for clarity. Use a straightedge or rule to complete the sketch.

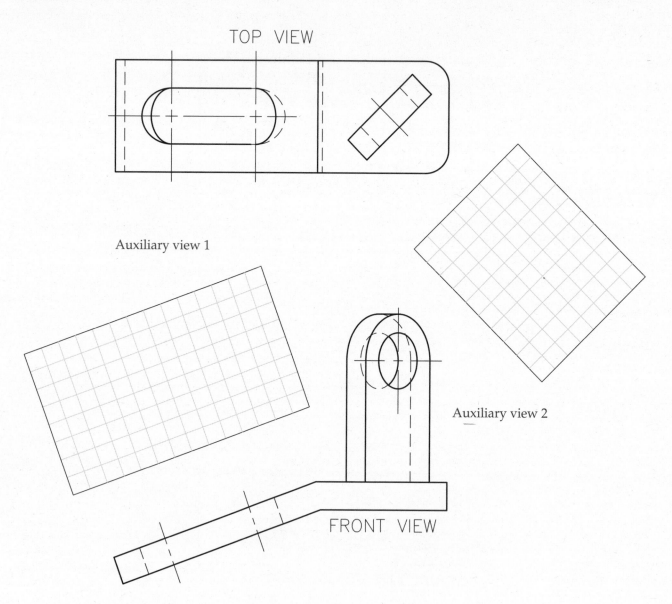

TOP VIEW

Auxiliary view 1

Auxiliary view 2

FRONT VIEW

Notes

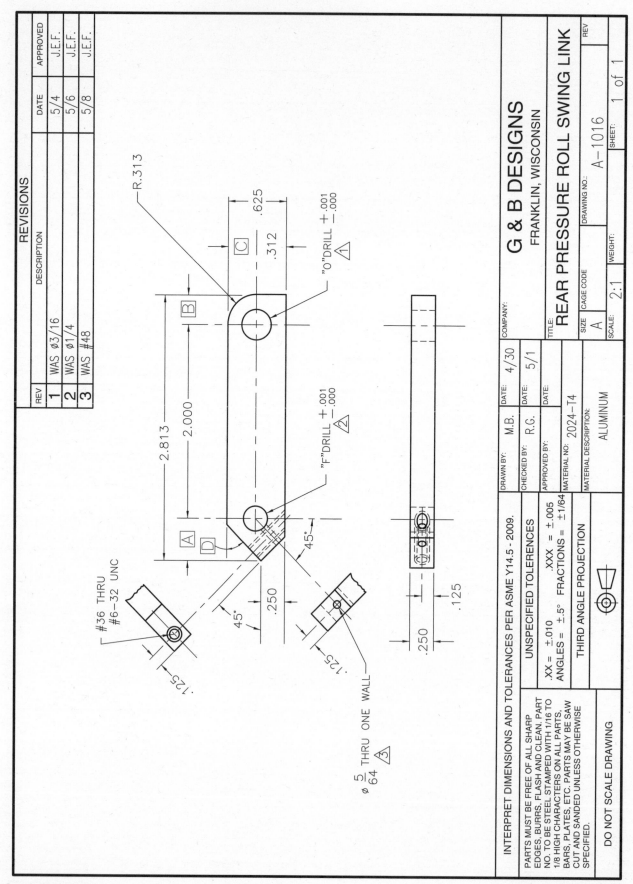

REVISIONS			
REV	DESCRIPTION	DATE	APPROVED
1	WAS ⌀3/16	5/4	J.E.F.
2	WAS ⌀1/4	5/6	J.E.F.
3	WAS #48	5/8	J.E.F.

R.313

.625

C

.312

"O"DRILL +.001 −.000 △1

B

2.813

2.000

A   D

"F"DRILL +.001 −.000 △2

45°

.250

45°

.125

⌀ 5/64 THRU ONE WALL △3

#36 THRU #6-32 UNC

.125

.250

.125

.250

G & B DESIGNS

FRANKLIN, WISCONSIN

REV

REAR PRESSURE ROLL SWING LINK

COMPANY:

TITLE:

SIZE	CAGE CODE	DRAWING NO.:	REV
A		A–1016	
SCALE: **2:1**	WEIGHT:	SHEET: 1 of 1	

DRAWN BY: M.B.	DATE: 4/30
CHECKED BY: R.G.	DATE: 5/1
APPROVED BY:	DATE:
MATERIAL NO: 2024–T4	
MATERIAL DESCRIPTION: ALUMINUM	

INTERPRET DIMENSIONS AND TOLERANCES PER ASME Y14.5 - 2009.

UNSPECIFIED TOLERANCES	
.XX = ±.010	.XXX = ±.005
ANGLES = ±.5°	FRACTIONS = ±1/64

THIRD ANGLE PROJECTION

PARTS MUST BE FREE OF ALL SHARP EDGES, BURRS, FLASH AND CLEAN. PART NO. TO BE STEEL STAMPED WITH 1/16 TO 1/8 HIGH CHARACTERS ON ALL PARTS, BARS, PLATES, ETC. PARTS MAY BE SAW CUT AND SANDED UNLESS OTHERWISE SPECIFIED.

DO NOT SCALE DRAWING

Activity 14-1 Rear Pressure Roll Swing Link.

Name _____ Date _____ Class _____

Activity 14-1
Rear Pressure Roll Swing Link

Refer to **Activity 14-1**. *Study the drawing and familiarize yourself with the views, dimensions, title block, and notes. Read the questions, refer to the print, and write your answers in the blanks provided.*

1. What is the decimal size of the "O" drill?

2. What kind of material is used to make the part?

3. How thick is the part?

4. How many auxiliary views are on the print?

5. Determine distance B.

6. How many holes are drilled into the part?

7. Determine distance A.

8. How many foreshortened holes are depicted in the bottom view?

9. What was Revision 1?

10. What tolerance is used on three-place decimal dimensions?

11. What decimal size is the No. 36 tap drill?

12. Determine distance C.

13. Who checked the print?

14. What tolerance is used on angular dimensions?

15. Is the Ø5/64 hole drilled through the center of the tapped hole?

16. What type(s) of auxiliary views are used on this print?

17. Determine angle D.

18. How far from the end of the part is the tapped hole located?

19. What is the center distance between the two largest drilled holes?

20. At what angle is the Ø5/64 hole drilled to the "F" hole?

1. _____

2. _____

3. _____

4. _____

5. _____

6. _____

7. _____

8. _____

9. _____

10. _____

11. _____

12. _____

13. _____

14. _____

15. _____

16. _____

17. _____

18. _____

19. _____

20. _____

Notes

Learning Objectives

After studying this unit, you will be able to:

✓ Describe the purposes of geometric dimensioning and tolerancing (GD&T).
✓ Define various applications of geometric tolerances.
✓ Identify various GD&T symbols.
✓ Define various terms relating to GD&T.
✓ Identify datum surfaces.
✓ Interpret the meaning of material condition modifiers on a print.
✓ Interpret the meaning of GD&T symbols on a print.

Key Terms

angularity	flatness	profile of a line
basic dimension	form tolerance	profile of a surface
circularity	geometric dimensioning and	profile tolerance
circular runout	tolerancing (GD&T)	regardless of feature size (RFS)
concentricity	least material condition (LMC)	runout tolerance
cylindricity	location tolerance	size feature
datum	material condition modifier	straightness
datum feature	maximum material condition (MMC)	symmetry
datum feature symbol	orientation tolerance	tolerance
datum target	parallelism	total runout
feature	perpendicularity	true position
feature control frame	position	

Advancements in technology have brought about greater control in the accuracy of machined parts. Still, it is almost impossible to manufacture a perfect part, so it becomes necessary to define the amount of variation permitted to a specific form of a part. These factors have led to the use of a drafting system known as *geometric dimensioning and tolerancing (GD&T)*.

The method of applying geometric dimensioning and tolerancing to conventional dimensioning is through the use of geometric symbols that have been recommended by the American Society of Mechanical Engineers (ASME).

This unit is a brief introduction to geometric dimensioning and tolerancing. Information contained in the unit is a partial, summary version of GD&T practices. For more complete and detailed information, refer to the publication ASME Y14.5-2009, *Dimensioning and Tolerancing*.

The purposes of GD&T include helping to reduce drawing changes, increasing productivity, and avoiding the accidental scrapping of functional parts. GD&T is used when operating automated equipment and working with functional gaging. It can also help clarify the features of parts which are critical to function or interchangeability.

GD&T Application

Applying geometric dimensioning and tolerancing involves specifying the allowable variation permitted on exact form or true position on part features. Geometric tolerances are applied to the following five areas of concern: form, orientation, location, profile, and runout tolerances.

- *Form tolerances* control the form or shape of various geometric figures. Form tolerances control straightness, flatness, circularity (roundness), and cylindricity. Form tolerances are not related to datums.
- *Orientation tolerances* control angularity, perpendicularity, and parallelism.
- *Location tolerances* define the allowable variation of a feature from the exact or true position shown on the drawing.
- *Profile tolerances* are used to control form, or combination of size, form, and orientation.

They specify a constant boundary along the true profile within which all points or elements of the surface must lie. Profile tolerances include line profile and surface profile.

- *Runout tolerances* control the relationships of one or more features of a part to its axis.

GD&T Symbols

Previous units described the use of symbols adopted by various ASME standards, including the diameter ($\varnothing$), radius (R), counterbore, spotface, countersink, and depth symbols. Geometric control characteristics have been established for form, orientation, location, profile, and runout tolerances. The accompanying chart, **Figure 15-1**, shows the symbols for the 14 geometric control characteristics used in GD&T.

Type of Tolerance	Characteristic	Symbol
Form	Straightness	—
	Flatness	▱
	Circularity (Roundness)	◯
	Cylindricity	⌭
Orientation	Angularity	∠
	Perpendicularity	⊥
	Parallelism	∥
Location	Position	⊕
	Concentricity	◎
	Symmetry	≡
Profile	Profile of a Line	⌒
	Profile of a Surface	⌓
Runout	Circular Runout	↗ or ↗
	Total Runout	↗↗ or ↗↗

The American Society of Mechanical Engineers

Figure 15-1 Geometric characteristic symbols.

Geometric characteristic symbols are enclosed in a rectangular box containing the allowed tolerance at datum with a leader pointing to the feature to which it applies, **Figure 15-2**.

The *datum feature symbol* shown in **Figure 15-3** consists of a frame containing the datum reference letter. An equilateral triangle connected to the square identifies the feature.

All geometric symbols require a *feature control frame*. It consists of a rectangular box containing the geometric characteristic symbol, followed by the allowable tolerance. See **Figure 15-4**.

All orientation, location, and runout tolerances must be related to a datum reference. This relationship is stated in the feature control frame. The datum reference letter is placed after the geometric symbol and the tolerance, as shown in **Figure 15-5**. A single geometric control can have multiple datum references. Datum references are read from left to right.

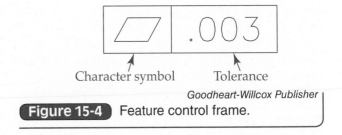

Character symbol   Tolerance

Goodheart-Willcox Publisher

Figure 15-4 Feature control frame.

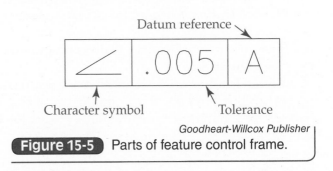

Datum reference

Character symbol   Tolerance

Goodheart-Willcox Publisher

Figure 15-5 Parts of feature control frame.

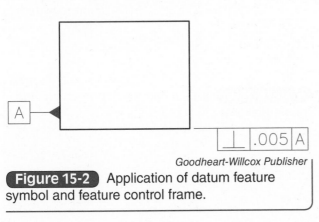

Goodheart-Willcox Publisher

Figure 15-2 Application of datum feature symbol and feature control frame.

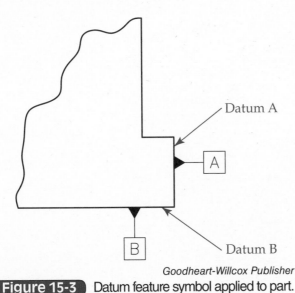

Datum A

A

B   Datum B

Goodheart-Willcox Publisher

Figure 15-3 Datum feature symbol applied to part.

GD&T Terminology

The following are several terms appearing in this unit that should be reviewed prior to diving into GD&T any further. Various other new terms will be defined as they appear within this unit.

A *datum* is defined as an exact plane, point, line, or axis of a part. A datum is used as a reference base or origin position from which location or geometric characteristics of features of a part are derived.

In reality, a dimension can never be exact because of numerous variables that can cause variations in the manufacturing process. A *basic dimension* is the theoretical exact value of a feature or datum's size, shape, or location. A basic dimension has no allowances or tolerances applied to its value. In addition, title block tolerances do not apply to basic dimensions. A rectangular frame placed around a dimension's value indicates a basic dimension. A basic dimension indicates an important dimension for a feature. Basic dimensions are used for the following:

■ Specifying the true location of a feature
■ Defining the profile of a line or surface
■ Locating datum targets

A *feature* is a universal term applied to an actual portion of a part, such as a surface, hole, thread, or groove. A *size feature* is a feature with a center plane or center axis. The theoretically exact location of the center plane or center axis of a size feature is known as the *true position*.

A *tolerance* is the total amount a dimension is allowed to vary or the difference between its maximum and minimum limits. Geometric tolerances can be two-dimensional or three-dimensional.

Datums

Datum planes are theoretically exact reference bases. *Datum features* are actual surfaces or features of an object used to establish datums that include any surface and feature irregularities. See **Figure 15-6**.

Datum features can be controlled through the use of various symbols. For example, to control flatness for a datum feature, the flatness symbol is used, **Figure 15-7**. Also, note that the difference in height between the high points and low points of the surface must be within a tolerance of .005″. This is the flatness tolerance of the surface.

Besides the flatness tolerance required for the lower surface, a size tolerance of ±.010″ must be held between the upper and lower surfaces. The size between the high points of the surfaces must not exceed the high limit of .635″. Also, the size between the low points of the surfaces must not be less than the low limit of .615″.

In summary, when machining both the upper and lower surfaces, two concerns must be taken into account: the lower surface must be held to a flatness tolerance of .005″ and still be within a size tolerance of ±.010″ with the upper surface.

Often, an entire surface does not need to be machined. In these cases, only specific points or areas on the surface require machining. These points, lines, and areas are called *datum targets*.

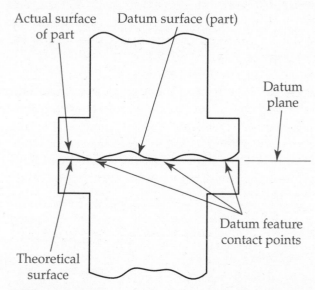

Goodheart-Willcox Publisher

Figure 15-6 Actual surface of part contacting theoretical (datum) surface.

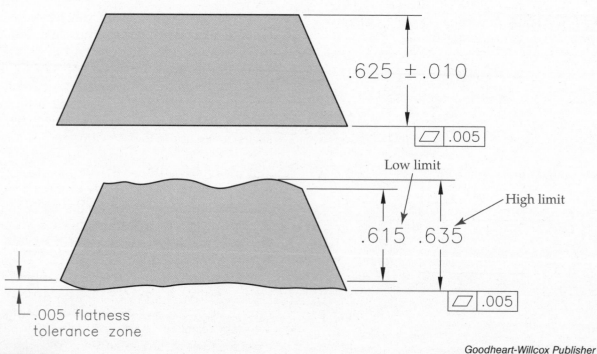

Goodheart-Willcox Publisher

Figure 15-7 Specifying flatness of a surface.

A datum target symbol is used to identify a datum target.

The datum target symbol is a circle divided by a horizontal line. In the lower part of the circle is the identification of the datum point. This identification includes the datum feature letter, followed by the number of the point. The upper half of the circle contains the diameter of the datum target area. This area is left blank for datum target points and datum target lines.

A datum target point is shown on the surface as an "X". A leader connects the point and the datum target symbol. See **Figure 15-8**. The details of the datum target symbol are shown in **Figure 15-9**.

A datum target area is marked with diagonal lines. The outer edge of the area is marked with a phantom line. See **Figure 15-10**.

A centerline is used as a datum axis on cylindrical and symmetrical parts. See **Figures 15-11** and **15-12**. Datum B is the datum axis of the part. The datum axis shown is perpendicular to datum feature A.

Material Condition Modifiers

Material condition modifiers indicate a bonus tolerance is to be applied to a feature control. Material condition modifiers are applied at maximum material condition (MMC), least material condition (LMC), or regardless of feature size (RFS). Both MMC and LMC use modifier symbols, an encircled M for MMC and an encircled L for LMC. Regardless of feature size is assumed when no other material condition modifier is specified. The symbol for RFS, an encircled S, is no longer used, but may be present on older prints. **Figure 15-13** shows the MMC, LMC, RFS, and other modifying symbols. These modifiers should appear in the feature control frame following the tolerance, as shown in **Figure 15-14**.

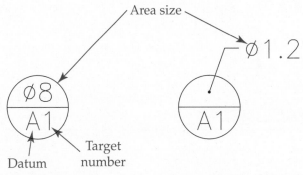

Goodheart-Willcox Publisher

Figure 15-9   Datum target symbol.

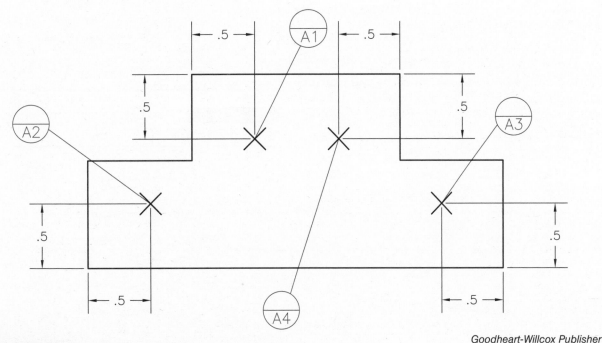

Goodheart-Willcox Publisher

Figure 15-8   Datum target points on a surface.

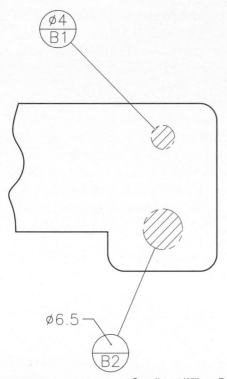

Goodheart-Willcox Publisher

Figure 15-10 Identifying datum target areas.

Maximum material condition (MMC) means that internal features such as holes and slots would be at their low limit (minimum) size, whereas external features such as a shaft would be at their high limit (maximum) size.

Example:
MMC of a ∅.750/.752 hole is .750″.
MMC of a ∅.750/.752 shaft is .752″.

When tolerance is specified at MMC, the allowable tolerance is determined by the feature size. If the actual feature size has varied from MMC, an increase in the tolerance equal to that amount of variance is allowed.

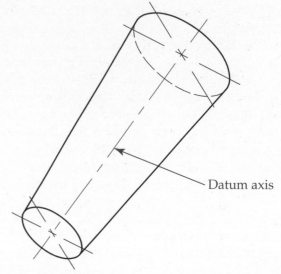

Goodheart-Willcox Publisher

Figure 15-11 Datum axis on a tapered cylinder.

GD&T Symbols	
Ⓜ	Maximum material condition modifier
Ⓛ	Least material condition modifier
Ⓢ	Regardless of feature size (old standard; obsolete)
Ⓟ	Projected tolerance zone
Ⓣ	Tangent plane
Ⓕ	Free state
⟨ST⟩	Statistical tolerance
↔	Between

Goodheart-Willcox Publisher

Figure 15-13 Material condition modifiers and other modifying symbols.

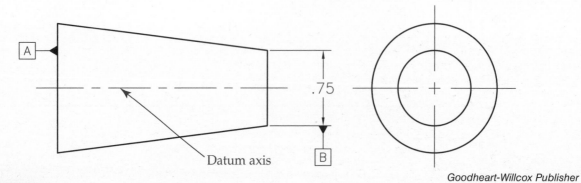

Figure 15-12 Centerline as a datum axis.

Goodheart-Willcox Publisher

Figure 15-14 Feature control frames with material condition modifiers.

Goodheart-Willcox Publisher

Example:

If a .010" position tolerance zone is allowed for a .750/.752" hole at MMC, **Figure 15-15**, the

tolerance zone would increase to .011" for a .751" hole and .012" for a .752" hole. Tolerance increases are equal to amount of hole diameter change.

Least material condition (LMC) means that the internal features such as holes and slots would be at their high limit (maximum) size. External features such as a shaft would be at their low limit (minimum) size.

Example:
LMC of a Ø.750/.752 hole is .752".
LMC of a Ø.750/.752 shaft is .750".

When tolerance is specified at LMC, the allowable tolerance is determined by the feature size. If the actual feature size has varied from

Drawing

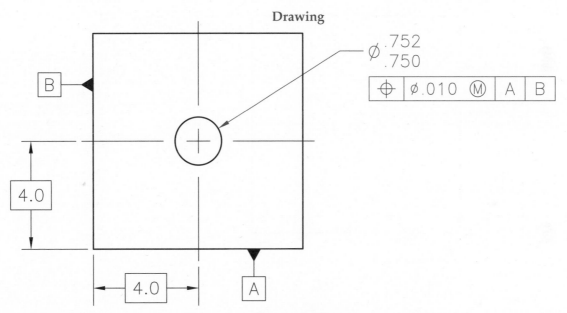

Interpretation

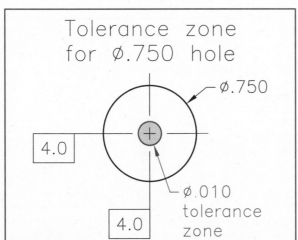
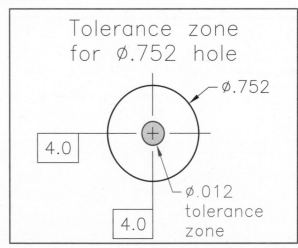

Goodheart-Willcox Publisher

Figure 15-15 Position tolerance applied to a hole at MMC. The diameter symbol (Ø) indicates that the tolerance zone is round. The centerline axis of the hole must be located in the tolerance zone.

LMC, increase in the tolerance equal to that amount of variance is allowed.

Example:

If a .010″ position tolerance zone is allowed for a .750/.752 hole at LMC, **Figure 15-16**, the tolerance zone would increase to .011″ for a .751 hole and .012″ for a .750 hole. Tolerance change is equal to hole diameter change.

Regardless of feature size (RFS) means that the tolerance zone is limited to the specified value and does not change regardless of the actual size of the feature.

Example:

If a .010″ tolerance is allowed for a .750/.752 hole, the tolerance zone remains the same, **Figure 15-17**, regardless of whether the hole is .750, .751, or .752″ in diameter. Tolerance at RFS is independent of feature size.

Other modifying symbols include the projected tolerance zone and free state symbols. The projected

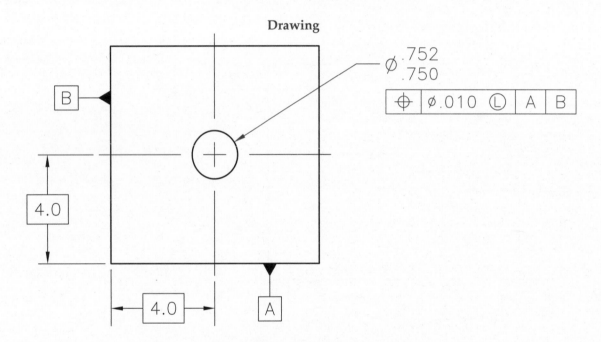

Drawing

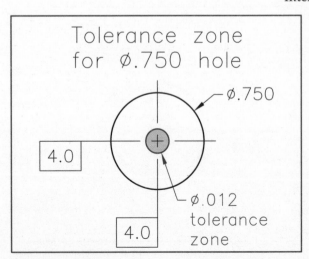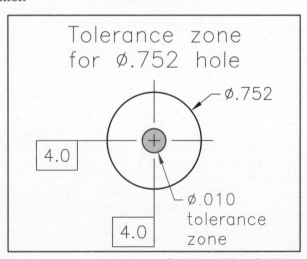

Interpretation

Figure 15-16 Position tolerance applied to a hole at LMC.

tolerance zone symbol is used to indicate that the tolerance zone is to be projected outside the object. The free state symbol indicates the tolerance is to be applied to an object in a free state (objects that bend or flex, such as a rubber gasket).

Form Tolerances

Form tolerance symbols depict circularity, cylindricity, straightness, and flatness.

Circularity

Circularity applies to a surface of revolution making all points of a surface intersected by a cutting plane perpendicular to a common axis and equidistant (same distance) from axis, **Figure 15-18**. Circularity is also known as *roundness*.

Circularity tolerance states the maximum variation allowed between maximum and minimum radii in a single cross section, **Figure 15-19**. The .005″ wide tolerance zone for circularity is between two concentric circles. The actual round surface of the cylinder must lie within the .005″ wide tolerance zone.

Cylindricity

Cylindricity applies to a surface of revolution making all points on the surface equidistant along a common axis. The cylindricity tolerance states a tolerance zone between two concentric cylinders within which all points on a surface must lie. See **Figure 15-20**. Cylindricity takes into consideration circularity, straightness, and taper of a cylinder.

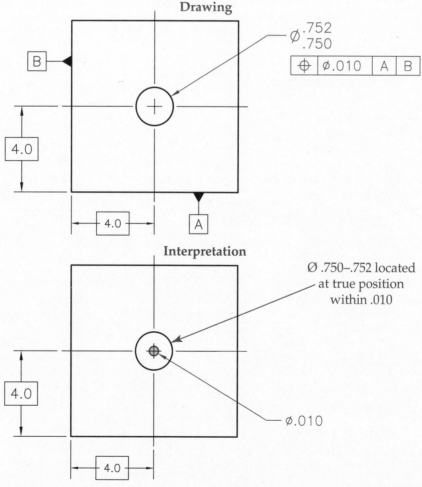

Figure 15-17   Location tolerance applied to a hole at RFS.

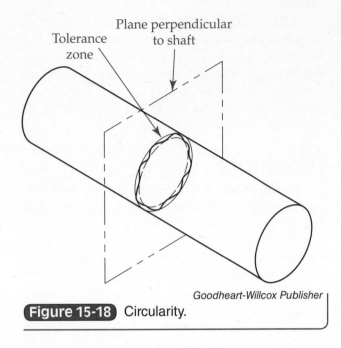

Figure 15-18 Circularity.

Straightness

Straightness is a condition where all points on any surface or axis must lie in a straight line. Straightness tolerance creates a tolerance zone (two parallel planes) in which all points on a surface must lie, as shown in **Figures 15-21** and **15-22**.

Flatness

Flatness requires all points on a surface to lie in one plane. Flatness tolerance creates a tolerance zone formed by two parallel planes between which the entire surface must lie. See **Figure 15-23**.

Orientation Tolerances

Orientation tolerance symbols depict angularity, perpendicularity, and parallelism.

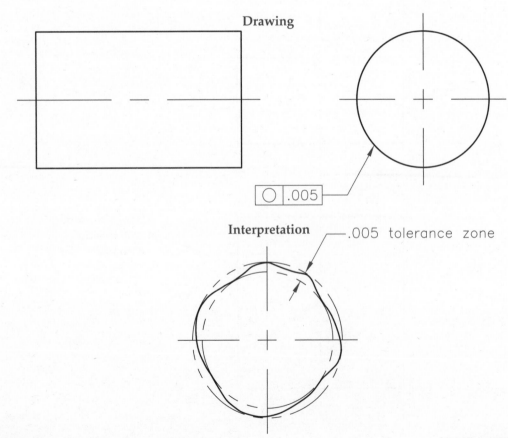

Figure 15-19 Circularity tolerance zone.

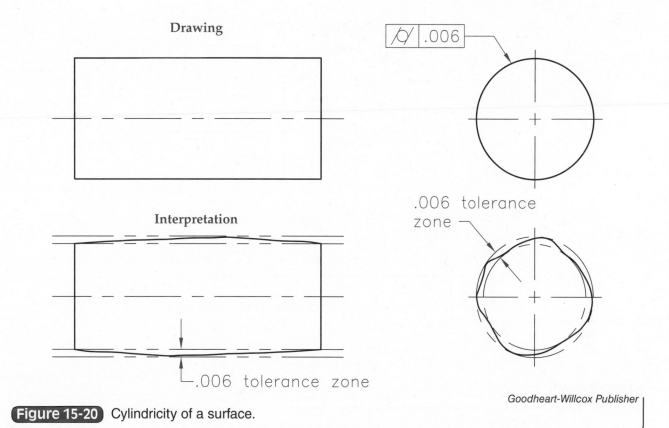

Figure 15-20   Cylindricity of a surface.

Goodheart-Willcox Publisher

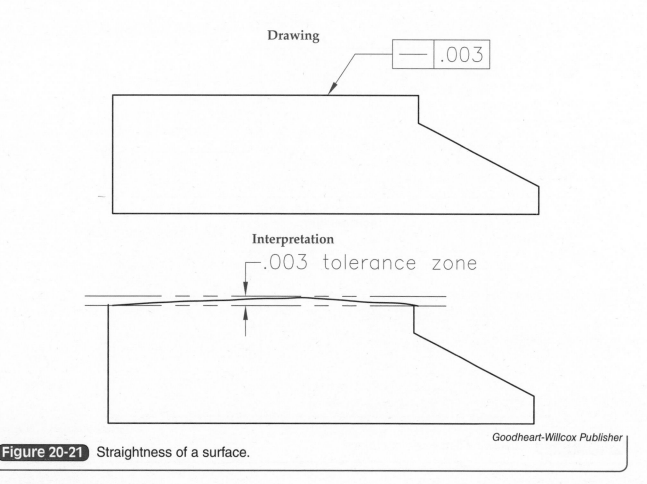

Figure 20-21   Straightness of a surface.

Goodheart-Willcox Publisher

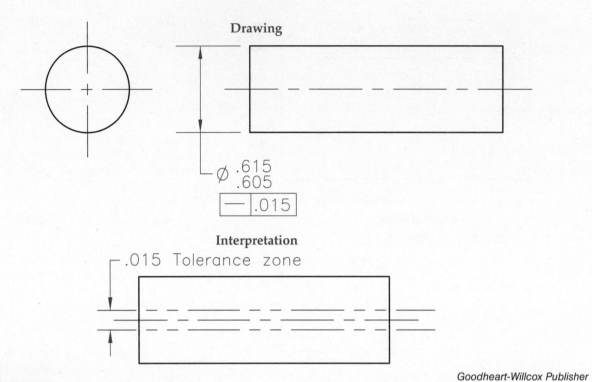

Goodheart-Willcox Publisher

Figure 15-22 Specifying axis straightness.

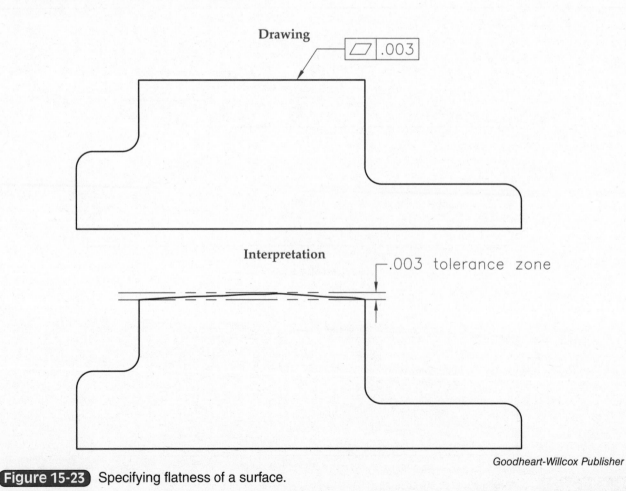

Goodheart-Willcox Publisher

Figure 15-23 Specifying flatness of a surface.

Angularity

Angularity applies to a surface or axis which is at a specified angle (other than 90°) to a datum plane or axis. See **Figure 15-24**. A basic dimension is used, and no tolerance of degrees is needed.

Perpendicularity

Perpendicularity requires a surface or axis to be at a right angle (90°) to a datum plane or axis, **Figure 15-25**.

Parallelism

Parallelism requires that a surface or axis remains the same distance at all points from a datum plane or axis, as shown in **Figure 15-26**.

Location Tolerances

Location tolerance symbols include position, concentricity, and symmetry.

Position

Position defines a tolerance zone within which the axis or center plane of a feature is allowed to vary from the true (exact) position. Position tolerancing is applied to a feature control on MMC, LMC, or RFS basis.

Coordinate tolerancing of a position provides a square tolerance zone. This zone results from coordinate plus and minus tolerancing, as shown in **Figure 15-27**.

Position tolerancing increases a tolerance zone by 57% more area by using a round tolerance zone.

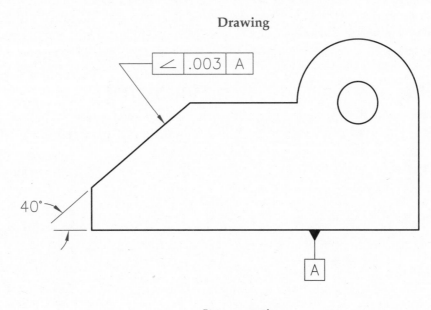

Drawing

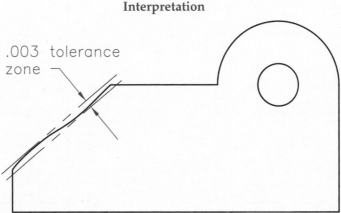

Interpretation

Figure 15-24   Angularity of a surface.

Goodheart-Willcox Publisher

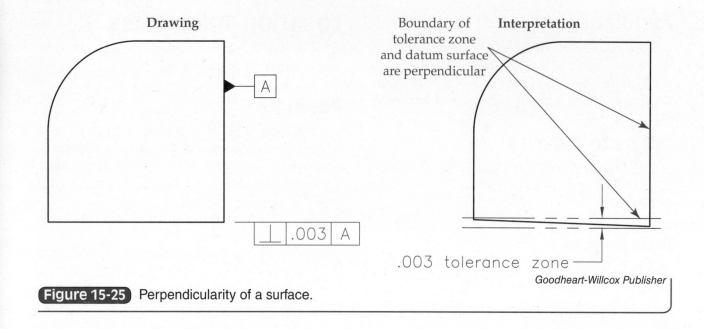

Figure 15-25 Perpendicularity of a surface.

This occurs by using basic dimensions and specifying a position tolerance. The round tolerance zone, **Figure 15-28**, is centered at the intersection of these basic dimensions.

to the datum axis. Concentricity maintains equal midpoint distance of all elements on each side of a datum axis, regardless of feature shape or size. See **Figure 15-29**.

Concentricity

Concentricity exists when the axes of each part feature's surface of revolution are common

Symmetry

Symmetry exists when a part feature is symmetrically arranged about the center plane

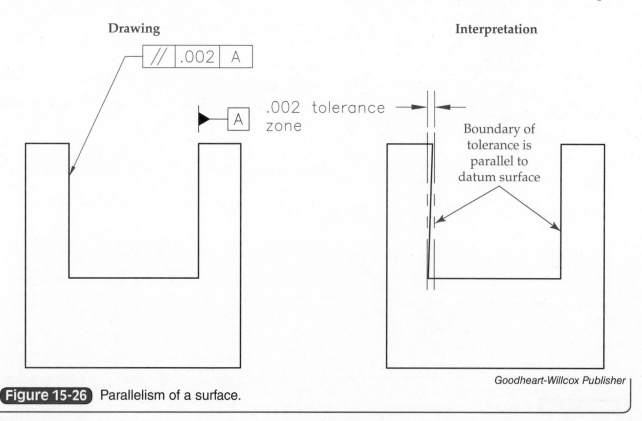

Figure 15-26 Parallelism of a surface.

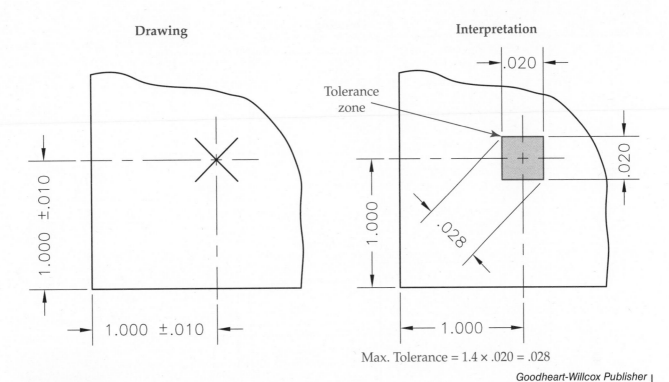

Max. Tolerance = 1.4 × .020 = .028

Goodheart-Willcox Publisher

Figure 15-27 Coordinate tolerance and tolerance zone.

of a datum feature. The elements controlled by a symmetry tolerance are located equidistant about a center plane. See **Figure 15-30**.

Profile Tolerances

Profile tolerance symbols include profile of a line and profile of a surface.

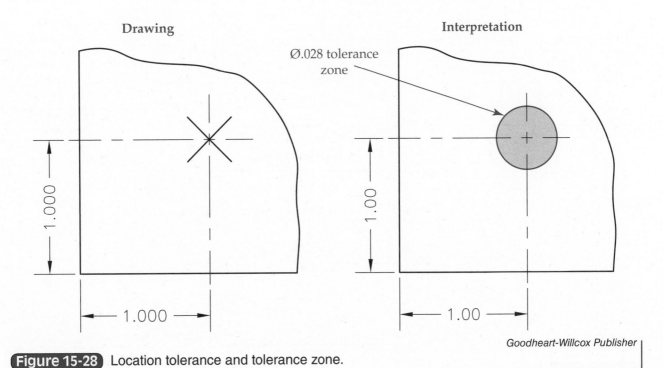

Goodheart-Willcox Publisher

Figure 15-28 Location tolerance and tolerance zone.

Drawing

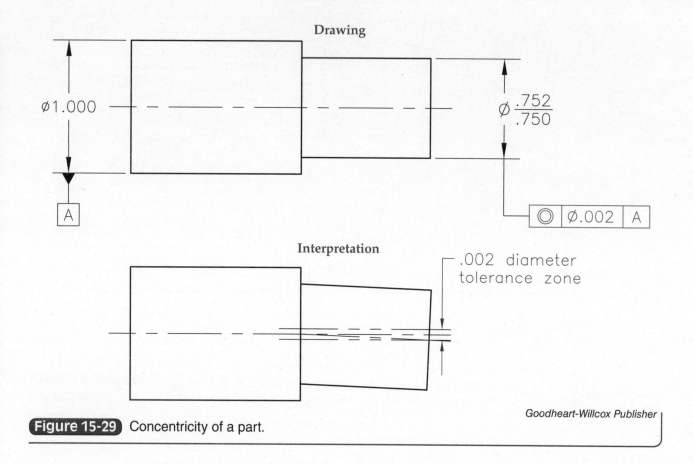

Interpretation

Figure 15-29   Concentricity of a part.

Goodheart-Willcox Publisher

Drawing

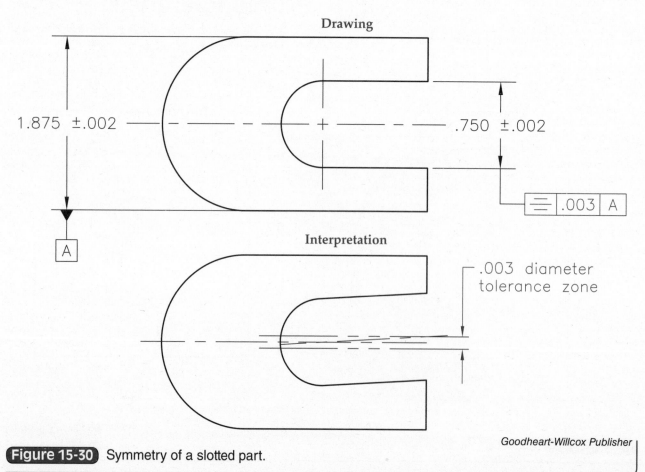

Interpretation

Figure 15-30   Symmetry of a slotted part.

Goodheart-Willcox Publisher

Profile of a Line

Profile of a line states all points on a line must lie within the specified tolerance zone of the true profile. The between symbol, **Figure 15-31**, can be used to specify a profile tolerance between two points, such as BETWEEN X & Y.

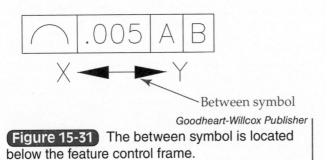

Between symbol

Goodheart-Willcox Publisher

Figure 15-31 The between symbol is located below the feature control frame.

Profile of a line is used when the entire surface control is not required. It is also used on shapes with varying cross-sectional dimensions along their length or width, as shown in **Figure 15-32**.

Profile of a Surface

Profile of a surface states all points along an entire surface must lie within the specified tolerance zone of the true profile. See **Figure 15-33**.

Runout Tolerances

Runout is applied to surfaces at right angles to a datum axis and to surfaces existing around a

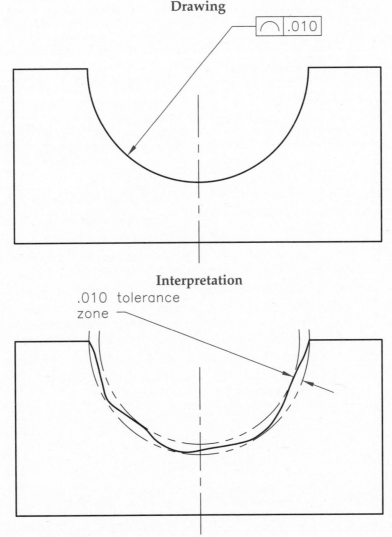

Drawing

Interpretation

.010 tolerance zone

Goodheart-Willcox Publisher

Figure 15-32 Profile of a line.

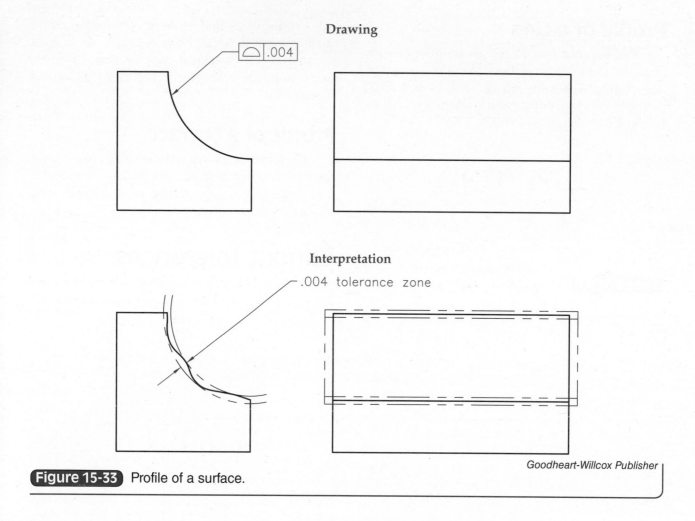

Goodheart-Willcox Publisher

Figure 15-33 Profile of a surface.

datum axis. Runout tolerance symbols represent circular runout and total runout.

Circular Runout

Circular runout provides control of circular elements of a surface applied independently at any measuring location as the part is rotated 360°. See **Figure 15-34**.

Circular runout applied to surfaces existing around a datum axis controls cumulative form and location characteristics (roundness, concentricity). Circular runout applied to a surface at right angles to a datum axis controls wobble.

Total Runout

Total runout provides control of all circular elements of a surface applied simultaneously to all measuring locations as the part is rotated 360°. See **Figure 15-35**.

Total runout applied to surfaces existing around a datum axis controls cumulative form and location characteristics such as circularity, concentricity, angularity, straightness, taper, and profile of a surface. Total runout applied to a surface at right angles to a datum axis controls wobble and flatness.

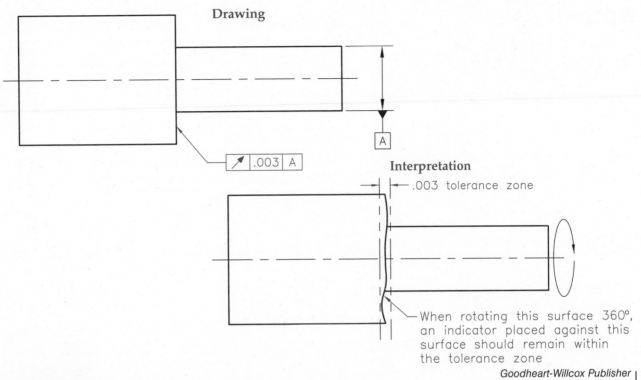

Figure 15-34 Circular runout (wobble) on a shouldered surface.

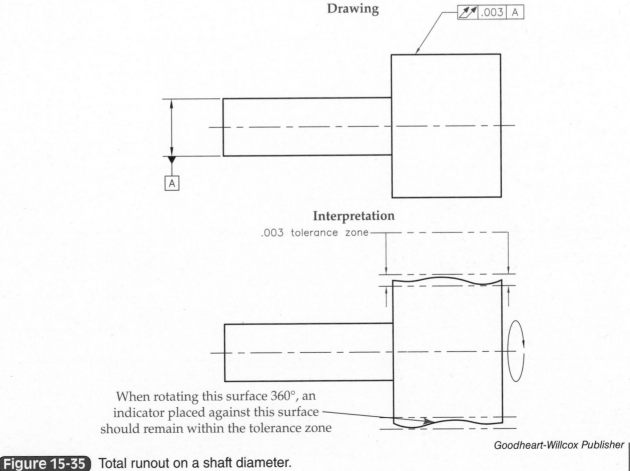

Figure 15-35 Total runout on a shaft diameter.

Notes

Name _____ Date _____ Class _____

Activity 15-1
GD&T Symbols

Match each term or description with the correct symbol by placing the appropriate letter in the blank provided.

_____ 1.   Datum identification symbol.

_____ 2.   Geometric symbol that specifies a tolerance zone between two concentric circles.

_____ 3.   Geometric symbol denoting a tolerance zone between two concentric cylinders.

_____ 4.   Symmetry.

_____ 5.   Datum target symbol.

_____ 6.   Flatness.

_____ 7.   Symbol requiring axes of each part features' surface of revolution to be common to the datum axis.

_____ 8.   Internal features (holes, slots) are at their low limit.

_____ 9.   Total runout.

_____ 10. External features (shafts) are at their low limit.

A.

B.

C.

D.

E.

F.

G.

H.

I.

J.

Name _____ Date _____ Class _____

Activity 15-2
GD&T Symbols

Match each term or description with the correct symbol by placing the appropriate letter in the blank provided.

_____ 1. Straightness.

_____ 2. Geometric symbol used to identify a tolerance of location.

_____ 3. Geometric symbol specifying a surface or axis at an angle other than 90°.

_____ 4. Symbol requiring a surface or axis to be at right angle to a datum plane or axis.

_____ 5. Profile of a line.

_____ 6. Symbol provides control of circular elements of a surface.

_____ 7. Parallelism.

_____ 8. Obsolete symbol specifying a tolerance zone limited to specified value.

_____ 9. Profile of a surface.

_____ 10. Symbol located below the feature control frame.

A. ——

B. Ⓢ

C. ⌖

D. ⊥

E. ↔

F. ∠

G. //

H. ⌒

I. ↗

J. ◠

Name _____ Date _____ Class _____

Activity 15-3
Position Tolerancing

Complete the chart for the part below by providing position tolerance values for the hole shown as the machined hole size changes.

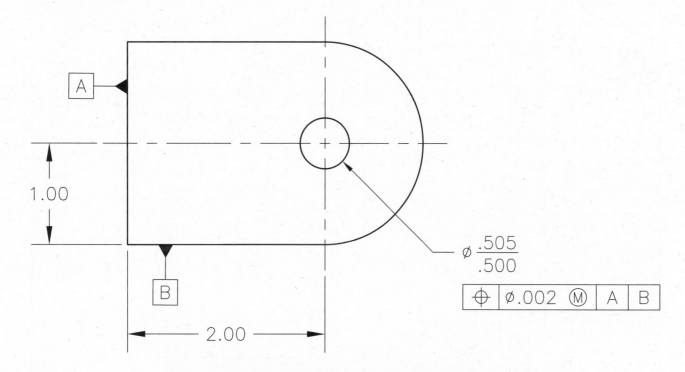

Hole Size	Position Tolerance
.505	
.504	
.503	
.502	
.501	
.500	

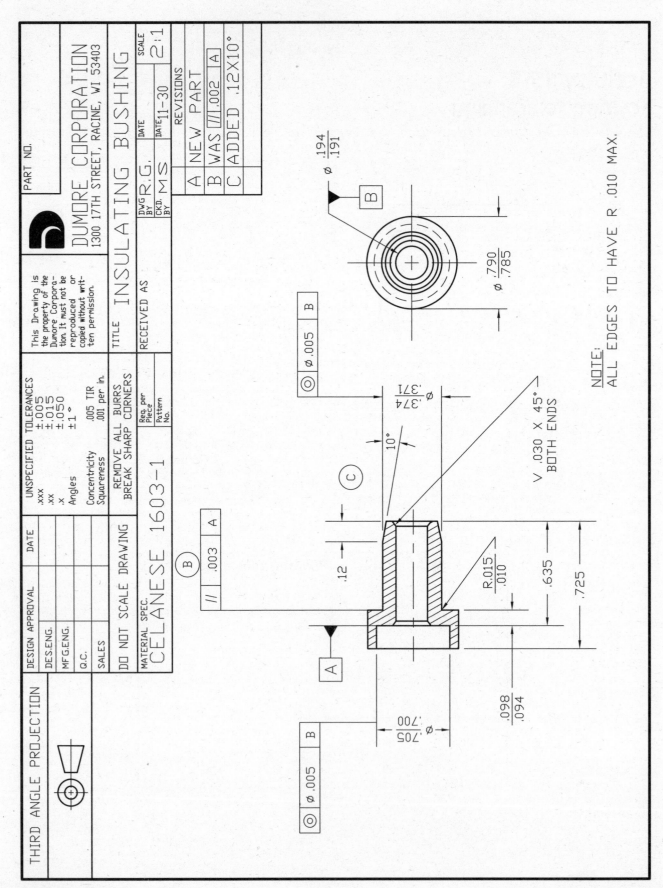

THIRD ANGLE PROJECTION

DUMORE CORPORATION
1300 17TH STREET, RACINE, WI 53403

TITLE   INSULATING BUSHING

SCALE
2:1

PART NO.

DWG BY R.G.
CKD. BY MS
DATE 11-30

REVISIONS
A   NEW PART
B   WAS ///|.002 |A|
C   ADDED .12X10°

This Drawing is the property of the Dumore Corpora- tion. It must not be reproduced or copied without writ- ten permission.

RECEIVED AS

DESIGN APPROVAL
DES.ENG.
MFG.ENG.
Q.C.
SALES
DATE

UNSPECIFIED TOLERANCES
.xxx   ±.005
.xx   ±.015
.x   ±.050
Angles ±1°
Concentricity   .005 TIR
Squareness   .001 per in.

REMOVE ALL BURRS
BREAK SHARP CORNERS

Req. per Piece
Pattern No.

DO NOT SCALE DRAWING

MATERIAL SPEC.
CELANESE 1603-1

Ø .194/.191

[B]

Ø .790/.785

⌀ .005 | B

Ø .374/.371

10°

[C]

|| | .003 | A

[B]

V .030 X 45°
BOTH ENDS

R.015/.010

.12

.635

.725

[A]

Ø .705/.700

.098/.094

⌀ .005 | B

NOTE:
ALL EDGES TO HAVE R .010 MAX.

Activity 15-4   Insulating Bushing.

Name _____ Date _____ Class _____

Activity 15-4
Insulating Bushing

Refer to **Activity 15-4**. *Study the drawing and familiarize yourself with the views, dimensions, title block, and notes. Read the questions, refer to the print, and write your answers in the blanks provided.*

1. How deep is the counterbore?   1. _____

2. How long is the .790/.785 diameter?   2. _____

3. What is the high limit of the total length of the part?   3. _____

4. Must the largest outside diameter be concentric to the counterbore diameter?   4. _____

5. State the size of the external chamfer.   5. _____

6. What feature is Datum B?   6. _____

7. What was revision C?   7. _____

8. What material is the bushing?   8. _____

9. What is the low limit of the large outside diameter?   9. _____

10. How long is the .374/.371 diameter?   10. _____

11. What is the diameter of the second largest circle shown in the right side view?   11. _____

12. What feature is Datum A?   12. _____

13. How far apart are the two surfaces that must be parallel?   13. _____

14. What is the minimum and maximum length of dimension 14 when the overall length is at the maximum tolerance?   14. _____

15. What tolerance is allowed on concentricity?   15. _____

16. What is the maximum radius allowable on all edges?   16. _____

17. If the total length of the bushing is at low limit and the .635 length is at its high limit, what would be the counterbore depth?   17. _____

18. What tolerance is allowed on angles?   18. _____

19. What tolerance is allowed on two place decimals?   19. _____

20. What scale is the print?   20. _____

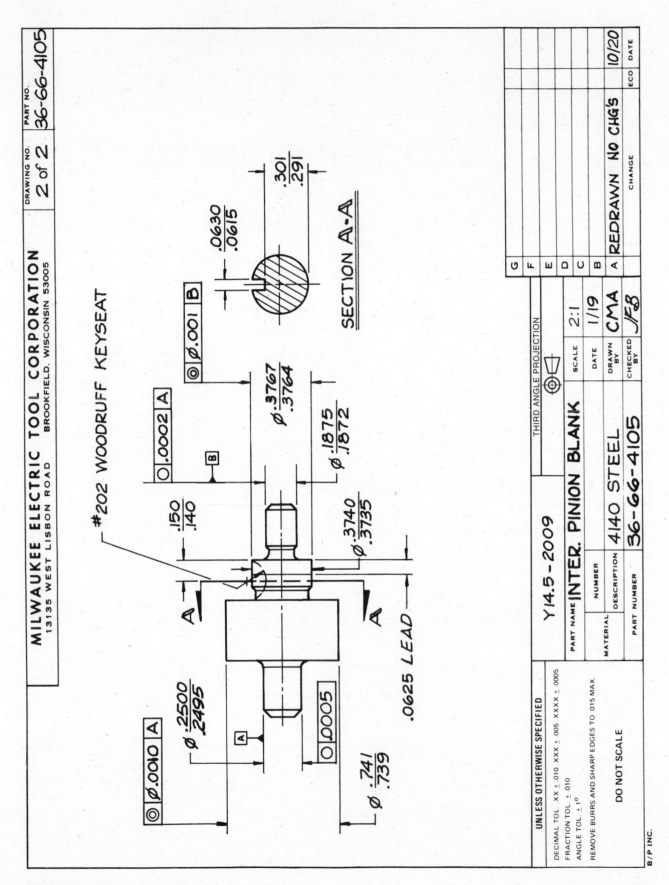

MILWAUKEE ELECTRIC TOOL CORPORATION
13135 WEST LISBON ROAD   BROOKFIELD, WISCONSIN 53005

DRAWING NO. **2 of 2**
PART NO. **36-66-4105**

#202 WOODRUFF KEYSEAT

⊚ Ø .001 B

◯ .0002 A

Ø .3767 / .3764

Ø .1875 / .1872

B

.150 / .140

Ø .3740 / .3735

.0625 LEAD

Ⓐ

Ⓐ

⊚ Ø .0010 A

Ø .2500 / .2495

◯ .0005

◯ .0005

Ⓐ

Ø .741 / .739

.0630 / .0615

.301 / .291

SECTION A-A

G						
F						
E						
D						
C						
B						
A	REDRAWN NO CHG'S			10/20	
	CHANGE			ECO	DATE

THIRD ANGLE PROJECTION ⊕

Y14.5 - 2009

PART NAME INTER. PINION BLANK

	NUMBER		SCALE **2:1**	DATE **1/19**
MATERIAL	DESCRIPTION 4140 STEEL		DRAWN BY **CMA**	
	PART NUMBER 36-66-4105		CHECKED BY **JFB**	

UNLESS OTHERWISE SPECIFIED

DECIMAL TOL. XX ± .010 .XXX ± .005 .XXXX ± .0005
FRACTION TOL. ± .010
ANGLE TOL. ± 1°
REMOVE BURRS AND SHARP EDGES TO .015 MAX.

DO NOT SCALE

B/P INC.

Activity 15-5 Inter. Pinion Blank.

Name _____ Date _____ Class _____

Activity 15-5
Inter. Pinion Blank

Refer to **Activity 15-5**. *Study the drawing and familiarize yourself with the views, dimensions, title block, and notes. Read the questions, refer to the print, and write your answers in the blanks provided.*

1. How wide is the keyseat?

2. List all the low limit dimensions shown for diameters.

3. Why are length dimensions missing on this print?

4. Why are datums not identified in the feature control frames for roundness?

5. What dimension gives the location of the keyway?

6. Which diameters are used as datums?

7. What diameter woodruff key is used on this part?

8. Do the views shown represent the actual size of the part?

9. What other dimensions besides lengths are missing on this print?

10. What is the approximate depth of the keyseat?

11. Which diameter requiring a roundness specification is more critical in tolerance allowed?

12. What type of tolerance is concentricity?

13. Which diameter must be concentric to the .3767/.3764 diameter?

14. What tolerance is allowed on the smallest diameter shown on this print?

15. Which diameter is shown in the sectional view?

16. How many centerlines are shown on this print?

17. How many feature control frames are shown on this print?

18. What is the size of the lead diameter?

19. What type of material is used to make the part?

20. What type of section view is Section A-A?

1. _____

2. _____

3. _____

4. _____

5. _____

6. _____

7. _____

8. _____

9. _____

10. _____

11. _____

12. _____

13. _____

14. _____

15. _____

16. _____

17. _____

18. _____

19. _____

20. _____

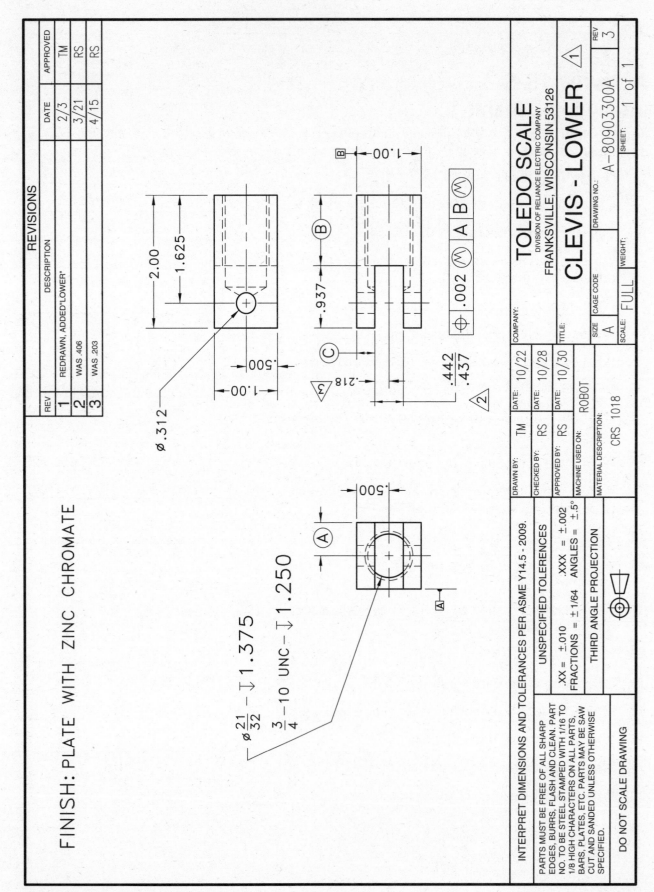

FINISH: PLATE WITH ZINC CHROMATE

∅ .312

2.00
1.625
.500
1.00

$\varnothing\frac{21}{32}$ − ↧ 1.375
$\frac{3}{4}$ −10 UNC − ↧ 1.250

.500
Ⓐ

Ⓑ 1.00
Ⓑ
.937
Ⓒ
.218 ⬡3

| ⌖ | .002 Ⓜ | Ⓐ | Ⓑ | Ⓜ |

.442
.437
⬡2

REVISIONS

REV	DESCRIPTION	DATE	APPROVED
1	REDRAWN, ADDED"LOWER"	2/3	TM
2	WAS .406	3/21	RS
3	WAS .203	4/15	RS

INTERPRET DIMENSIONS AND TOLERANCES PER ASME Y14.5 - 2009.

UNSPECIFIED TOLERENCES		
.XX = ±.010	.XXX = ±.002	
FRACTIONS = ±1/64	ANGLES = ±.5°	

THIRD ANGLE PROJECTION

DO NOT SCALE DRAWING

PARTS MUST BE FREE OF ALL SHARP EDGES, BURRS, FLASH AND CLEAN. PART NO. TO BE STEEL STAMPED WITH 1/16 TO 1/8 HIGH CHARACTERS ON ALL PARTS, BARS, PLATES, ETC. PARTS MAY BE SAW CUT AND SANDED UNLESS OTHERWISE SPECIFIED.

DRAWN BY: TM	DATE: 10/22	COMPANY:
CHECKED BY: RS	DATE: 10/28	**TOLEDO SCALE**
APPROVED BY: RS	DATE: 10/30	DIVISION OF RELIANCE ELECTRIC COMPANY

FRANKSVILLE, WISCONSIN 53126

MACHINE USED ON: ROBOT

MATERIAL DESCRIPTION: CRS 1018

TITLE:

CLEVIS - LOWER ⬡1

SIZE A	CAGE CODE	DRAWING NO.: A−80903300A	REV 3

| SCALE: FULL | WEIGHT: | SHEET: 1 of 1 |

Name _____ Date _____ Class _____

Activity 15-6
Clevis-Lower

Refer to **Activity 15-6**. *Study the drawing and familiarize yourself with the views, dimensions, title block, and notes. Read the questions, refer to the print, and write your answers in the blanks provided.*

1. How far is the 5/16 drilled hole from the left end of the part?

2. What material is the part?

3. What does "encircled M" symbol mean in the feature control frame?

4. How far can the centerline (plane) of the slot vary to the centerline of the part?

5. The position tolerance refers to what part characteristic?

6. Datum feature B must have what relationship to datum feature A?

7. What is dimension A?

8. How wide is the slot?

9. How deep is the 21/32 hole?

10. What feature is datum B?

11. What size tap drill is used on this print?

12. What was the original width of the slot?

13. How deep is the slot?

14. What is dimension B?

15. What finish is required on this part?

16. What machine is used to make this part?

17. What is the high limit on the width of the part?

18. What is dimension C?

19. Which dimension on the print was originally .203?

20. What thread series is the tapped hole?

1. _____

2. _____

3. _____

4. _____

5. _____

6. _____

7. _____

8. _____

9. _____

10. _____

11. _____

12. _____

13. _____

14. _____

15. _____

16. _____

17. _____

18. _____

19. _____

20. _____

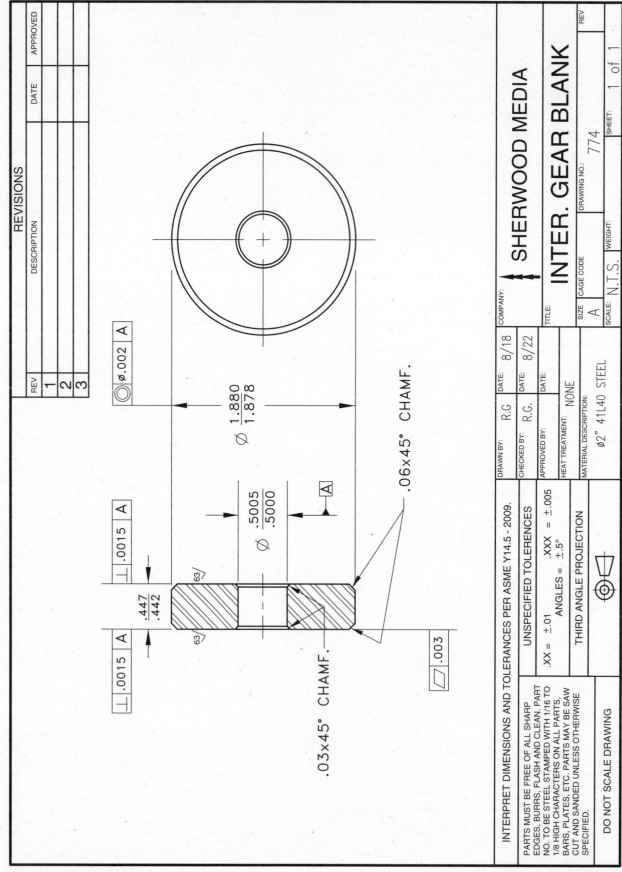

REVISIONS

REV	DESCRIPTION	DATE	APPROVED
1			
2			
3			

⌖ ⌀.002 A

⌀ 1.880 / 1.878

⊥ .0015 A

⊥ .0015 A

.447 / .442

63

63

⌀ .5005 / .5000

A

A

.06×45° CHAMF.

.03×45° CHAMF.

⬦ .003

INTERPRET DIMENSIONS AND TOLERANCES PER ASME Y14.5 - 2009.

PARTS MUST BE FREE OF ALL SHARP EDGES, BURRS, FLASH AND CLEAN. PART NO. TO BE STEEL STAMPED WITH 1/16 TO 1/8 HIGH CHARACTERS ON ALL PARTS. BARS, PLATES, ETC. PARTS MAY BE SAW CUT AND SANDED UNLESS OTHERWISE SPECIFIED.

UNSPECIFIED TOLERENCES	
.XX = ±.01	.XXX = ±.005
ANGLES = ±.5°	

THIRD ANGLE PROJECTION

DO NOT SCALE DRAWING

DRAWN BY: R.G	DATE: 8/18	COMPANY:
CHECKED BY: R.G.	DATE: 8/22	
APPROVED BY:	DATE:	
HEAT TREATMENT: NONE		
MATERIAL DESCRIPTION: ⌀2" 41L40 STEEL		

◀ **SHERWOOD MEDIA**

TITLE: **INTER. GEAR BLANK**

SIZE A	CAGE CODE	DRAWING NO.: 774	REV
SCALE: N.T.S.	WEIGHT:	SHEET: 1 of 1	

Activity 15-7 Inter. Gear Blank.

Name _____ Date _____ Class _____

Activity 15-7
Inter. Gear Blank

Refer to **Activity 15-7**. *Study the drawing and familiarize yourself with the views, dimensions, title block, and notes. Read the questions, refer to the print, and write your answers in the blanks provided.*

1. What is the maximum thickness dimension allowed on the part?

2. How many datums are given on the print?

3. What kind of material is used for the part?

4. What finish is required on the ends of the gear blank?

5. What limits of flatness are given for the part?

6. How concentric must the outside diameter be?

7. What feature is Datum A?

8. How many surfaces are given a tolerance for flatness?

9. Does the cross hatching in the section view match the material given on the print?

10. What does the abbreviation N.T.S. mean?

11. Determine the tolerance allowed on the hole of the part.

12. What does the symbol in the two topmost feature control frames mean?

13. Is this part heat treated?

14. What size internal chamfer is required for the part?

15. Why does the flatness tolerance not reference a datum?

1. _____

2. _____

3. _____

4. _____

5. _____

6. _____

7. _____

8. _____

9. _____

10. _____

11. _____

12. _____

13. _____

14. _____

15. _____

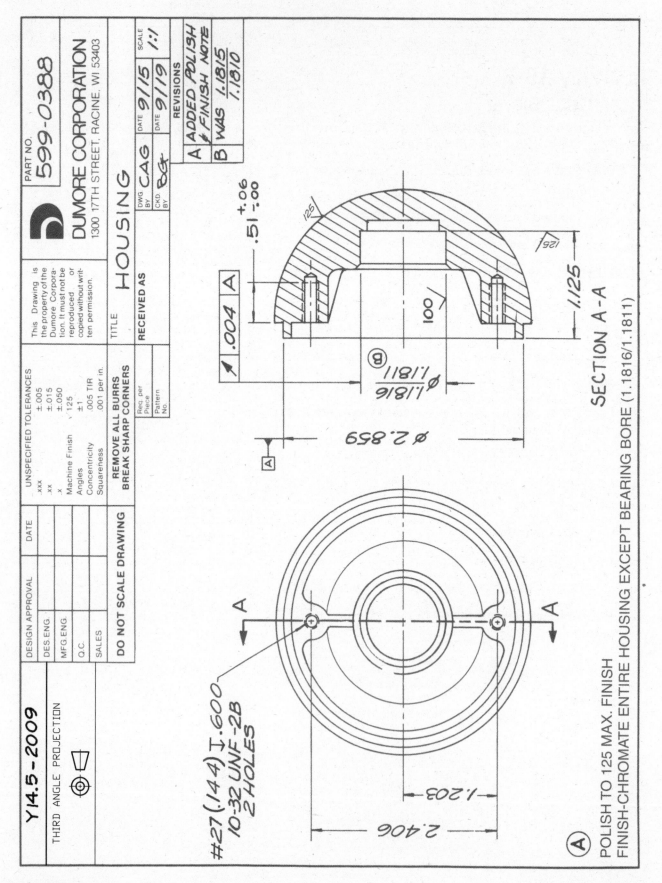

SECTION A-A

POLISH TO 125 MAX. FINISH
FINISH-CHROMATE ENTIRE HOUSING EXCEPT BEARING BORE (1.1816/1.1811)

Y14.5-2009

THIRD ANGLE PROJECTION

	DESIGN APPROVAL	DATE
	DES. ENG.	
	MFG. ENG.	
	Q.C.	
	SALES	

DO NOT SCALE DRAWING

UNSPECIFIED TOLERANCES	
.xxx	±.005
.xx	±.015
.x	±.050
Machine Finish	√ 125
Angles	±1
Concentricity	.005 TIR
Squareness	.001 per in.

REMOVE ALL BURRS
BREAK SHARP CORNERS

Req. per Piece	
Pattern No	

RECEIVED AS

TITLE **HOUSING**

This Drawing is the property of the Dumore Corporation. It must not be reproduced or copied without written permission.

DUMORE CORPORATION
1300 17TH STREET, RACINE, WI 53403

DWG BY	CAG	SCALE
CKD BY	BG	1:1

PART NO.
599-0388

REVISIONS

A	ADDED POLISH & FINISH NOTE	DATE 9/15
B	WAS 1.1815 / 1.1810	DATE 9/19

#27(.144) ↧.600
10-32 UNF-2B
2 HOLES

2.406

1.203

⌀ 2.859

1.125

.51 +.06 -.00

.004 | A

⌀ 1.1816 / 1.1811 Ⓑ

Name _____ Date _____ Class _____

Activity 15-8
Housing

Refer to **Activity 15-8**. *Study the drawing and familiarize yourself with the views, dimensions, title block, and notes. Read the questions, refer to the print, and write your answers in the blanks provided.*

1. How many tapped holes are in the part?

   1. _____

2. What dimensions are used in the location of the tapped holes?

   2. _____

3. What size is the bearing bore in the housing?

   3. _____

4. What type of view is on the right side of the drawing?

   4. _____

5. What is the maximum depth allowed for the 10-32 UNF thread?

   5. _____

6. What is the minimum depth allowed for the 10-32 UNF thread?

   6. _____

7. How much tolerance is permitted on the bearing bore?

   7. _____

8. What kind of feature is Datum A?

   8. _____

9. What does the symbol in the feature control frame represent?

   9. _____

10. Is the outside of the housing finished? If yes, to what finish?

    10. _____

11. What was revision B?

    11. _____

12. How much runout is permitted between the bore and the outside diameter?

    12. _____

13. How many ribs does this part possess?

    13. _____

14. How far apart are the tapped holes?

    14. _____

15. How is the housing finished?

    15. _____

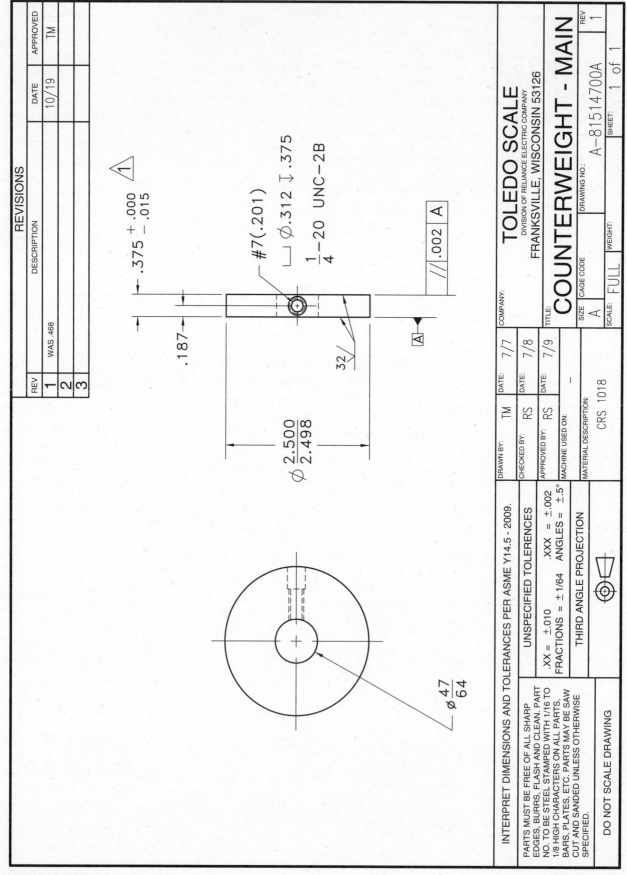

Activity 15-9 Counterweight.

Name _____ Date _____ Class _____

Activity 15-9
Counterweight

Refer to **Activity 15-9**. *Study the drawing and familiarize yourself with the views, dimensions, title block, and notes. Read the questions, refer to the print, and write your answers in the blanks provided.*

1. What size is the counterbore diameter?

   1. _____

2. What is the minimum thickness allowed on the counterweight?

   2. _____

3. What tolerance is allowed on the outside diameter of the part?

   3. _____

4. How many datums are shown on the print?

   4. _____

5. What three symbols are used to describe the small hole?

   5. _____

6. What does UNC mean?

   6. _____

7. How thick was the part originally?

   7. _____

8. What tolerance is allowed on the counterbore diameter?

   8. _____

9. How deep is the counterbore?

   9. _____

10. Approximately how long is the tapped hole?

   10. _____

11. What does the B mean in the thread designation?

   11. _____

12. How much wall thickness is there between the counterbore and the sides of the counterweight?

   12. _____

13. What tolerance is allowed for fractional dimensions?

   13. _____

14. What type of material is 1018?

   14. _____

15. What surface finish is required on the sides of the counterweight?

   15. _____

Notes

Detail Drawings

Learning Objectives

After studying this unit, you will be able to:

✓ Interpret a one-view detail drawing.
✓ Interpret a two-view detail drawing.
✓ Interpret a simple three-view detail drawing.

Key Terms

detail drawing
multiview drawing

The units up to this point were designed in stages to explain various methods of print reading and skill-building techniques. The last three units of this text will explore different types of drawings and provide a review of the text. A review of the units that illustrate the various views, title block information, and drawing examples may be helpful at this point.

Detail Drawings

Specific information for manufacturing the part must be clearly and carefully located on the drawing. A *detail drawing* provides all the information necessary for the production of a part. The information should include the correct views, dimensions, tolerances, materials, part number, and other specifications for manufacturing. The detail drawing is the actual print used in a shop to manufacture that individual part. A concise drawing helps the print reader understand what is required to make the part properly.

Detail drawings generally are single-view or multiview drawings. *Multiview drawings* are drawings involving more than one projection plane. Basic parts may require only one or two views. Many prints need three views to fully describe the part. However, complex parts may require additional views, including sectional views, auxiliary views, and any or all of the six principal views.

One-View Detail Drawing

Objects that are uniform in cross section, such as cylindrical parts and thin, flat parts appear best on prints as one-view drawings. See **Figure 16-1**. A centerline through a cylindrical part indicates that the part is symmetrical.

A one-view drawing of a thin, flat object usually has its thickness listed in the materials specification area of the title block. The thickness may also be stated as a special note. See **Figure 16-2**.

Two-View Detail Drawing

Two-view drawings are commonly used to describe machined parts. Two-view drawings generally are not complex, but fulfill certain requirements:

■ The part shown is simply shaped, such as cylindrical, rectangular, or stamped pieces.

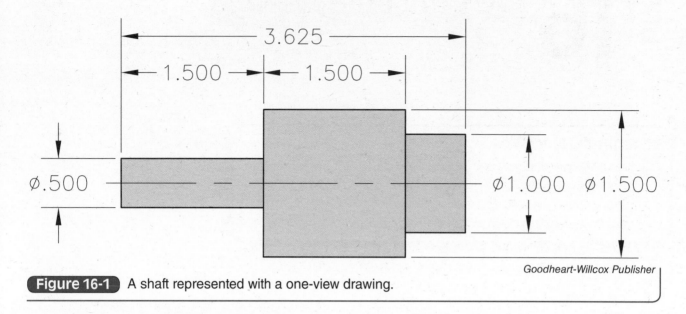

Goodheart-Willcox Publisher

Figure 16-1 A shaft represented with a one-view drawing.

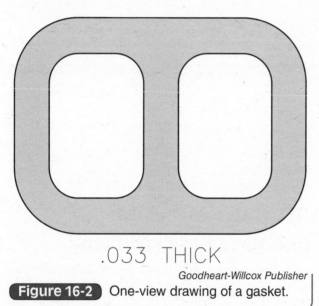

.033 THICK

Goodheart-Willcox Publisher

Figure 16-2 One-view drawing of a gasket.

- A third view would not add specific information, such as dimensions.
- A third view would show no significant contours of the object.
- Two of the three views would be the same.

The designer or drafter decides which two views best represent the details of the part. The two views selected will differ depending on the part. A common arrangement for two-view drawings is the top and front views, as shown in **Figure 16-3**. Notice that the top view shows the hole by use of a circle and two centerlines. The circle in the top view correctly indicates true size and shape of the hole. If only the front and right side views were shown, the hole could be easily misread.

Likewise, the front view describes the beveled ends more clearly than the top view and right side view (not shown). As mentioned, the selection of the two views is most important to the print reader for detail clarification.

In **Figure 16-4**, the front view of a bearing block shows a small circle within a larger circle. This is the best view to represent the hole. In addition, the front view shows two angular supports on both sides of the large circle. The top view represents both angular supports and the two holes. The top and front views supply the necessary details to describe the bearing block without the addition of a third view.

Three-View Detail Drawing

Three-view drawings are generally more complex than one- and two-view drawings. A three-view drawing may use any of the six principal views to illustrate the part. However, when drawing three views, it is general practice to include the front view as one of the selected views.

When selecting an additional view such as the third view, place it in the correct position in relationship to the other two selected views. **Figure 16-5** shows the correct position of the top view in relationship with the front view. The right side view provides the additional information needed to understand the part fully. The addition of the third view gives a clearer representation of the double yoke.

TOP

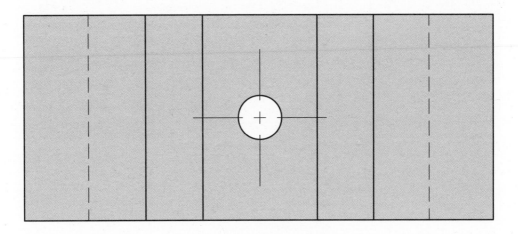

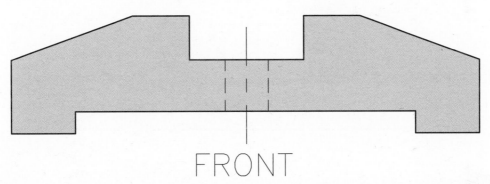

FRONT

Goodheart-Willcox Publisher

Figure 16-3 Only two views are needed to represent this guide.

The next illustration, **Figure 16-6**, uses similar view arrangements: top, front, and right side. The top and front views of the fork contain the majority of the necessary details. The right side view, however, is key to describing the contour of the fork. Notice the slightly curved outer edges as shown in the right side view. The top and front views cannot show these curves. The left side view (not shown) is the only other view that would show the same contour information, but would include numerous hidden lines. Therefore, the right side view is the best option because it shows the features most clearly.

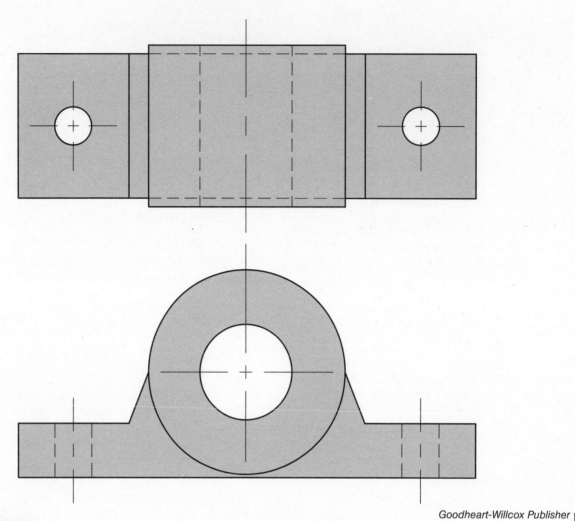

Figure 16-4 Bearing block represented with a two-view drawing.

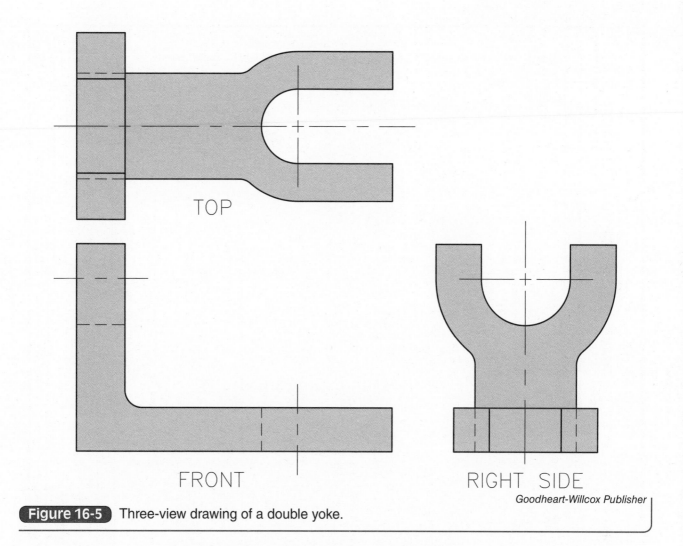

Goodheart-Willcox Publisher

Figure 16-5   Three-view drawing of a double yoke.

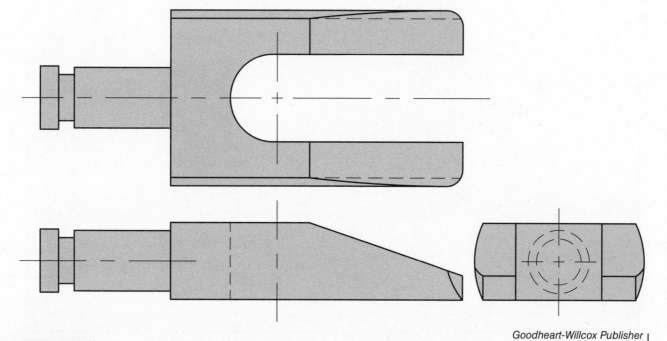

Goodheart-Willcox Publisher

Figure 16-6   The drawing of this fork requires three views to explain the part.

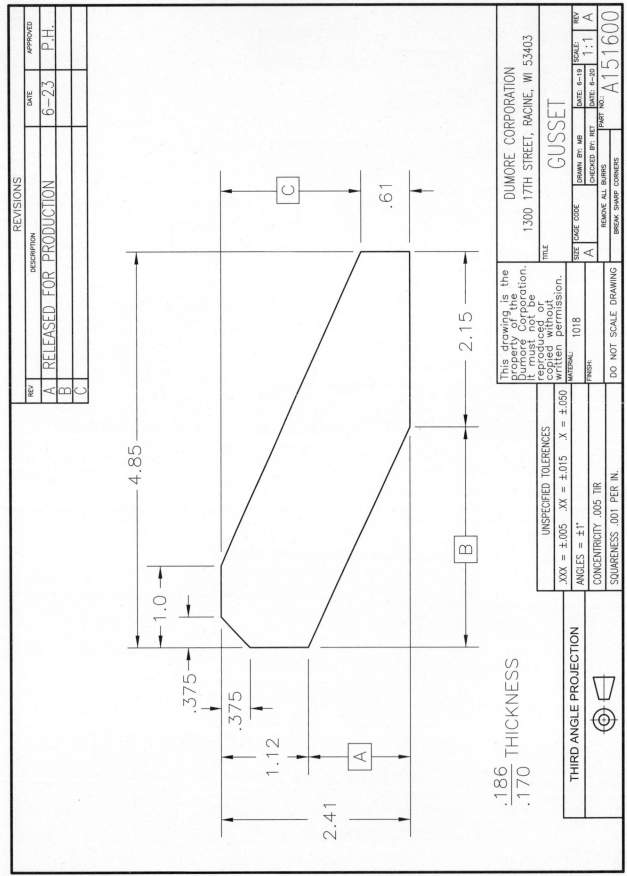

Goodheart-Willcox Publisher

Copyright Goodheart-Willcox Co., Inc.

Name _____ Date _____ Class _____

Activity 16-1
Gusset

Refer to **Activity 16-1**. *Study the drawing and familiarize yourself with the views, dimensions, title block, and notes. Read the questions, refer to the print, and write your answers in the blanks provided.*

1. How thick is the part?

2. What material specifications are given?

3. What is the minimum overall length allowed for the part?

4. How long is dimension A?

5. What is the high limit on the 2.15 dimension?

6. What is the high limit on the .375 dimension?

7. How long is dimension B?

8. By whom was the print drawn?

9. Are there any revisions made on the print?

10. Who checked the print?

11. What tolerance is used on one-place decimal dimensions?

12. How long is dimension C?

13. What scale is the print?

14. What directions are given on the print in regard to sharp corners?

15. What minimum part thickness is allowed?

1. _____

2. _____

3. _____

4. _____

5. _____

6. _____

7. _____

8. _____

9. _____

10. _____

11. _____

12. _____

13. _____

14. _____

15. _____

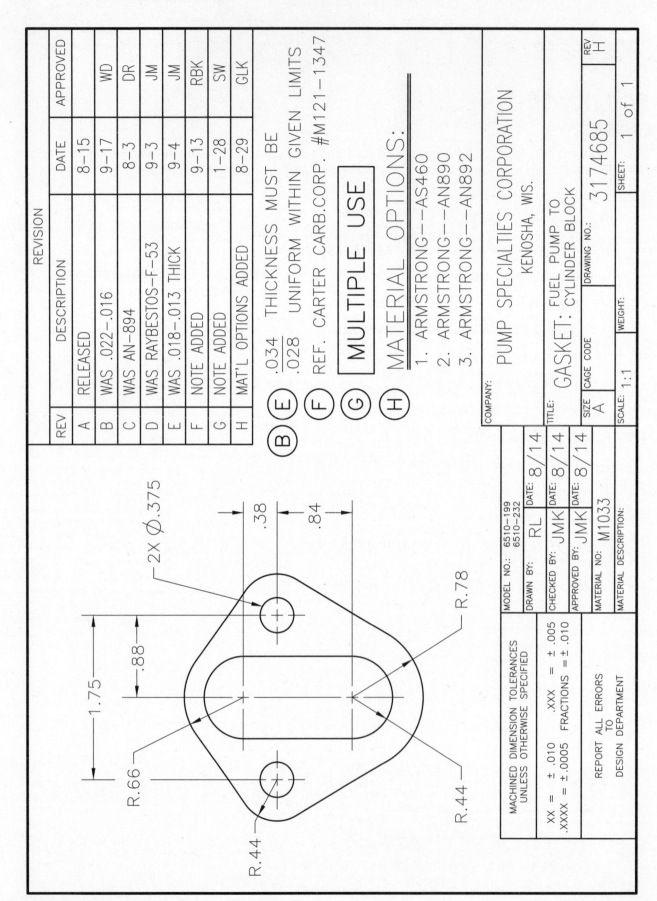

Activity 16-2 Gasket.

Name _____ Date _____ Class _____

Activity 16-2
Gasket

Refer to **Activity 16-2**. *Study the drawing and familiarize yourself with the views, dimensions, title block, and notes. Read the questions, refer to the print, and write your answers in the blanks provided.*

1. What maximum thickness is allowed on the part? 1. _____

2. What is the full length of the slot? 2. _____

3. Determine the center distance between the two holes? 3. _____

4. Give the high limit on the hole sizes. 4. _____

5. List the radii shown on the part print. 5. _____

6. By whom was the print drawn? 6. _____

7. What note is given regarding part thickness? 7. _____

8. How far is the right hole from the vertical centerline of the part? 8. _____

9. List model numbers using the gasket. 9. _____

10. How wide is the gasket at its maximum width? 10. _____

11. Give the part description. 11. _____

12. What scale is the print? 12. _____

13. How wide is the slot? 13. _____

14. How thick was the gasket when originally manufactured? 14. _____

15. List the materials currently used to make the gasket. 15. _____

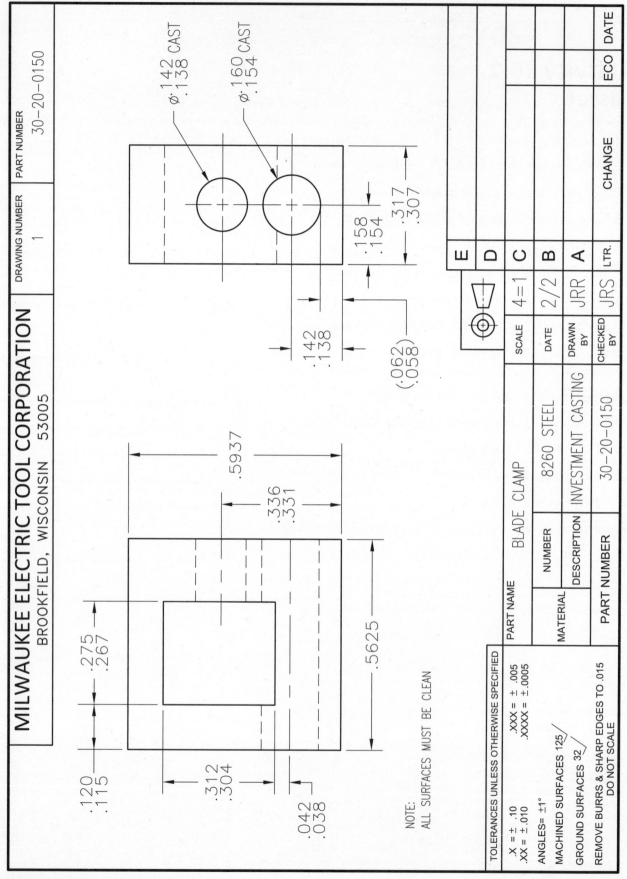

ø.142 CAST
.138

ø.160 CAST
.154

.317
.307

.158
.154

.142
.138

(.062)
(.058)

.5937

.336
.331

.5625

.275
.267

.312
.304

.120
.115

.042
.038

NOTE:
ALL SURFACES MUST BE CLEAN

MILWAUKEE ELECTRIC TOOL CORPORATION
BROOKFIELD, WISCONSIN   53005

DRAWING NUMBER
1

PART NUMBER
30-20-0150

		DATE
E		
D		
C		
B	2/2	
A	JRR	
LTR.	CHANGE	ECO DATE

SCALE
4=1

DATE
2/2

DRAWN BY
JRR

CHECKED BY
JRS

PART NAME   BLADE CLAMP

MATERIAL	NUMBER	8260 STEEL
	DESCRIPTION	INVESTMENT CASTING

PART NUMBER   30-20-0150

TOLERANCES UNLESS OTHERWISE SPECIFIED

.X = ± .10   .XXX = ± .005
.XX = ±.010   .XXXX = ±.0005

ANGLES= ±1°

MACHINED SURFACES 125 ╱

GROUND SURFACES 32 ╱

REMOVE BURRS & SHARP EDGES TO .015
DO NOT SCALE

Activity 16-3 Blade Clamp.

Name _____ Date _____ Class _____

Activity 16-3
Blade Clamp

Refer to **Activity 16-3**. *Study the drawing and familiarize yourself with the views, dimensions, title block, and notes. Read the questions, refer to the print, and write your answers in the blanks provided.*

1. What size is the rectangular hole in the part?

1. _____

2. What is the length of the part?

2. _____

3. What is the height of the part?

3. _____

4. What tolerance is used on four-place decimal dimensions?

4. _____

5. Which hole does not go through the entire part?

5. _____

6. What is the low limit of the diameter of the large cast hole?

6. _____

7. What is the high limit of the diameter of the small cast hole?

7. _____

8. What are the minimum and maximum acceptable distances between the centers of the two cast holes?

8. _____

9. What scale is the print?

9. _____

10. Give the description of the material used to make the part.

10. _____

11. What is the maximum allowable width of the part?

11. _____

12. How far does the rectangular hole go into the large cast diameter hole?

12. _____

13. Give the material number of the part.

13. _____

14. What type of line shows the rectangular hole in the side view?

14. _____

15. What is the minimum amount of material allowable between the two cast holes? Be sure to consider both hole position variation and hole size variation.

15. _____

16. If all dimensions are at their maximums, would the small hole be in the center of the rectangular hole?

16. _____

17. What are the minimum and maximum allowable dimensions for the long side of the rectangular hole?

17. _____

18. In the side view, are the cast holes in the center of the part?

18. _____

19. What is the high limit of the 9/16 dimension?

19. _____

20. Is the shape of the part square or rectangular?

20. _____

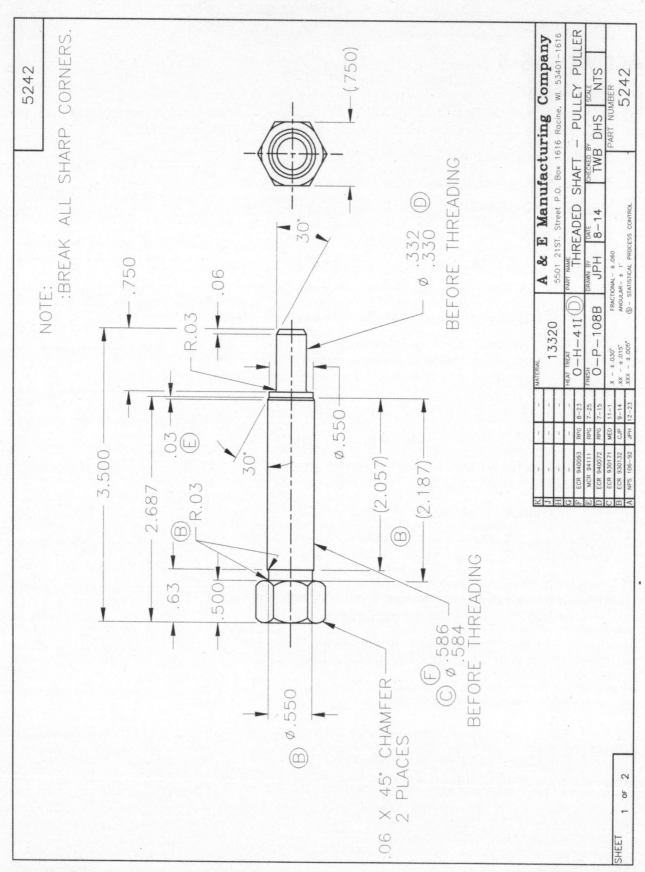

NOTE:
:BREAK ALL SHARP CORNERS.

5242

.750

R.03

.06

30°

∅ .332
∅ .330 Ⓓ

BEFORE THREADING

3.500

2.687

.03 Ⓔ

30°

∅ .550

(2.057)

(2.187)

Ⓑ

Ⓑ

R.03

Ⓑ R.03

.63

.500

∅ .586
∅ .584 Ⓕ

BEFORE THREADING

∅ .550

.06 X 45° CHAMFER
2 PLACES

Ⓑ

(.750)

				A & E Manufacturing Company		
				5501 21ST. Street P.O. Box 1616 Racine, WI. 53401-1616		
MATERIAL				PART NAME		
13320				THREADED SHAFT – PULLEY PULLER		
HEAT TREAT				DRAWN BY	DATE	CHECKED BY
O-H-41I Ⓓ				JPH	8-14	TWB DHS
FINISH						SCALE
O-P-108B						NTS
				FRACTIONAL – ± .060		PART NUMBER
				ANGULAR – ± 1°		5242
				X – ± .030″		
				XX – ± .015″		
				XXX – ± .005″		
				Ⓢ – STATISTICAL PROCESS CONTROL		

K	–	–	–
J	–	–	–
H	–	–	–
G	–	–	–
F	ECR 940093	RPG	8-23
E	MCR 94111	RPG	7-25
D	ECR 940072	RPG	7-15
C	ECR 930171	MED	11-1
B	ECR 930132	CJP	9-14
A	NPS 106-92	JPH	12-23

SHEET 1 OF 2

Activity 16-4 Threaded Shaft–Pulley Puller.

Name _____ Date _____ Class _____

Activity 16-4
Threaded Shaft-Pulley Puller

Refer to **Activity 16-4**. *Study the drawing and familiarize yourself with the views, dimensions, title block, and notes. Read the questions, refer to the print, and write your answers in the blanks provided.*

1. What tolerance is required for two-place decimals?

1. _____

2. What is the length of the 30° chamfer on the small threaded diameter?

2. _____

3. What is the scale of the print?

3. _____

4. What is the minimum overall length of the part allowed?

4. _____

5. How long is the chamfer used for the large threaded diameter?

5. _____

6. How long is the .550 diameter section found between the two threaded diameters?

6. _____

7. What size is the hex stock?

7. _____

8. What is the diameter of the thread relief neck required for the large diameter thread?

8. _____

9. What is the length of the thread relief neck for the large diameter thread?

9. _____

10. What is the length of the large threaded diameter including chamfer?

10. _____

11. What is the minimum outside diameter of the small diameter thread?

11. _____

12. What is the high limit on the large threaded diameter?

12. _____

13. When was the last print revision?

13. _____

14. When was the print originally drawn?

14. _____

15. What tolerance is specified for angular dimensions?

15. _____

16. How many radii does the part contain?

16. _____

17. What does the symbol comprised of an encircled letter "S" mean?

17. _____

18. What is the distance from the shoulder of the small thread to the hex head shoulder?

18. _____

19. Give the heat treatment specification for the part.

19. _____

20. Which change revision was required by the manufacturing department?

20. _____

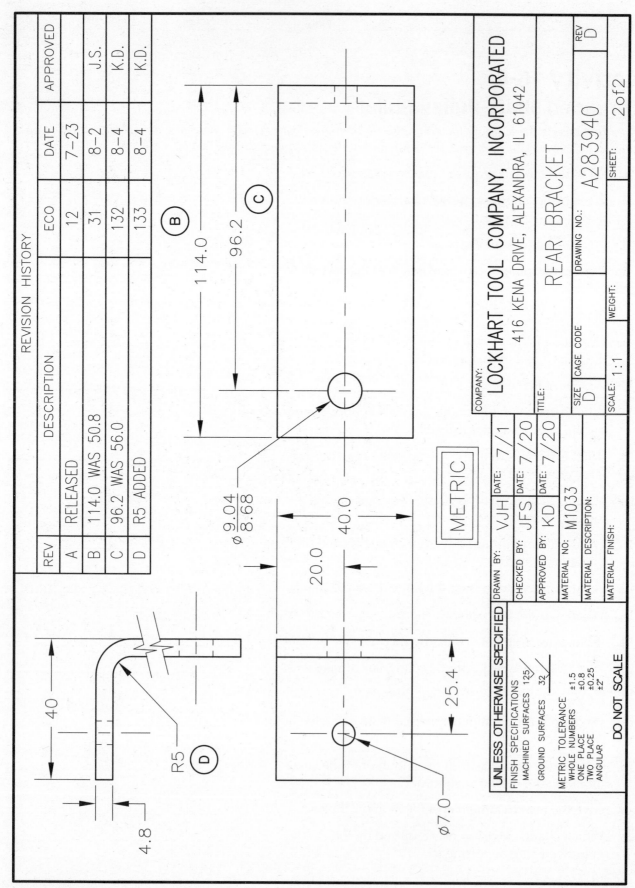

REV	DESCRIPTION	ECO	DATE	APPROVED
A	RELEASED	12	7–23	J.S.
B	114.0 WAS 50.8	31	8–2	K.D.
C	96.2 WAS 56.0	132	8–4	K.D.
D	R5 ADDED	133	8–4	K.D.

REVISION HISTORY

114.0

96.2

Ⓒ

Ⓑ

Ø 9.04 / 8.68

40.0

20.0

25.4

40

R5

Ⓓ

4.8

Ø7.0

METRIC

COMPANY:

LOCKHART TOOL COMPANY, INCORPORATED

416 KENA DRIVE, ALEXANDRA, IL 61042

TITLE:

REAR BRACKET

DRAWING NO.: A283940

REV D

SHEET: 2 of 2

SIZE D	CAGE CODE	WEIGHT:

SCALE: 1:1

UNLESS OTHERMSE SPECIFIED

FINISH SPECIFICATIONS
MACHINED SURFACES 125/
GROUND SURFACES 32/

METRIC TOLERANCE
WHOLE NUMBERS   ±1.5
ONE PLACE   ±0.8
TWO PLACE   ±0.25
ANGULAR   ±2°

DO NOT SCALE

DRAWN BY: VJH   DATE: 7/1
CHECKED BY: JFS DATE: 7/20
APPROVED BY: KD DATE: 7/20
MATERIAL NO.: M1033
MATERIAL DESCRIPTION:
MATERIAL FINISH:

Activity 16-5 Rear Bracket.

Name _____ Date _____ Class _____

Activity 16-5
Rear Bracket

Refer to **Activity 16-5**. *Study the drawing and familiarize yourself with the views, dimensions, title block, and notes. Read the questions, refer to the print, and write your answers in the blanks provided.*

1. How many holes does the part have?

1. _____

2. What is the system of measurement for the print?

2. _____

3. What size is the smallest hole?

3. _____

4. How long is the bracket?

4. _____

5. What is the material number?

5. _____

6. How thick is the bracket?

6. _____

7. What is the largest allowable diameter of the small hole?

7. _____

8. What was revision C?

8. _____

9. How far is the ⌀9.04 hole located from the end of the short leg of the bracket?

9. _____

10. What type of break line is shown in the top view?

10. _____

11. What is the tolerance for a two-place metric dimension?

11. _____

12. How long is the short leg of the bracket?

12. _____

13. What is the tolerance on the ⌀9.04 hole?

13. _____

14. What new information was added in revision D?

14. _____

15. What does the symbol ⌀ represent?

15. _____

Notes

Learning Objectives

After studying this unit, you will be able to:

✓ Identify and read an assembly drawing.
✓ Explain the use of assembly drawings in industry.
✓ Identify a subassembly and exploded assembly drawing.
✓ Identify components in a parts list on an assembly drawing.

Key Terms

assembly   balloon   parts list
assembly drawing   exploded assembly drawing   subassembly

Assembly Drawings

Assembly drawings are a special application of pictorial drawings that are common in the manufacturing world. An *assembly* is a collection of parts fitted together to form a machine or structure. Therefore, an *assembly drawing* is a drawing that uses two-dimensional or pictorial views to show how individual parts are fitted together to form a machine or structure. See **Figure 17-1**.

Understanding detail drawings, as described in the previous unit, is important to reading an assembly drawing. The assembly drawing may have a pictorial representation showing where the individual part is placed in the completed product. On less complex assemblies, the assembly drawing may be a simple two-dimensional representation, illustrating the necessary component parts. See **Figure 17-2**.

Subassembly

Many of the parts shown in an assembly drawing have a corresponding part drawing. Thus, the part is produced based on the part drawing. After all parts have been produced, the assembly drawing is used to build the assembly.

In many cases, one assembly serves as a "part" for a larger assembly. These intermediate assemblies are called subassemblies. A *subassembly* is an assembly that is intended to be used as part of a larger assembly. The assembly shown in **Figure 17-1** is a subassembly. Complex machines, such as automobiles or airplanes, can include thousands of subassemblies.

Exploded Assembly

In the pictorial drawing shown in **Figure 17-1**, the individual components are listed and placed in their respective positions, but spread away from each other. This type of drawing is known as an *exploded assembly drawing*. Exploded assembly drawings help show the proper order and manner of assembling a machine or structure.

Reading Assembly Drawings

Hidden lines are often omitted from assembly drawings. Hidden lines in limited use can aid in the clarification of edges and contours. Hidden lines may also be necessary to the placement

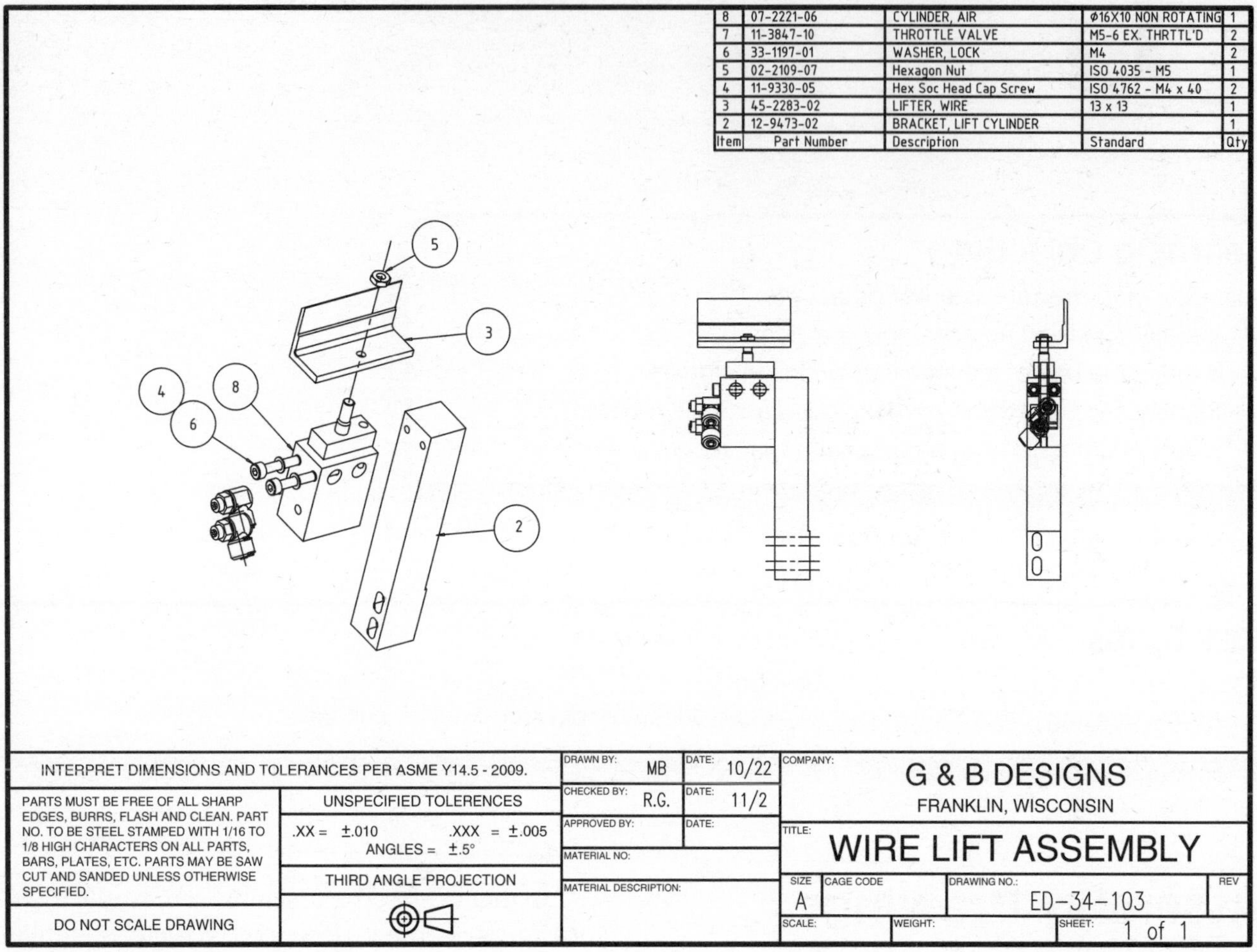

8 07-2221-06 CYLINDER, AIR ⌀16X10 NON ROTATING 1
7 11-3847-10 THROTTLE VALVE M5-6 EX. THRTTL'D 2
6 33-1197-01 WASHER, LOCK M4 2
5 02-2109-07 Hexagon Nut ISO 4035 - M5 1
4 11-9330-05 Hex Soc Head Cap Screw ISO 4762 - M4 x 40 2
3 45-2283-02 LIFTER, WIRE 13 x 13 1
2 12-9473-02 BRACKET, LIFT CYLINDER 1
Item Part Number Description Standard Qty
5
3
4
8
6
2
INTERPRET DIMENSIONS AND TOLERANCES PER ASME Y14.5 - 2009.
PARTS MUST BE FREE OF ALL SHARP EDGES, BURRS, FLASH AND CLEAN. PART NO. TO BE STEEL STAMPED WITH 1/16 TO 1/8 HIGH CHARACTERS ON ALL PARTS, BARS, PLATES, ETC. PARTS MAY BE SAW CUT AND SANDED UNLESS OTHERWISE SPECIFIED.
DO NOT SCALE DRAWING
UNSPECIFIED TOLERENCES
.XX = ±.010 .XXX = ±.005
ANGLES = ±.5°
THIRD ANGLE PROJECTION
DRAWN BY: MB DATE: 10/22
CHECKED BY: R.G. DATE: 11/2
APPROVED BY: DATE:
MATERIAL NO:
MATERIAL DESCRIPTION:
COMPANY: G & B DESIGNS
FRANKLIN, WISCONSIN
TITLE: WIRE LIFT ASSEMBLY
SIZE A
CAGE CODE
DRAWING NO.: ED-34-103
REV
SCALE:
WEIGHT:
SHEET: 1 of 1

Goodheart-Willcox Publisher

Figure 17-1 A simple assembly drawing. Some assembly drawings include only a pictorial view. This drawing includes both pictorial and orthographic views.

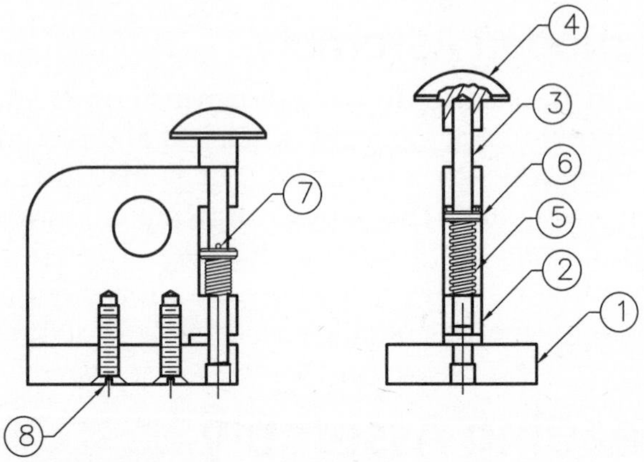

4
3
7
6
5
2
1
8

8	1/4-20UNC-2A CAP SCREW	STEEL	2
7	⌀1/16 X 3/8 LG. COTTER PIN	STEEL	1
6	WASHER	BRASS	1
5	SPRING	MUSIC WIRE	1
4	HEAD	BRASS	1
3	PLUNGER	DRILL ROD	1
2	BODY	ALUMINUM	1
1	BASE	ALUMINUM	1
ITEM	DESCRIPTION	MATERIAL	QTY.
PARTS LIST			

Goodheart-Willcox Publisher

Figure 17-2 This assembly drawing includes only two-dimensional orthographic views. No pictorial view is provided.

of the parts in the assembly. In **Figure 17-3** the hidden lines are used for the positioning of parts as well as the identification of fasteners.

Sectional views are common on assembly drawings. Section lines with different spacing or set at different angles can be used to indicate the individual components of the assembly. Section line patterns representing part materials can also be used. Thin parts like gaskets and seals may be filled solid. See **Figure 17-4**.

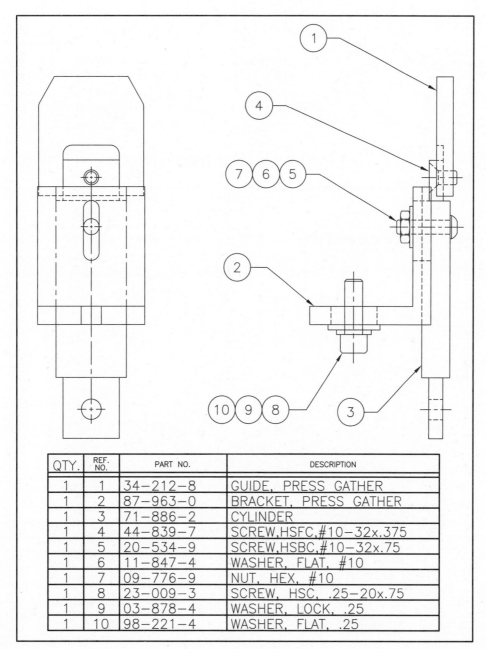

QTY.	REF. NO.	PART NO.	DESCRIPTION
1	1	34-212-8	GUIDE, PRESS GATHER
1	2	87-963-0	BRACKET, PRESS GATHER
1	3	71-886-2	CYLINDER
1	4	44-839-7	SCREW,HSFC,#10-32x.375
1	5	20-534-9	SCREW,HSBC,#10-32x.75
1	6	11-847-4	WASHER, FLAT, #10
1	7	09-776-9	NUT, HEX, #10
1	8	23-009-3	SCREW, HSC, .25-20x.75
1	9	03-878-4	WASHER, LOCK, .25
1	10	98-221-4	WASHER, FLAT, .25

Goodheart-Willcox Publisher

Figure 17-3 Hidden lines help show how this mechanism is assembled.

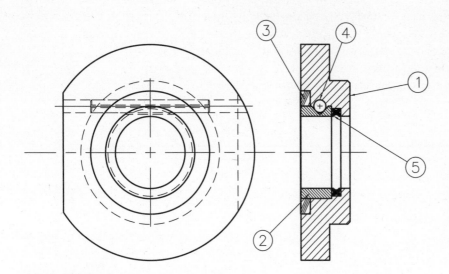

ITEM NO.	QTY.	PART NO.	DESCRIPTION
1	1	39385	CAP
2	1	83480	BUSHING
3	1	49349	CUSHION
4	1	98474	PIN, SPRING
5	1	19348	SEAL, QUAD RING

Goodheart-Willcox Publisher

Figure 17-4 The sectional view features three different patterns of section lines. Also note how the gasket is filled solid.

Parts Lists in Assembly Drawings

The component parts should be clearly marked and identified. Many assembly drawings use leader lines attached to *balloons*, circles containing identifying numbers or letters. The letters or numbers reference individual parts on a corresponding chart called a *parts list*. See **Figure 17-5**. Other names for a parts list include

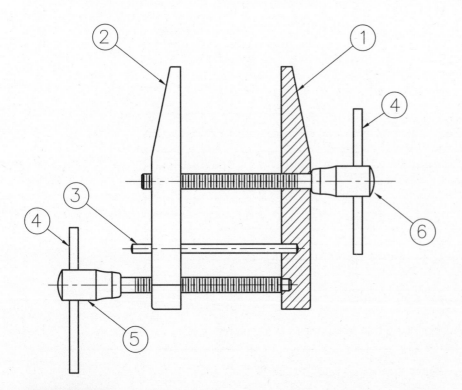

No.	PART	QTY.
	PARTS LIST	
1	JAW—RIGHT	1
2	JAW—LEFT	1
3	GUIDE ROD	1
4	HANDLE	2
5	ADJUSTING SCREW—LEFT	1
6	ADJUSTING SCREW—RIGHT	1

Goodheart-Willcox Publisher

Figure 17-5 The numbers in the balloons correspond to the numbers on the parts list.

materials list, bill of materials, and parts schedule. Parts list often list the item or reference number, part name or number, a description of the item, and the quantity required for the assembly.

Assembly drawings often contain fasteners such as pins, nuts, bolts, washers, keys, springs, fittings, and machine screws. Fasteners are generally not subject to sectioning techniques, as shown in **Figure 17-6**. By eliminating the "fastener family" from sectioning on the assembly drawing, the parts are more readily identified. See the *Machinery's Handbook* for more examples of fasteners.

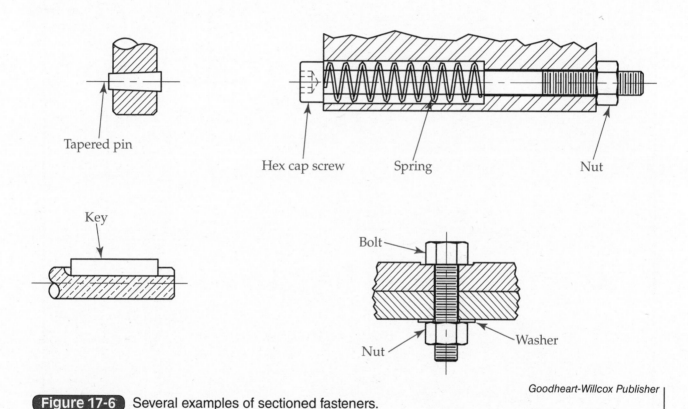

Tapered pin

Hex cap screw   Spring   Nut

Key

Bolt

Nut   Washer

Figure 17-6   Several examples of sectioned fasteners.

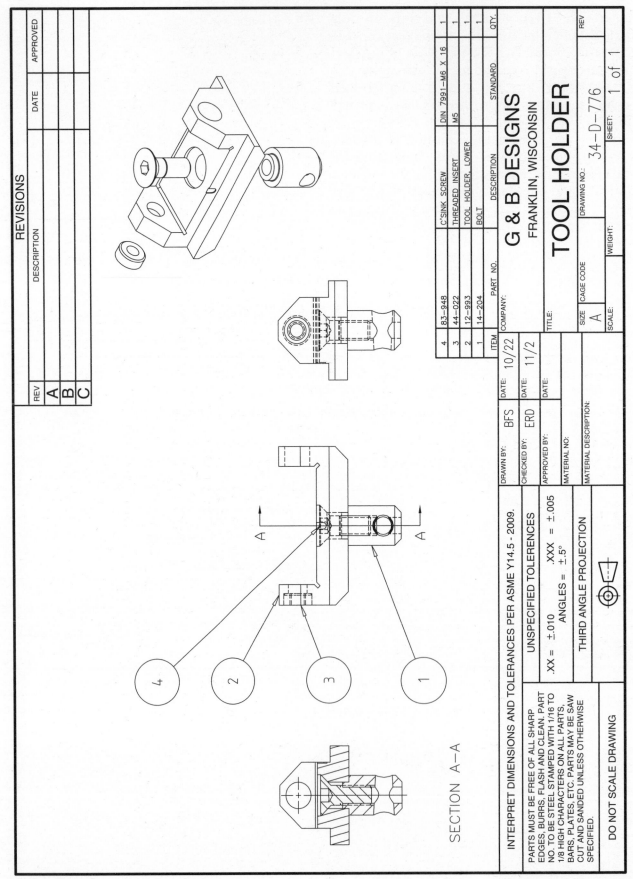

SECTION A–A

REV	REVISIONS		
	DESCRIPTION	DATE	APPROVED
A			
B			
C			

ITEM	PART NO.	DESCRIPTION	STANDARD	QTY.
4	83–948	C'SINK SCREW	DIN 7991–M6 X 16	1
3	44–022	THREADED INSERT	M5	1
2	12–993	TOOL HOLDER, LOWER		1
1	14–204	BOLT		1

COMPANY:
G & B DESIGNS
FRANKLIN, WISCONSIN

TITLE:
TOOL HOLDER

DRAWN BY:	BFS	DATE:	10/22
CHECKED BY:	ERD	DATE:	11/2
APPROVED BY:		DATE:	
MATERIAL NO:			
MATERIAL DESCRIPTION:			

SIZE	CAGE CODE	DRAWING NO.:	REV
A		34–D–776	

WEIGHT: SCALE: SHEET: 1 of 1

INTERPRET DIMENSIONS AND TOLERANCES PER ASME Y14.5 - 2009.

UNSPECIFIED TOLERENCES	
.XX = ±.010	.XXX = ±.005
ANGLES = ±.5°	

THIRD ANGLE PROJECTION

PARTS MUST BE FREE OF ALL SHARP EDGES, BURRS, FLASH AND CLEAN. PART NO. TO BE STEEL STAMPED WITH 1/16 TO 1/8 HIGH CHARACTERS ON ALL PARTS, BARS, PLATES, ETC. PARTS MAY BE SAW CUT AND SANDED UNLESS OTHERWISE SPECIFIED.

DO NOT SCALE DRAWING

Activity 17-1 Tool Holder.

Name _____ Date _____ Class _____

Activity 17-1
Tool Holder

Refer to **Activity 17-1**. *Study the drawing and familiarize yourself with the views, dimensions, title block, and notes. Read the questions, refer to the print, and write your answers in the blanks provided.*

1. How many holes are in item 2?

2. How many threads per inch does item 4 have?

3. What is the description of item 3?

4. What standard refers to item 4?

5. Does M5 refer to a coarse or fine thread?

6. What quantity of item 2 is required for the assembly?

7. How are items 1 and 2 assembled?

8. What type of sectional view is Section A-A?

9. What paper size was the original version of this print?

10. *True or False?* The pictorial view is an exploded assembly drawing.

11. What angle of projection are the two-dimensional views shown in?

12. Is item 4 threaded the entire length of the shaft?

13. What type of screw thread representation is used on this drawing?

14. What is the part number of the threaded insert?

15. *True or False?* Item 1 is a subassembly.

1. _____

2. _____

3. _____

4. _____

5. _____

6. _____

7. _____

8. _____

9. _____

10. _____

11. _____

12. _____

13. _____

14. _____

15. _____

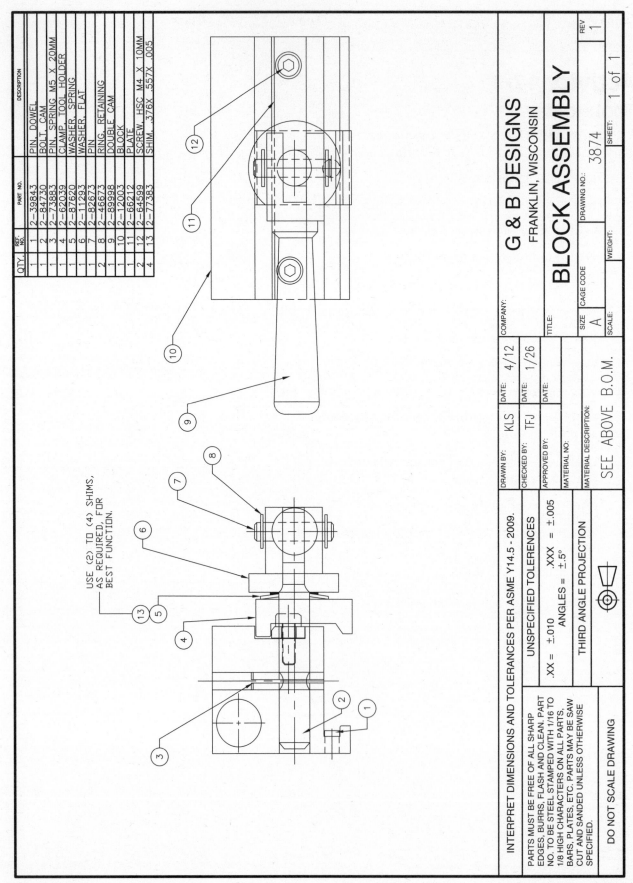

QTY.	REF. NO.	PART NO.	DESCRIPTION
1	1	2-39843	PIN, DOWEL
1	2	2-84730	BOLT, CAM
1	3	2-73883	PIN, SPRING M5 X 20MM
1	4	2-62039	CLAMP, TOOL HOLDER
1	5	2-87620	WASHER, SPRING
1	6	2-11293	WASHER, FLAT
1	7	2-82673	PIN
2	8	2-46673	RING, RETAINING
1	9	2-89998	DOUBLE CAM
1	10	2-12003	BLOCK
1	11	2-66212	PLATE
2	12	2-64599	SCREW, HSC M4 X 10MM
4	13	2-77383	SHIM .376X .557X .005

USE (2) TO (4) SHIMS,
AS REQUIRED, FOR
BEST FUNCTION.

G & B DESIGNS
FRANKLIN, WISCONSIN

BLOCK ASSEMBLY

COMPANY:		
DRAWN BY: KLS	DATE: 4/12	
CHECKED BY: TFJ	DATE: 1/26	
APPROVED BY:	DATE:	
MATERIAL NO:		
MATERIAL DESCRIPTION:		

SEE ABOVE B.O.M.

SIZE	CAGE CODE	DRAWING NO:	REV
A		3874	1
	WEIGHT:	SCALE:	SHEET: 1 of 1

INTERPRET DIMENSIONS AND TOLERANCES PER ASME Y14.5 - 2009.

UNSPECIFIED TOLERANCES

.XX = ±.010   .XXX = ±.005
ANGLES = ±.5°

THIRD ANGLE PROJECTION

PARTS MUST BE FREE OF ALL SHARP
EDGES, BURRS, FLASH AND CLEAN. PART
NO. TO BE STEEL STAMPED WITH 1/16 TO
1/8 HIGH CHARACTERS ON ALL PARTS,
BARS, PLATES, ETC. PARTS MAY BE SAW
CUT AND SANDED UNLESS OTHERWISE
SPECIFIED.

DO NOT SCALE DRAWING

Activity 17-2 Block Assembly.

Name _____ Date _____ Class _____

Activity 17-2
Block Assembly

Refer to **Activity 17-2**. *Study the drawing and familiarize yourself with the views, dimensions, title block, and notes. Read the questions, refer to the print, and write your answers in the blanks provided.*

1. What does B.O.M. mean?

2. What is part 2-12003?

3. What size are the shims used in the assembly?

4. How many retaining rings are required?

5. What revision is this print?

6. How many shims are required for this assembly?

7. What size is the spring pin?

8. What is the item number and description for the part held in place by the retaining rings?

9. How many threaded fasteners are used in the assembly?

10. What is the description for item 10?

11. Which item is grooved?

12. What size is item 12?

13. Who was the print drawn by?

14. List the two types of washers used in this assembly.

15. Why is the large hole not shown in both views?

1. _____

2. _____

3. _____

4. _____

5. _____

6. _____

7. _____

8. _____

9. _____

10. _____

11. _____

12. _____

13. _____

14. _____

15. _____

Notes

This final unit of *Machine Trades Print Reading* contains a series of review activities. These activities review the topics covered throughout the text. These topics include line usage, multiview drawings, machining details, and tolerances.

When completing these review questions, use the following procedure:

1. Begin by studying the drawing. Imagine that you are making the part that is shown in the print. Try to visualize the piece by asking yourself these questions:
    - What type of material is used? What color will the finished product be? Will the edges be smooth or rough?
    - How large is the piece? Could you easily hold it in one hand or would several people be needed to lift it?
    - Which machining processes are used to create the piece? Do holes require reaming or tapping? Which surfaces need to be machined?
    - For whom is the piece being manufactured? Why do they need this piece?
    - Are there any unusual notes or tolerances that will require additional work or time?

2. Once you have visualized the piece, answer the questions.
3. After completing the questions, recheck your answers. Use common sense to make sure that your answers are reasonable.

There is a significant difference between simply reading numbers from a print and understanding what those numbers mean. By learning to visualize what is being made and knowing the steps required to make it, you will gain a better understanding of the views, dimensions, tolerances, and symbols discussed in this text.

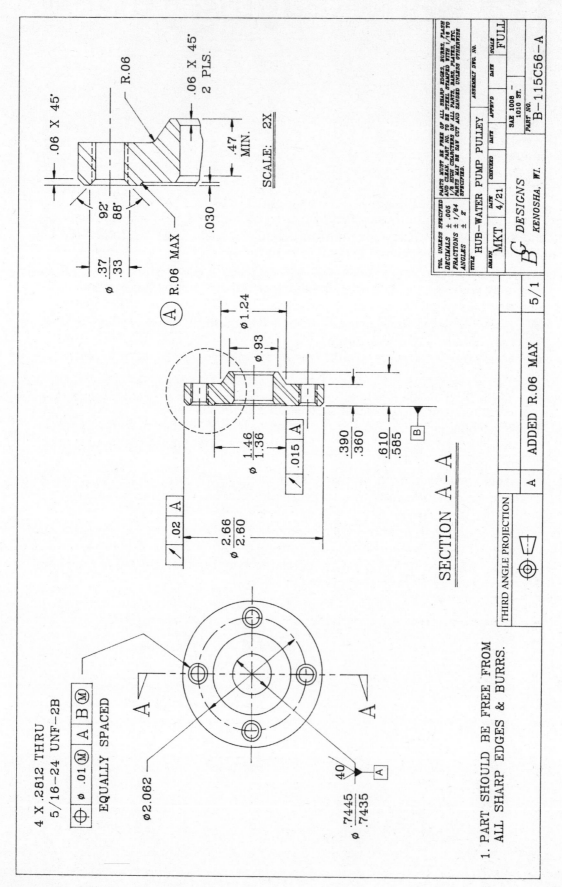

SCALE: 2X

.06 X 45°

R.06

.06 X 45°
2 PLS.

.47
MIN.

92°
88°

.030

$\phi \dfrac{.37}{.33}$

Ⓐ R.06 MAX

ϕ1.24

ϕ.93

$\phi \dfrac{1.46}{1.36}$

.015 | A

.390 / .360

.610 / .585

B

.02 | A

$\phi \dfrac{2.66}{2.60}$

SECTION A–A

4 X .2812 THRU
5/16–24 UNF–2B

⊕ | ϕ .01 Ⓜ | A | B Ⓜ

EQUALLY SPACED

A

A

ϕ2.062

$\phi \dfrac{.7445}{.7435}$

40

A

1. PART SHOULD BE FREE FROM
   ALL SHARP EDGES & BURRS.

THIRD ANGLE PROJECTION

5/1

ADDED R.06 MAX

A

TOL. UNLESS SPECIFIED	PARTS MUST BE FREE OF ALL SHARP EDGES, BURRS, FLASH AND CLEAN. PART NO. TO BE STEEL STAMPED WITH 1/16 TO 3 HIGH CHARACTERS ON ALL PARTS. PARTS MAT BE SAW CUT AND SANDED UNLESS OTHERWISE SPECIFIED.		ASSEMBLY DWG. NO.		
DECIMALS ± .005				SCALE	
FRACTIONS ± 1/64				FULL	
ANGLES ± 2°					
	TITLE HUB–WATER PUMP PULLEY				
DRAFT	DATE	CHECKED	DATE	APP'VD	DATE
MKT	4/21				
Ⓑ DESIGNS					
KENOSHA, WI.	SAE 1008 – 1010 ST.				
				PART NO. B–115C56–A	

SECTION A–A

GB Designs

Name _____ Date _____ Class _____

Activity 18-1
Water Pump Pulley Hub

Refer to **Activity 18-1**. *Study the drawing and familiarize yourself with the views, dimensions, title block, and notes. Read the questions, refer to the print, and write your answers in the blanks provided.*

1. What is the maximum outside diameter of the pulley?   1. _____

2. What is the maximum length of the .744 diameter bore?   2. _____

3. How long is the tapered section of the pulley? Calculate using high limit dimensions.   3. _____

4. What fit are the threaded holes?   4. _____

5. What is the diameter of the bolt circle?   5. _____

6. How many chamfers does the part have?   6. _____

7. How much does the taper differ in diameter?   7. _____

8. What is the minimum thickness of the part?   8. _____

9. What is the maximum chamfer diameter allowed on the threaded holes?   9. _____

10. Does the section line pattern match the material specified?   10. _____

11. What is the minimum runout shown on the print?   11. _____

12. What is the high limit on the depth of the shallow counterbore?   12. _____

13. How many revisions have been made on the part?   13. _____

14. What feature is datum A?   14. _____

15. What type of sectional view is shown?   15. _____

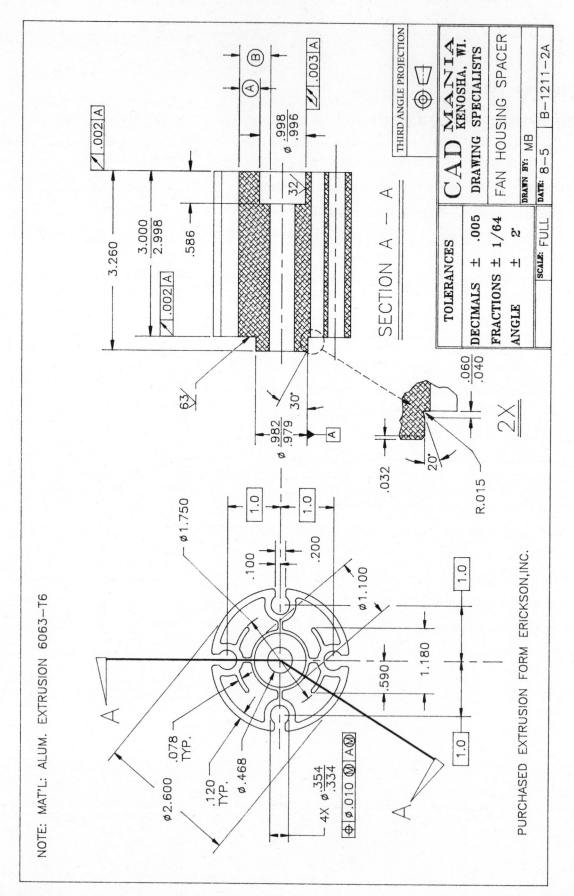

SECTION A – A

CAD MANIA		
KENOSHA, WI.		
DRAWING SPECIALISTS		
FAN HOUSING SPACER		
DRAWN BY: MB		
DATE: 8–5	B–1211–2A	

THIRD ANGLE PROJECTION

TOLERANCES		
DECIMALS ± .005		
FRACTIONS ± 1/64		
ANGLE ± 2°		
SCALE: FULL		

NOTE: MAT'L: ALUM. EXTRUSION 6063–T6

PURCHASED EXTRUSION FORM ERICKSON,INC.

Name _____ Date _____ Class _____

Activity 18-2
Fan Housing Spacer

Refer to **Activity 18-2**. *Study the drawing and familiarize yourself with the views, dimensions, title block, and notes. Read the questions, refer to the print, and write your answers in the blanks provided.*

1. What is the minimum overall length of the spacer?   1. _____

2. What is the minimum depth allowed on the counterbored hole?   2. _____

3. What is the largest diameter on the spacer?   3. _____

4. What size is the feature referred to as datum A?   4. _____

5. What is the thinnest wall thickness on the spacer?   5. _____

6. What is the maximum length of the 2.600 diameter section?   6. _____

7. What is the maximum length for the .982/.979 diameter?   7. _____

8. What tolerance is given for total runout?   8. _____

9. What positional tolerance is specified for the ⌀.354/.334 holes?   9. _____

10. What tolerance is specified for circular runout?   10. _____

11. What is the width of the 4 slots in the housing spacer?   11. _____

12. What size is the chamfer on the .982/.979 diameter?   12. _____

13. What material is used to manufacture the spacer?   13. _____

14. Determine distance A when holes are at the low limit.   14. _____

15. Determine distance B when holes are at the high limit.   15. _____

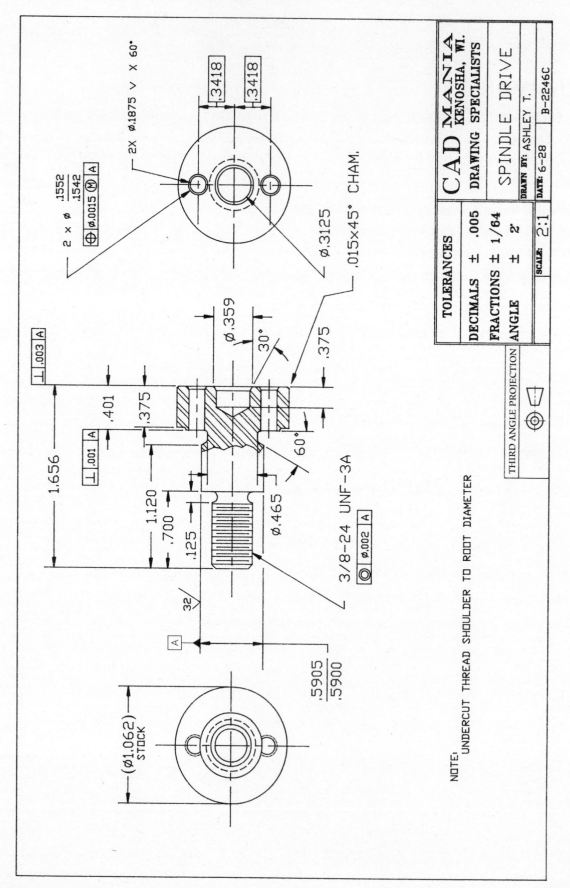

Activity 18-3 Spindle Drive.

CAD Mania

Name _____ Date _____ Class _____

Activity 18-3
Spindle Drive

Refer to **Activity 18-3**. *Study the drawing and familiarize yourself with the views, dimensions, title block, and notes. Read the questions, refer to the print, and write your answers in the blanks provided.*

1. What size is the large internal chamfer?

2. What is the length of the .5900 diameter?

3. How deep is the .3125 diameter hole?

4. Which dimension requires the smoothest machined surface?

5. What is the maximum length of the head of the spindle drive?

6. The 60° shoulder relief should be machined to what diameter?

7. What feature is datum A?

8. List the dimension(s) referred to as basic dimensions.

9. What diameter are the chamfers on the ∅.1552/.1542 holes?

10. What geometric characteristics are used to define this part?

11. How wide is the 60° groove at the opening?

12. What tolerance is given for concentricity of the thread?

13. What size must the two ∅.1552/.1542 holes be to constitute MMC?

14. What is the smallest tolerance shown on the print?

15. What is the minimum allowable wall thickness between the .3125 and .1552/.1542 diameter holes if they are at the low limit?

1. _____

2. _____

3. _____

4. _____

5. _____

6. _____

7. _____

8. _____

9. _____

10. _____

11. _____

12. _____

13. _____

14. _____

15. _____

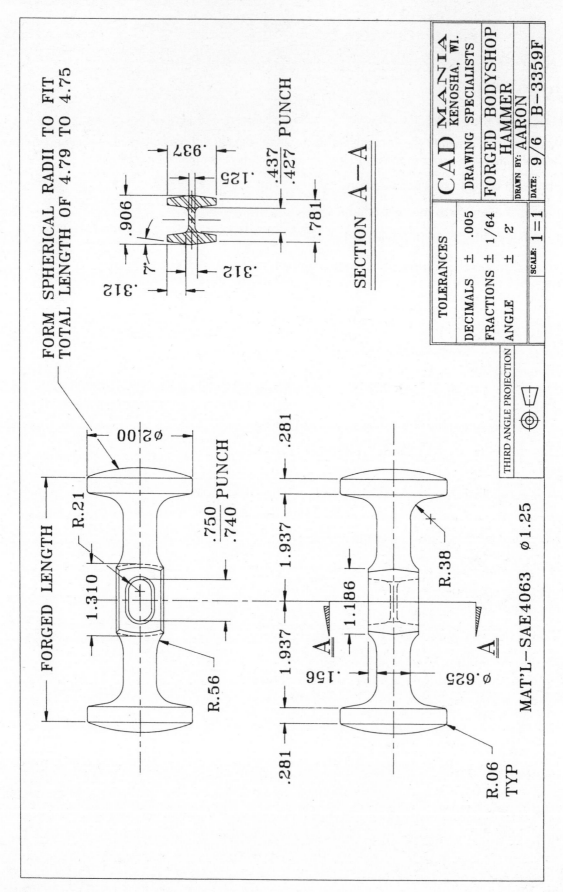

SECTION A—A

FORM SPHERICAL RADII TO FIT
TOTAL LENGTH OF 4.79 TO 4.75

TOLERANCES		CAD MANIA KENOSHA, WI.
		DRAWING SPECIALISTS
DECIMALS ± .005		FORGED BODYSHOP
FRACTIONS ± 1/64		HAMMER
ANGLE ± 2°		DRAWN BY: AARON
		DATE: 9/6 | B-3359F
		SCALE: 1=1

THIRD ANGLE PROJECTION

MAT'L—SAE4063 Ø1.25

Activity 18-4   Forged Bodyshop Hammer.

Name _____ Date _____ Class _____

Activity 18-4
Forged Bodyshop Hammer

Refer to **Activity 18-4**. *Study the drawing and familiarize yourself with the views, dimensions, title block, and notes. Read the questions, refer to the print, and write your answers in the blanks provided.*

1. What diameter are the hammer head ends?

2. What diameters are the necks?

3. How high is the middle portion of the hammer?

4. What is the width of the handle slot?

5. What is the maximum length of the handle slot?

6. Referring to the top view, determine the minimum sidewall thickness of the handle slot.

7. What material is used for the forging?

8. What is the minimum total length of the hammer?

9. What is the maximum forged length of the hammer?

10. What size are the smallest radii on the forging?

1. _____

2. _____

3. _____

4. _____

5. _____

6. _____

7. _____

8. _____

9. _____

10. _____

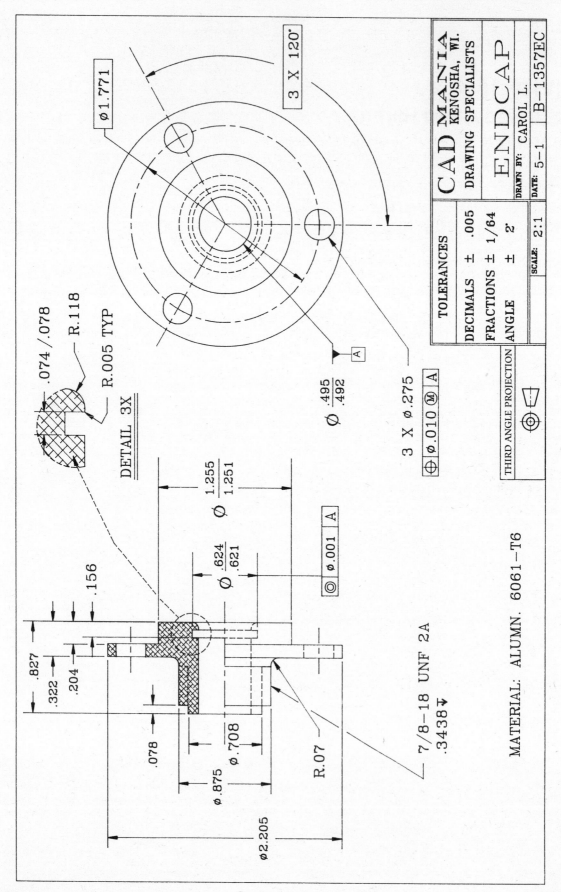

Activity 18-5 Endcap

Name _____ Date _____ Class _____

Activity 18-5
Endcap

Refer to **Activity 18-5**. *Study the drawing and familiarize yourself with the views, dimensions, title block, and notes. Read the questions, refer to the print, and write your answers in the blanks provided.*

1. How far apart, in degrees, are the three .275 diameter holes from each other?

2. What is the low limit of the large hole?

3. What is the maximum diameter of the internal groove?

4. What is the maximum length of the endcap?

5. How long is the 1.255/1.251 diameter section?

6. What is the high limit on the largest diameter?

7. What is the wall thickness between the .275 diameter holes and the large outside diameter?

8. What is the minimum width of the slot?

9. What feature is datum A?

10. What geometric characteristic is used to define the three equally spaced holes?

11. With the bore at high limit and the slot diameter at low limit, how deep is the slot?

12. What dimension locates the internal groove?

13. How thick is the flange on the endcap?

14. What is the concentricity tolerance?

1. _____

2. _____

3. _____

4. _____

5. _____

6. _____

7. _____

8. _____

9. _____

10. _____

11. _____

12. _____

13. _____

14. _____

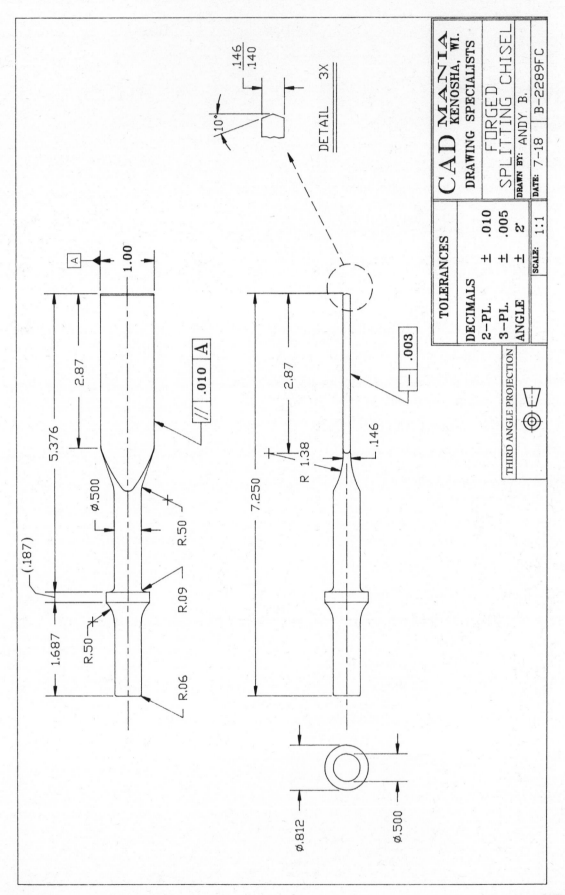

DETAIL   3X

$\dfrac{.146}{.140}$

10°

// | .010 | A

A

A

1.00

2.87

5.376

Ø.500

R.50

R.09

R.50

R.06

(.187)

1.687

— | .003

2.87

7.250

R 1.38

.146

Ø.812

Ø.500

TOLERANCES

DECIMALS

2–PL. ± .010

3–PL. ± .005

ANGLE ± 2°

SCALE: 1:1

CAD MANIA
KENOSHA, WI.
DRAWING SPECIALISTS

FORGED
SPLITTING CHISEL

DRAWN BY: ANDY B.

DATE: 7–18   B–2289FC

THIRD ANGLE PROJECTION

Activity 18-6 Forged Splitting Chisel.

Name _____ Date _____ Class _____

Activity 18-6
Forged Splitting Chisel

Refer to **Activity 18-6**. *Study the drawing and familiarize yourself with the views, dimensions, title block, and notes. Read the questions, refer to the print, and write your answers in the blanks provided.*

1. What is the scale of the print?

2. What is the maximum length allowed for the chisel?

3. What is the shank diameter?

4. What is the shank length?

5. What is the total included angle of the chisel point?

6. What geometric symbol is shown in the front view?

7. What type of tolerance is this geometric symbol?

8. What geometric symbol is shown in the top view?

9. What tolerance is specified for this symbol?

10. What is the scale of the detail?

11. What is the high limit on the width of the chisel blade?

12. What is the maximum thickness allowed on the chisel blade?

13. What size is the largest fillet shown on the print?

14. What is the smallest tolerance shown on the print?

15. What is the print drawing number?

1. _____

2. _____

3. _____

4. _____

5. _____

6. _____

7. _____

8. _____

9. _____

10. _____

11. _____

12. _____

13. _____

14. _____

15. _____

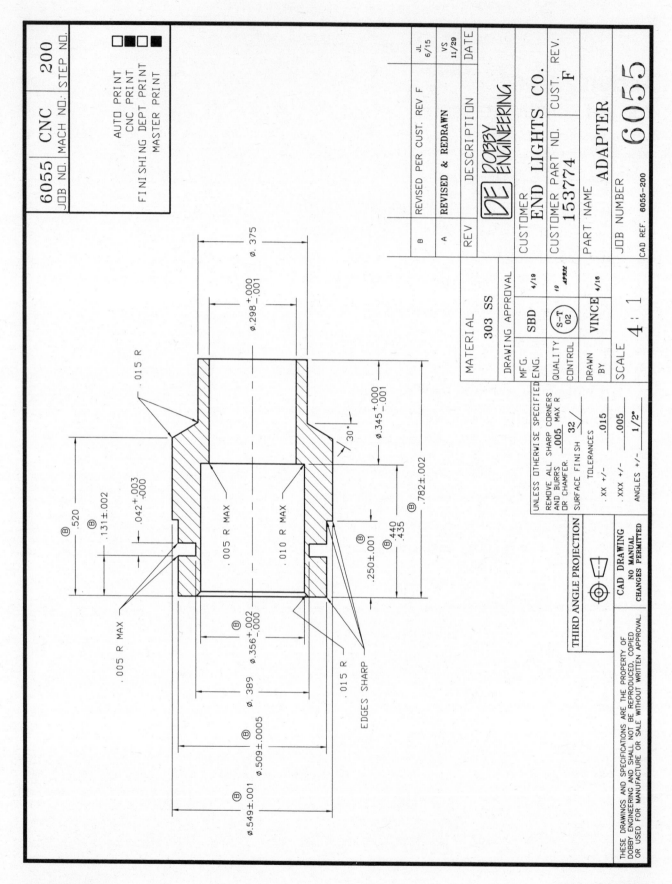

Activity 18-7 Adapter.

Name _____ Date _____ Class _____

Activity 18-7
Adapter

Refer to **Activity 18-7**. *Study the drawing and familiarize yourself with the views, dimensions, title block, and notes. Read the questions, refer to the print, and write your answers in the blanks provided.*

1. Identify the customer part number.

2. What is the minimum total length of the workpiece?

3. What is the maximum diameter of the external groove?

4. What is the minimum width allowed on the external groove?

5. What is the low limit on the .375 diameter?

6. Identify the scale of the print.

7. What is the minimum length of the .509 diameter?

8. What is the largest outside diameter?

9. What is the low limit on the large internal diameter?

10. How many revisions were made on June 15th?

11. What dimension is used to locate the external groove?

12. Which dimension on the print has the smallest tolerance?

13. What surface finish is required?

14. How many edges are required to remain sharp?

15. How many corners require a .015 radius?

16. If both the overall length and the small internal bore are at maximum length, what would be the length of the large internal bore?

17. With the .509 diameter at its high limit and the external groove diameter at its low limit, determine the depth of the groove.

18. What type of sectional view is shown on the print?

19. What specified dimension has the largest tolerance?

20. With the .509 diameter at its maximum length, what is the length of the .549 diameter if the .520 dimension is held?

1. _____

2. _____

3. _____

4. _____

5. _____

6. _____

7. _____

8. _____

9. _____

10. _____

11. _____

12. _____

13. _____

14. _____

15. _____

16. _____

17. _____

18. _____

19. _____

20. _____

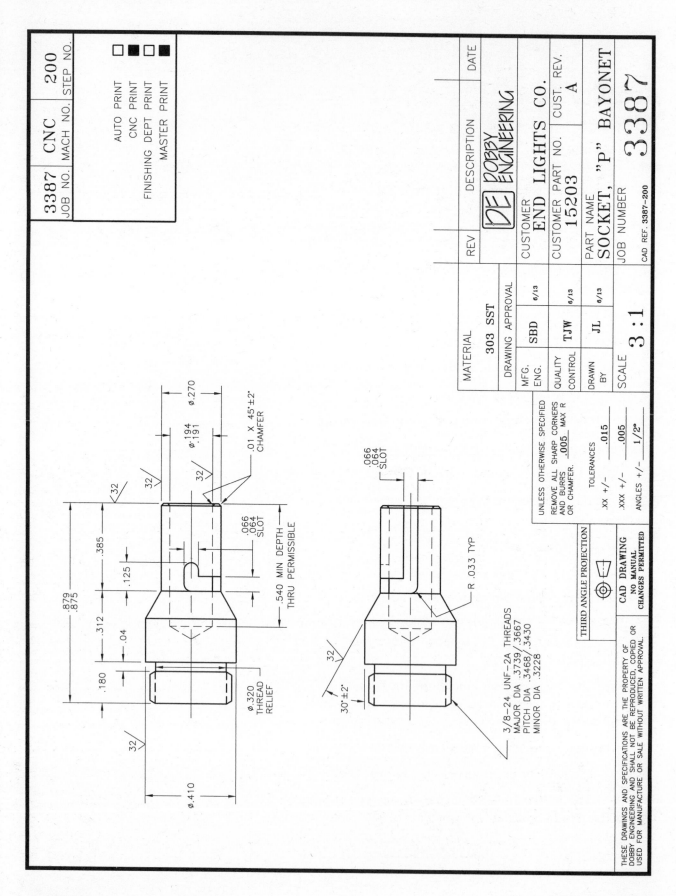

Name _____ Date _____ Class _____

Activity 18-8
"P" Bayonet Socket

Refer to **Activity 18-8**. *Study the drawing and familiarize yourself with the views, dimensions, title block, and notes. Read the questions, refer to the print, and write your answers in the blanks provided.*

1. What is the minimum overall length of the workpiece?

2. How long is the section that has a .270 diameter?

3. What size is the diameter with the angular shoulder shown in the top view?

4. What length are the 45° chamfers?

5. What is the thread diameter?

6. If the major diameter of the thread is at the high limit, what is the depth of the thread?

7. How wide is the slot?

8. How long is the short leg of the slot?

9. What is the high limit on the section with the .410 diameter?

10. What operation or step number is this part?

11. If the overall length is at maximum, what is the length of the thread?

12. How many degrees does the slot revolve?

13. If the thread relief diameter were at high limit, would the relief interfere with the minor diameter?

14. Does the slot go through the wall of the hole and outside diameter?

15. How far is the end of the slot from the right end of the workpiece?

16. If the overall length is at the minimum, how far is the short end of the slot from the left end of the workplace?

17. What is the CAD reference number?

18. What does SST mean?

19. What is the maximum wall thickness of the slotted section of the part?

20. What is the maximum chamfer allowed on unspecified corners?

1. _____

2. _____

3. _____

4. _____

5. _____

6. _____

7. _____

8. _____

9. _____

10. _____

11. _____

12. _____

13. _____

14. _____

15. _____

16. _____

17. _____

18. _____

19. _____

20. _____

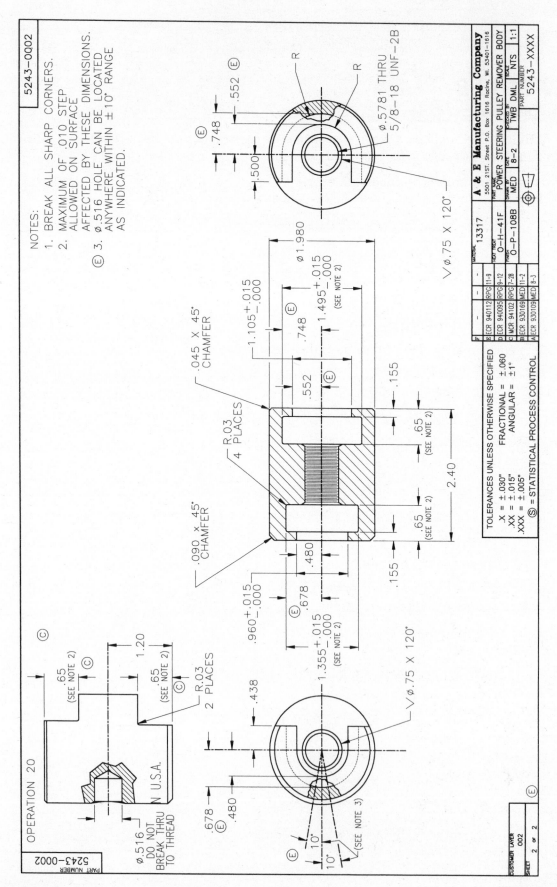

NOTES:
1. BREAK ALL SHARP CORNERS.
2. MAXIMUM OF .010 STEP ALLOWED ON SURFACE AFFECTED BY THESE DIMENSIONS.
Ⓔ 3. Ø.516 HOLE CAN BE LOCATED ANYWHERE WITHIN ±10° RANGE AS INDICATED.

Name _____ Date _____ Class _____

Activity 18-9
Power Steering Pulley Remover Body

Refer to **Activity 18-9**. *Study the drawing and familiarize yourself with the views, dimensions, title block, and notes. Read the questions, refer to the print, and write your answers in the blanks provided.*

1. What is the maximum overall length of the workplace?

   1. _____

2. What is the length of the 1.355 internal dimension?

   2. _____

3. What is the length of the thread?

   3. _____

4. What size is the external chamfer on the left end of the part?

   4. _____

   5. _____

5. What fit class is the thread?

6. What is the minimum size of the large internal circular groove?

   6. _____

7. What size chamfer is on the thread?

   7. _____

8. What is the maximum size of the .155 deep slot on the left end of the workpiece?

   8. _____

9. What size is the drilled hole in the top view?

   9. _____

10. What is the minimum outside diameter of the workpiece?

   10. _____

11. What type of sectional view is shown in the upper-left portion of the print?

   11. _____

12. When was revision E made?

   12. _____

13. What material is used to machine the workplace?

   13. _____

14. If the 1.495 wide circular groove is at maximum size and the outside diameter is at minimum size, what would be the distance from the groove to the outer surface?

   14. _____

15. What dimension does the hidden line in the right side view represent?

   15. _____

16. Within what total range can the .516 diameter hole be located?

   16. _____

17. What is the part number of the workpiece?

   17. _____

18. How many places call for a .03 radius?

   18. _____

19. When was the print first released?

   19. _____

20. Referring to the top view, what is the length of the protrusion from the right end to the lower shoulder (disregarding the fillet)?

   20. _____

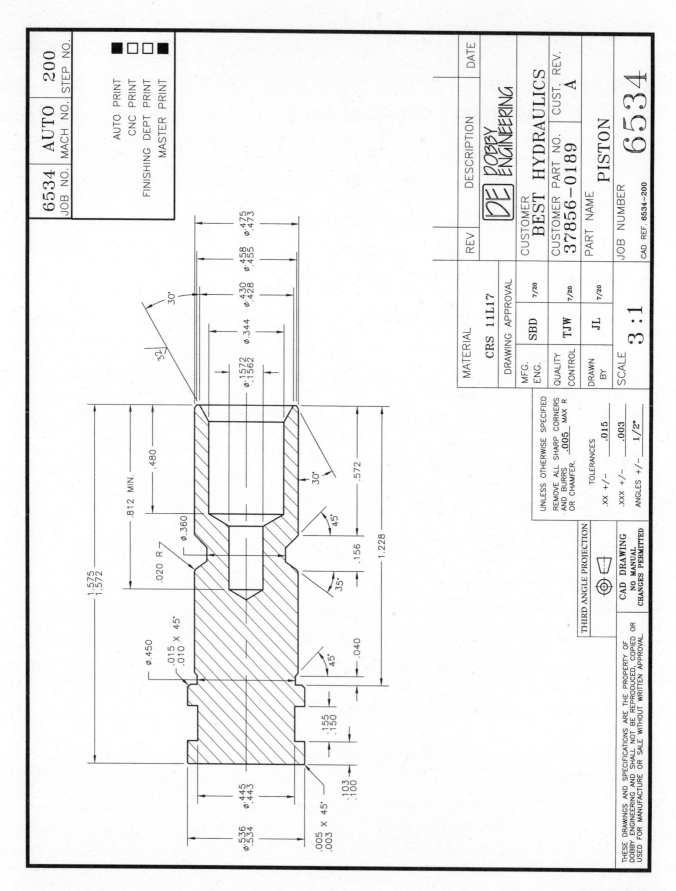

Activity 18-10 Piston.

Name _____ Date _____ Class _____

Activity 18-10
Piston

Refer to **Activity 18-10**. *Study the drawing and familiarize yourself with the views, dimensions, title block, and notes. Read the questions, refer to the print, and write your answers in the blanks provided.*

1. What is the smallest outside diameter?

2. What is the maximum overall length allowed for the workplace?

3. What diameter is the .040 wide groove?

4. How deep is the .1572 diameter hole?

5. What radius is given on the print?

6. What scale is the print?

7. What tolerance is used on the .450 diameter?

8. At its widest, what diameter is the 30° internal chamfer?

9. How deep is the .344 diameter hole?

10. What specific type of material is used to make the workplace?

11. What finish is required on the 30° internal chamfer?

12. What dimension specifies the width of the .360 diameter external groove?

13. What is the distance from the left end of the workpiece to the shoulder formed between the .536 and .450 diameters?

14. By whom was the print drawn?

15. What is the depth of the .155 wide groove if both diameters involved were at the high limit?

16. What is the tolerance for two place decimals?

17. How many 45° chamfers are on the workpiece?

18. What is the smallest tolerance shown on the print?

19. What dimension positions the .360 diameter groove?

20. What wall thickness exists if the large diameter hole and the outside diameter are at low limits?

1. _____

2. _____

3. _____

4. _____

5. _____

6. _____

7. _____

8. _____

9. _____

10. _____

11. _____

12. _____

13. _____

14. _____

15. _____

16. _____

17. _____

18. _____

19. _____

20. _____

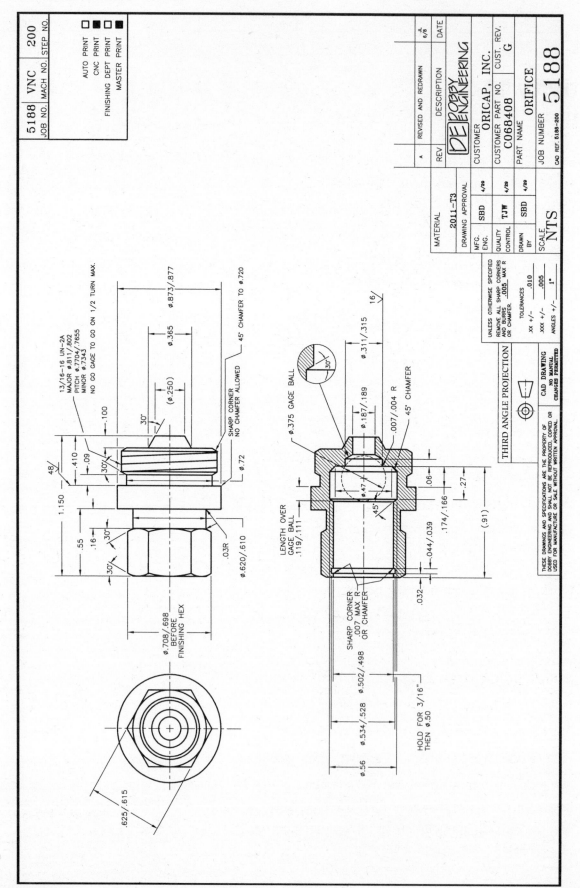

JOB NO.	MACH NO.	STEP NO.
5188	VNC.	200

☐ AUTO PRINT
☑ CNC PRINT
☐ FINISHING DEPT PRINT
☑ MASTER PRINT

REV	DESCRIPTION	DATE
A	REVISED AND REDRAWN	JL 6/8

DE DOBBY ENGINEERING

CUSTOMER **ORICAP, INC.**

CUSTOMER PART NO. **C068408**   CUST. REV. **G**

PART NAME **ORIFICE**

JOB NUMBER **5188**

CAD REF. **5188-200**

MATERIAL **2011-T3**

DRAWING APPROVAL

MFG. ENG.	SBD	4/20
QUALITY CONTROL	TJW	4/20
DRAWN BY	SBD	4/20

SCALE **NTS**

UNLESS OTHERWISE SPECIFIED
REMOVE ALL SHARP CORNERS AND BURRS .005 MAX R
OR CHAMFER.

TOLERANCES
.XX +/- .010
.XXX +/- .005
ANGLES +/- 1°

THIRD ANGLE PROJECTION

CAD DRAWING
NO MANUAL CHANGES PERMITTED

13/16-16 UN-2A
MAJOR ∅.811/.802
PITCH ∅.7704/.7655
MINOR ∅.7343
NO GO GAGE TO GO ON 1/2 TURN MAX.

∅.873/.877
∅.365
(∅.250)
45° CHAMFER TO ∅.720
SHARP CORNER NO CHAMFER ALLOWED
∅.72
.100
30°
30°
.410
.09
48°
1.150
.55
.16
30°
30°
.03R
∅.620/.610
∅.708/.698 BEFORE FINISHING HEX
.625/.615

∅.375 GAGE BALL
30°
∅.311/.315
16
∅.187/.189
.007/.004 R
45° CHAMFER
LENGTH OVER GAGE BALL .119/.111
∅.47
45°
.06
.27
.174/.166
.044/.039
.032
SHARP CORNER .007 MAX R OR CHAMFER
∅.502/.498
(.91)
∅.534/.528
HOLD FOR 3/16" THEN ∅.50
∅.56

Name _____ Date _____ Class _____

Activity 18-11
Orifice

Refer to **Activity 18-11**. *Study the drawing and familiarize yourself with the views, dimensions, title block, and notes. Read the questions, refer to the print, and write your answers in the blanks provided.*

1. When was the print approved by quality control?

2. What is the maximum overall length allowed on the part?

3. What is the maximum overall diameter of the workpiece?

4. What is the low limit on the size of the finished hex?

5. What is the length of the .873/.877 diameter?

6. What is the minimum allowable diameter of the .044 wide internal groove?

7. What is the low limit on the thread relief diameter?

8. What is the length of the thread, including chamfers?

9. What is the maximum major diameter of the thread?

10. Is this part produced on a CNC machine?

11. How wide is the .620 diameter groove?

12. What is the length of the hex portion of the orifice?

13. When was the print last revised?

14. What is the maximum unspecified radius allowed on the part?

15. How far is the narrow internal groove from the right end of the part if the width of the slot is at its high limit?

16. How long is the .311/.315 diameter?

17. What is the length of the rightmost internal 45° chamfer?

18. What is the large diameter of the short 30° angle at the tip of the orifice?

19. What is the length of the .502 internal diameter?

20. What is the length of the .56 diameter, including the 45° chamfers?

1. _____

2. _____

3. _____

4. _____

5. _____

6. _____

7. _____

8. _____

9. _____

10. _____

11. _____

12. _____

13. _____

14. _____

15. _____

16. _____

17. _____

18. _____

19. _____

20. _____

Notes

Reference Section

Care of Drawings and Prints

When working with drawings and prints, care must be taken not to damage them. Original drawings should be safely stored and removed from storage only when needed. Particular care should be taken when using hand drawings. It can be very time-consuming to redraw a damaged original or create a new digital drawing.

Drawings and prints should be kept as clean as possible. Never set food or a drink on a drawing or print. Always protect prints from damaging weather, such as rain and snow.

When rolling drawings and prints, always roll with the printed side turned out and the title block at the end of the roll. This will allow the title blocks to be viewed without unrolling the entire set. See **Figure R-1**. Also, by rolling with the printed side out, the print will not "reroll" itself when it is laid flat on a table. If the printed side is turned in when the drawing is rolled, it will be difficult to lay flat.

When folding prints, always fold them so that the title block can be viewed. This will prevent the need to unfold the drawing to see what it is. See **Figure R-2**. Sharp creases can be made by folding the drawing and then quickly pressing the crease (from middle to edges) with a scale or other straightedge.

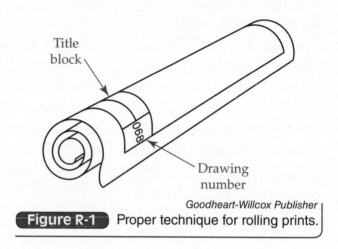

Goodheart-Willcox Publisher

Figure R-1   Proper technique for rolling prints.

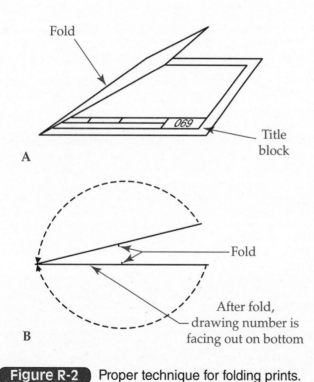

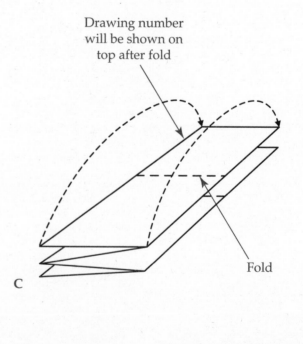

Goodheart-Willcox Publisher

Figure R-2   Proper technique for folding prints.

Standard Abbreviations

A

ADD / Addendum
ADJ / Adjust
ALIGN / Alignment
ALLOW / Allowance
ALT / Alteration
ALUM / Aluminum
ALY / Alloy
ANL / Anneal
ANOD / Anodize
APPD / Approved
APPROX / Approximate
ASSY / Assembly
AUTO / Automatic
AUX / Auxiliary
AWG / American Wire Gage

B

BC / Bolt circle
B/M / Bill of material
BEV / Bevel
BHN / Brinnel hardness number
BNH / Burnish
BRG / Bearing
BRKT / Bracket
BRS / Brass
BRZ / Bronze
BRZG / Brazing
BUSH / Bushing

C

C TO C / Center to center
C'BORE / Counterbore
C'SINK / Countersink
CARB / Carburize
CDS / Cold-drawn steel
CH / Case harden
CHAM / Chamfer
CI / Cast iron
CIR / Circular
CIRC / Circumference
CL / Clearance
CONC / Concentric
COND / Condition
CONT / Control
COP / Copper
CPLG / Coupling
CR VAN / Chrome vanadium
CRS / Cold-rolled steel
CSTG / Casting

CTD / Coated
CTR / Center
CTR / Contour
CYL / Cylinder

D

DAT / Datum
DCN / Drawing change notice
DF / Drop forge
DIA / Diameter
DIAG / Diagonal
DIM / Dimension
DR / Drill
DWG / Drawing
DWL / Dowel

E

EA / Each
ECC / Eccentric
ECO / Engineering change order
ECR / Engineering change revision
EQ / Equal
EQUIV / Equivalent
ES / Engineering specifications
EST / Estimate

F

FAB / Fabricate
FAO / Finish all over
FIL / Filler
FIM / Full indicator movement
FIN / Finish
FLG / Flange
FORG / Forging
FST / Forged steel
FTG / Fitting
FURN / Furnish

G

GA / Gage
GALV / Galvanized
GRD / Grind
GSKT / Gasket

H

HCS / High carbon steel
HDN / Harden
HEX / Hexagonal
HOR / Horizontal
HS / High speed
HSG / Housing
HT TR / Heat treat

I

ID / Inside diameter
INSTL / Installation

K

KWY / Keyway

L

LAM / Laminate
LC / Low carbon
LG / Length
LH / Left-hand

M

MACH / Machine
MAG / Magnesium
MATL / Material
MAX / Maximum
MCR / Manufacturing change revision
MECH / Mechanical
MI / Malleable iron
MIL / Military
MIN / Minimum
MISC / Miscellaneous
MOD / Modification
MTG / Mounting

N

NO. / Number
NOM / Nominal
NORM / Normal
NS / Nickel steel
NTS / Not to scale

O

OBS / Obsolete
OD / Outside diameter

P

P / Pitch
PC / Piece
PROC / Process

Q

QTY / Quantity
QUAL / Quality

R

R / Radius
RD / Round

REF

REF / Reference
REQD / Required
REV / Revision
RH / Right-hand
RH / Rockwell hardness
RIV / Rivet

S

SCH / Schedule
SCR / Screw
SECT / Section
SEQ / Sequence
SERR / Serrate
SF / Spotface
SH / Sheet
SPC / Statistical process control
SPEC / Specification
SPL / Special
SQ / Square
SST / Stainless steel
STD / Standard
STK / Stock
STL / Steel
SYM / Symmetrical

T

TAP / Tapping
TEM / Temper
THD / Thread
THK / Thick
TIF / True involute form
TIR / Total indicator reading
TOL / Tolerance
TS / Tensile strength
TS / Tool steel
TYP / Typical

U

UNC / Unified Screw Thread Coarse
UNF / Unified Screw Thread Fine

V

VAR / Variable
VERT / Vertical

W

W / Width
WI / Wrought iron
WT / Weight

Decimal and Metric Equivalents

INCHES Fractions	INCHES Decimals	MILLIMETERS	INCHES Fractions	INCHES Decimals	MILLIMETERS
	.00394	.1	15/32	.46875	11.9063
	.00787	.2		.47244	12.00
	.01181	.3	31/64	.484375	12.3031
1/64	.015625	.3969	1/2	.5000	12.70
	.01575	.4		.51181	13.00
	.01969	.5	33/64	.515625	13.0969
	.02362	.6	17/32	.53125	13.4938
	.02756	.7	35/64	.546875	13.8907
1/32	.03125	.7938		.55118	14.00
	.0315	.8	9/16	.5625	14.2875
	.03543	.9	37/64	.578125	14.6844
	.03937	1.00		.59055	15.00
3/64	.046875	1.1906	19/32	.59375	15.0813
1/16	.0625	1.5875	39/64	.609375	15.4782
5/64	.078125	1.9844	5/8	.625	15.875
	.07874	2.00		.62992	16.00
3/32	.09375	2.3813	41/64	.640625	16.2719
7/64	.109375	2.7781	21/32	.65625	16.6688
	.11811	3.00		.66929	17.00
1/8	.125	3.175	43/64	.671875	17.0657
9/64	.140625	3.5719	11/16	.6875	17.4625
5/32	.15625	3.9688	45/64	.703125	17.8594
	.15748	4.00		.70866	18.00
11/64	.171875	4.3656	23/32	.71875	18.2563
3/16	.1875	4.7625	47/64	.734375	18.6532
	.19685	5.00		.74803	19.00
13/64	.203125	5.1594	3/4	.7500	19.05
7/32	.21875	5.5563	49/64	.765625	19.4469
15/64	.234375	5.9531	25/32	.78125	19.8438
	.23622	6.00		.7874	20.00
1/4	.2500	6.35	51/64	.796875	20.2407
17/64	.265625	6.7469	13/16	.8125	20.6375
	.27559	7.00		.82677	21.00
9/32	.28125	7.1438	53/64	.828125	21.0344
19/64	.296875	7.5406	27/32	.84375	21.4313
5/16	.3125	7.9375	55/64	.859375	21.8282
	.31496	8.00		.86614	22.00
21/64	.328125	8.3344	7/8	.875	22.225
11/32	.34375	8.7313	57/64	.890625	22.6219
	.35433	9.00		.90551	23.00
23/64	.359375	9.1281	29/32	.90625	23.0188
3/8	.375	9.525	59/64	.921875	23.4157
25/64	.390625	9.9219	15/16	.9375	23.8125
	.3937	10.00		.94488	24.00
13/32	.40625	10.3188	61/64	.953125	24.2094
27/64	.421875	10.7156	31/32	.96875	24.6063
	.43307	11.00		.98425	25.00
7/16	.4375	11.1125	63/64	.984375	25.0032
29/64	.453125	11.5094	1	1.0000	25.4000

Number and Letter Size Drills Conversion Chart

Number and Letter Size Drills Conversion Chart

Drill No. or Letter	Inch	mm	Drill No. or Letter	Inch	mm	Drill No. or Letter	Inch	mm	Drill No. or Letter	Inch	mm	Drill No. or Letter	Inch	mm
	.001	0.0254		.101	2.5654	7	.201	5.1054		.301	7.6454		.401	10.1854
	.002	0.0508	38 .1015	.102	2.5908		.202	5.1308	N	.302	7.6708		.402	10.2108
	.003	0.0762		.103	2.6162		.203	5.1562		.303	7.6962		.403	10.2362
	.004	0.1016	37	.104	2.6416	13/64	.2031	5.1594		.304	7.7216	Y	.404	10.2616
	.005	0.1270		.105	2.6670	6	.204	5.1816		.305	7.7470		.405	10.2870
	.006	0.1524	36 1065	.106	2.6924	5 .2055	.205	5.2070		.306	7.7724		.406	10.3124
	.007	0.1778		.107	2.7178		.206	5.2324		.307	7.7978	13/32	.4062	10.3187
	.008	0.2032		.108	2.7432		.207	5.2578		.308	7.8232		.407	10.3378
	.009	0.2286		.109	2.7686		.208	5.2832		.309	7.8486		.408	10.3632
	.010	0.2540	7/64	.1094	2.7781	4	.209	5.3086		.310	7.8740		.409	10.3886
	.011	0.2794	35	.110	2.7940		.210	5.3340		.311	7.8994		.410	10.4140
	.012	0.3048	34	.111	2.8194		.211	5.3594		.312	7.9248		.411	10.4394
	.013	0.3302		.112	2.8448		.212	5.3848	5/16	.3125	7.9375		.412	10.4648
80 .0135	.014	0.3556	33	.113	2.8702	3	.213	5.4102		.313	7.9502	Z	.413	10.4902
79 .0145	.015	0.3810		.114	2.8956		.214	5.4356		.314	7.9756		.414	10.5156
1/64 .0156		0.3969		.115	2.9210		.215	5.4610		.3150	8.0000		.415	10.5410
78	.016	0.4064	32	.116	2.9464		.216	5.4864		.315	8.0010		.416	10.5664
	.017	0.4318		.117	2.9718		.217	5.5118	O	.316	8.0264		.417	10.5918
77	.018	0.4572		.118	2.9972		.218	5.5372		.317	8.0518		.418	10.6172
	.019	0.4826		.1181	3.0000	7/32	.2187	5.5562		.318	8.0772		.419	10.6426
76	.020	0.5080		.119	3.0226		.219	5.5626		.319	8.1026		.420	10.6680
75	.021	0.5334	31	.120	3.0480		.220	5.5880		.320	8.1280		.421	10.6934
	.022	0.5588		.121	3.0734	2	.221	5.6134		.321	8.1534	27/64	.4219	10.7156
74 .0225	.023	0.5842		.122	3.0988		.222	5.6388		.322	8.1788		.422	10.7188
73	.024	0.6096		.123	3.1242		.223	5.6642	P	.323	8.2042		.423	10.7442
72	.025	0.6350		.124	3.1496		.224	5.6896		.324	8.2296		.424	10.7696
71	.026	0.6604	1/8	.125	3.1750		.225	5.7150		.325	8.2550		.425	10.7950
	.027	0.6858		.126	3.2004		.226	5.7404		.326	8.2804		.426	10.8204
70	.028	0.7112		.127	3.2258		.227	5.7658		.327	8.3058		.427	10.8458
	.029	0.7366		.128	3.2512	1	.228	5.7912		.328	8.3312		.428	10.8712
69 .0292	.030	0.7620	30 .1285	.129	3.2766		.229	5.8166	21/64	.3281	8.3344		.429	10.8966
68	.031	0.7874		.130	3.3020		.230	5.8410		.329	8.3566		.430	10.9220
1/32 .0312		0.7937		.131	3.3274		.231	5.8674		.330	8.3820		.431	10.9474
67	.032	0.8128		.132	3.3528		.232	5.8928		.331	8.4074		.432	10.9728
66	.033	0.8382		.133	3.3782		.233	5.9182	Q	.332	8.4328		.433	10.9982
	.034	0.8636		.134	3.4036	A	.234	5.9436		.333	8.4582		.4331	11.0000
65	.035	0.8890		.135	3.4290	15/64	.2344	5.9531		.334	8.4836		.434	11.0236
64	.036	0.9144	29	.136	3.4544		.235	5.9690		.335	8.5090		.435	11.0490
63	.037	0.9398		.137	3.4798		.236	5.9944		.336	8.5344		.436	11.0744
62	.038	0.9652		.138	3.5052		.2362	6.0000		.337	8.5598		.437	11.0998
61	.039	0.9906		.139	3.5306		.237	6.0198		.338	8.5852	7/16	.4375	11.1125
	.0394	1.0000	28 .1405	.140	3.5560	B	.238	6.0452	R	.339	8.6106		.438	11.1252
60	.040	1.0160	9/64	.1406	3.5719		.239	6.0706		.340	8.6360		.439	11.1506
59	.041	1.0414		.141	3.5814		.240	6.0960		.341	8.6614		.440	11.1760
58	.042	1.0668		.142	3.6068		.241	6.1214		.342	8.6868		.441	11.2014
57	.043	1.0922		.143	3.6322	C	.242	6.1468		.343	8.7122		.442	11.2268
	.044	1.1176		.144	3.6576		.243	6.1722	11/32	.3437	8.7312		.443	11.2522
	.045	1.1430	27	.145	3.6830		.244	6.1976		.344	8.7376		.444	11.2776
56 .0465	.046	1.1684		.146	3.7084		.245	6.2230		.345	8.7630		.445	11.3030
3/64 .0469		1.1906	26	.147	3.7338	D	.246	6.2484		.346	8.7884		.446	11.3284
	.047	1.1938		.148	3.7592		.247	6.2738		.347	8.8138		.447	11.3538
	.048	1.2192	25 .1495	.149	3.7846		.248	6.2992	S	.348	8.8392		.448	11.3792
	.049	1.2446		.150	3.8100		.249	6.3246		.349	8.8646		.449	11.4046
	.050	1.2700		.151	3.8354	1/4	.250	6.3500		.350	8.8900		.450	11.4300
55	.051	1.2954	24	.152	3.8608		.251	6.3754		.351	8.9154		.451	11.4554
	.052	1.3208		.153	3.8862		.252	6.4008		.352	8.9408		.452	11.4808
	.053	1.3462	23	.154	3.9116		.253	6.4262		.353	8.9662		.453	11.5062
	.054	1.3716		.155	3.9370		.254	6.4516		.354	8.9916	29/64	.4531	11.5094
54	.055	1.3970		.156	3.9624		.255	6.4770		.3543	9.0000		.454	11.5316
	.056	1.4224	5/32	.1562	3.9687		.256	6.5024		.355	9.0170		.455	11.5570
	.057	1.4478	22	.157	3.9878	F	.257	6.5278		.356	9.0424		.456	11.5824
	.058	1.4732		.1575	4.0000		.258	6.5532		.357	9.0678		.457	11.6078
	.059	1.4986		.158	4.0132		.259	6.5786	T	.358	9.0932		.458	11.6332
53 .0595	.060	1.5240		.159	4.0386		.260	6.6040		.359	9.1186		.459	11.6586
	.061	1.5494	21	.160	4.0640	G	.261	6.6294	23/64	.3594	9.1281		.460	11.6840
	.062	1.5748		.161	4.0894		.262	6.6548		.360	9.1440		.461	11.7094
1/16 .0625		1.5875	20	.162	4.1148		.263	6.6802		.361	9.1694		.462	11.7348
52 .0635	.063	1.6002		.163	4.1402		.264	6.7056		.362	9.1948		.463	11.7602
	.064	1.6256		.164	4.1656		.265	6.7310		.363	9.2202		.464	11.7856
	.065	1.6510		.165	4.1910	17/64	.2656	6.7469		.364	9.2456		.465	11.8110
	.066	1.6764	19	.166	4.2164	H	.266	6.7564		.365	9.2710		.466	11.8364
51	.067	1.7018		.167	4.2418		.267	6.7818		.366	9.2964		.467	11.8618
	.068	1.7272		.168	4.2672		.268	6.8072		.367	9.3218		.468	11.8872
	.069	1.7526		.169	4.2926		.269	6.8326	U	.368	9.3472	15/32	.4687	11.9062
50	.070	1.7780	18 .1695	.170	4.3180		.270	6.8580		.369	9.3726		.469	11.9126
	.071	1.8034		.171	4.3434		.271	6.8834		.370	9.3980		.470	11.9380
	.072	1.8288	11/64	.1719	4.3656	I	.272	6.9088		.371	9.4234		.471	11.9634
49	.073	1.8542		.172	4.3688		.273	6.9342		.372	9.4488		.472	11.9888
	.074	1.8796	17	.173	4.3942		.274	6.9596		.373	9.4742		.4724	12.0000
	.075	1.9050		.174	4.4196		.275	6.9850		.374	9.4996		.473	12.0142
48	.076	1.9304		.175	4.4450		.2756	7.0000	3/8	.375	9.5250		.474	12.0396
	.077	1.9558		.176	4.4704		.276	7.0104		.376	9.5504		.475	12.0650
47 .0785	.078	1.9812	16	.177	4.4958	J	.277	7.0358	V	.377	9.5758		.476	12.0904
5/64 .0781		1.9844		.178	4.5212		.278	7.0612		.378	9.6012		.477	12.1158
	.0787	2.0000		.179	4.5466		.279	7.0866		.379	9.6266		.478	12.1412
	.079	2.0066		.180	4.5720		.280	7.1120		.380	9.6520		.479	12.1666
	.080	2.0320	15	.181	4.5974	K	.281	7.1374		.381	9.6774		.480	12.1920
46	.081	2.0574		.182	4.6228	9/32	.2812	7.1437		.382	9.7028		.481	12.2174
45	.082	2.0828	14	.183	4.6482		.282	7.1628		.383	9.7282		.482	12.2428
	.083	2.1082		.184	4.6736		.283	7.1882		.384	9.7536		.483	12.2682
	.084	2.1336		.185	4.6990		.284	7.2136		.385	9.7790		.484	12.2936
	.085	2.1590	13	.186	4.7244		.285	7.2390	W	.386	9.8044	31/64	.4844	12.3031
44	.086	2.1844		.187	4.7498		.286	7.2644		.387	9.8298		.485	12.3190
	.087	2.2098	3/16	.1875	4.7625		.287	7.2898		.388	9.8552		.486	12.3444
	.088	2.2352		.188	4.7752		.288	7.3152		.389	9.8806		.487	12.3698
43	.089	2.2606	12	.189	4.8006		.289	7.3406		.390	9.9060		.488	12.3952
	.090	2.2860		.190	4.8260	L	.290	7.3660	25/64	.3906	9.9219		.489	12.4206
	.091	2.3114	11	.191	4.8514		.291	7.3914		.391	9.9314		.490	12.4460
	.092	2.3368		.192	4.8768		.292	7.4168		.392	9.9568		.491	12.4714
42 .0935	.093	2.3622		.193	4.9022		.293	7.4422		.393	9.9822		.492	12.4968
3/32 .0937		2.3812	10 .1935	.194	4.9278		.294	7.4676		.3937	10.0000		.493	12.5222
	.094	2.3876		.195	4.9530	M	.295	7.4930		.394	10.0076		.494	12.5476
	.095	2.4130	9	.196	4.9784		.296	7.5184		.395	10.0330		.495	12.5730
41	.096	2.4384		.1969	5.0000	19/64	.2969	7.5406		.396	10.0584		.496	12.5984
	.097	2.4638		.197	5.0038		.297	7.5438	X	.397	10.0838		.497	12.6238
40	.098	2.4892		.198	5.0292		.298	7.5692		.398	10.1092		.498	12.6492
39 .0995	.099	2.5146	8	.199	5.0546		.299	7.5946		.399	10.1346		.499	12.6746
	.100	2.5400		.200	5.0800		.300	7.6200		.400	10.1600	1/2	.500	12.7000

Metric Twist Drills

Metric Twist Drill Sizes					
Metric Drill Sizes (mm) [1]		**Decimal Equivalent in Inches (Ref)**	**Metric Drill Sizes (mm)** [1]		**Decimal Equivalent in Inches (Ref)**
Preferred	**Available**		**Preferred**	**Available**	
	.40	.0157	1.70		.0669
	.42	.0165		1.75	.0689
	.45	.0177	1.80		.0709
	.48	.0189		1.85	.0728
.50		.0197	1.90		.0748
	.52	.0205		1.95	.0768
.55		.0217	2.00		.0787
	.58	.0228		2.05	.0807
.60		.0236	2.10		.0827
	.62	.0244		2.15	.0846
.65		.0256	2.20		.0866
	.68	.0268		2.30	.0906
.70		.0276	2.40		.0945
	.72	.0283	2.50		.0984
.75		.0295	2.60		.1024
	.78	.0307		2.70	.1063
.80		.0315	2.80		.1102
	.82	.0323		2.90	.1142
.85		.0335	3.00		.1181
	.88	.0346		3.10	.1220
.90		.0354	3.20		.1260
	.92	.0362		3.30	.1299
.95		.0374	3.40		.1339
	.98	.0386		3.50	.1378
1.00		.0394	3.60		.1417
	1.03	.0406		3.70	.1457
1.05		.0413	3.80		.1496
	1.08	.0425		3.90	.1535
1.10		.0433	4.00		.1575
	1.15	.0453		4.10	.1614
1.20		.0472	4.20		.1654
1.25		.0492		4.40	.1732
1.30		.0512	4.50		.1772
	1.35	.0531		4.60	.1811
1.40		.0551	4.80		.1890
	1.45	.0571	5.00		.1969
1.50		.0591		5.20	.2047
	1.55	.0610	5.30		.2087
1.60		.0630		5.40	.2126
	1.65	.0650	5.60		.2205
				5.80	.2283

[1] Metric drill sizes listed in the "Preferred" column are based on the R'40 series of preferred numbers shown in the ISO Standard R497. Those listed in the "Available" column are based on the R80 series from the same document.

(Continued)

Metric Twist Drills

Metric Twist Drill Sizes					
Metric Drill Sizes (mm) [1]		Decimal Equivalent in Inches (Ref)	Metric Drill Sizes (mm) [1]		Decimal Equivalent in Inches (Ref)
Preferred	Available		Preferred	Available	
6.00		.2362		19.50	.7677
	6.20	.2441	20.00		.7874
6.30		.2480		20.50	.8071
	6.50	.2559	21.00		.8268
6.70		.2638		21.50	.8465
	6.80 [2]	.2677	22.00		.8661
	6.90	.2717		23.00	.9055
7.10		.2795	24.00		.9449
	7.30	.2874	25.00		.9843
7.50		.2953	26.00		1.0236
	7.80	.3071		27.00	1.0630
8.00		.3150	28.00		1.1024
	8.20	.3228		29.00	1.1417
8.50		.3346	30.00		1.1811
	8.80	.3465		31.00	1.2205
9.00		.3543	32.00		1.2598
	9.20	.3622		33.00	1.2992
9.50		.3740	34.00		1.3386
	9.80	.3858		35.00	1.3780
10.00		.3937	36.00		1.4173
	10.30	.4055		37.00	1.4567
10.50		.4134	38.00		1.4961
	10.80	.4252		39.00	1.5354
11.00		.4331	40.00		1.5748
	11.50	.4528		41.00	1.6142
12.00		.4724	42.00		1.6535
12.50		.4921		43.50	1.7126
13.00		.5118	45.00		1.7717
	13.50	.5315		46.50	1.8307
14.00		.5512	48.00		1.8898
	14.50	.5709	50.00		1.9685
15.00		.5906		51.50	2.0276
	15.50	.6102	53.00		2.0866
16.00		.6299		54.00	2.1260
	16.50	.6496	56.00		2.2047
17.00		.6693		58.00	2.2835
	17.50	.6890	60.00		2.3622
18.00		.7087			
	18.50	.7283			
19.00		.7480			

[1] Metric drill sizes listed in the "Preferred" column are based on the R'40 series of preferred numbers shown in the ISO Standard R497. Those listed in the "Available" column are based on the R80 series from the same document.
[2] Recommended only for use as a tap drill size.

Unified Thread Series and Tap and Clearance Drills

Size	Threads Per Inch	Major Dia.	Pitch Dia.	Tap Drill (75% Max. Thread)	Decimal Equivalent	Clearance Drill	Decimal Equivalent
Unified National Coarse and Unified National Fine Thread Series and Tap and Clearance Drills							
2	56	.0860	.0744	50	.0700	42	.0935
	64	.0860	.0759	50	.0700	42	.0935
3	48	.099	.0855	47	.0785	36	.1065
	56	.099	.0874	45	.0820	36	.1065
4	40	.112	.0958	43	.0890	31	.1200
	48	.112	.0985	42	.0935	31	.1200
6	32	.138	.1177	36	.1065	26	.1470
	40	.138	.1218	33	.1130	26	.1470
8	32	.164	.1437	29	.1360	17	.1730
	36	.164	.1460	29	.1360	17	.1730
10	24	.190	.1629	25	.1495	8	.1990
	32	.190	.1697	21	.1590	8	.1990
12	24	.216	.1889	16	.1770	1	.2280
	28	.216	.1928	14	.1820	2	.2210
1/4	20	.250	.2175	7	.2010	G	.2610
	28	.250	.2268	3	.2130	G	.2610
5/16	18	.3125	.2764	F	.2570	21/64	.3281
	24	.3125	.2854	I	.2720	21/64	.3281
3/8	16	.3750	.3344	5/16	.3125	25/64	.3906
	24	.3750	.3479	Q	.3320	25/64	.3906
7/16	14	.4375	.3911	U	.3680	15/32	.4687
	20	.4375	.4050	25/64	.3906	29/64	.4531
1/2	13	.5000	.4500	27/64	.4219	17/32	.5312
	20	.5000	.4675	29/64	.4531	33/64	.5156
9/16	12	.5625	.5084	31/64	.4844	19/32	.5937
	18	.5625	.5264	33/64	.5156	37/64	.5781
5/8	11	.6250	.5660	17/32	.5312	21/32	.6562
	18	.6250	.5889	37/64	.5781	41/64	.6406
3/4	10	.7500	.6850	21/32	.6562	25/32	.7812
	16	.7500	.7094	11/16	.6875	49/64	.7656
7/8	9	.8750	.8028	49/64	.7656	29/32	.9062
	14	.8750	.8286	13/16	.8125	57/64	.8906
1	8	1.0000	.9188	7/8	.8750	1-1/32	1.0312
	14	1.0000	.9536	15/16	.9375	1-1/64	1.0156
1-1/8	7	1.1250	1.0322	63/64	.9844	1-5/32	1.1562
	12	1.1250	1.0709	1-3/64	1.0469	1-5/32	1.1562
1-1/4	7	1.2500	1.1572	1-7/64	1.1094	1-9/32	1.2812
	12	1.2500	1.1959	1-11/64	1.1719	1-9/32	1.2812
1-1/2	6	1.5000	1.3917	1-11/32	1.3437	1-17/32	1.5312
	12	1.5000	1.4459	1-27/64	1.4219	1-17/32	1.5312

Metric Tap Drills

Tap Drill Sizes for ISO Metric Threads

Nominal Size mm	Series			
	Coarse		Fine	
	Pitch mm	Tap Drill mm	Pitch mm	Tap Drill mm
1.4	0.3	1.1	—	—
1.6	0.35	1.25	—	—
2	0.4	1.6	—	—
2.5	0.45	2.05	—	—
3	0.5	2.5	—	—
4	0.7	3.3	—	—
5	0.8	4.2	—	—
6	1.0	5.0	—	—
8	1.25	6.75	1	7.0
10	1.5	8.5	1.25	8.75
12	1.75	10.25	1.25	10.50
14	2	12.00	1.5	12.50
16	2	14.00	1.5	14.50
18	2.5	15.50	1.5	16.50
20	2.5	17.50	1.5	18.50
22	2.5	19.50	1.5	20.50
24	3	21.00	2	22.00
27	3	24.00	2	25.00

Unified Standard Screw Thread Series

Sizes		Basic Major Diameter	Threads Per Inch											Sizes
			Series with Graded Pitches			Series with Constant Pitches								
Primary	Secondary		Coarse UNC	Fine UNF	Extra Fine UNEF	4UN	6UN	8UN	12UN	16UN	20UN	28UN	32UN	
0		0.0600	—	80	—	—	—	—	—	—	—	—	—	0
	1	0.0730	64	72	—	—	—	—	—	—	—	—	—	1
2		0.0860	56	64	—	—	—	—	—	—	—	—	—	2
	3	0.0990	48	56	—	—	—	—	—	—	—	—	—	3
4		0.1120	40	48	—	—	—	—	—	—	—	—	—	4
5		0.1250	40	44	—	—	—	—	—	—	—	—	—	5
6		0.1380	32	40	—	—	—	—	—	—	—	—	UNC	6
8		0.1640	32	36	—	—	—	—	—	—	—	—	UNC	8
10		0.1900	24	32	—	—	—	—	—	—	—	—	UNF	10
	12	0.2160	24	28	32	—	—	—	—	—	—	UNF	UNEF	12
¼		0.2500	20	28	32	—	—	—	—	—	UNC	UNF	UNEF	¼
⁵⁄₁₆		0.3125	18	24	32	—	—	—	—	—	20	28	UNEF	⁵⁄₁₆
⅜		0.3750	16	24	32	—	—	—	—	UNC	20	28	UNEF	⅜
⁷⁄₁₆		0.4375	14	20	28	—	—	—	—	16	UNF	UNEF	32	⁷⁄₁₆
½		0.5000	13	20	28	—	—	—	—	16	UNF	UNEF	32	½
⁹⁄₁₆		0.5625	12	18	24	—	—	—	UNC	16	20	28	32	⁹⁄₁₆
⅝		0.6250	11	18	24	—	—	—	12	16	20	28	32	⅝
	¹¹⁄₁₆	0.6875	—	—	24	—	—	—	12	16	20	28	32	¹¹⁄₁₆
¾		0.7500	10	16	20	—	—	—	12	UNF	UNEF	28	32	¾
	¹³⁄₁₆	0.8125	—	—	20	—	—	—	12	16	UNEF	28	32	¹³⁄₁₆
⅞		0.8750	9	14	20	—	—	—	12	16	UNEF	28	32	⅞
	¹⁵⁄₁₆	0.9375	—	—	20	—	—	—	12	16	UNEF	28	32	¹⁵⁄₁₆
1		1.0000	8	12	20	—	—	UNC	UNF	16	UNEF	28	32	1
	1¹⁄₁₆	1.0625	—	—	18	—	—	8	12	16	20	28	—	1¹⁄₁₆
1⅛		1.1250	7	12	18	—	—	8	UNF	16	20	28	—	1⅛
	1³⁄₁₆	1.1875	—	—	18	—	—	8	12	16	20	28	—	1³⁄₁₆
1¼		1.2500	7	12	18	—	—	8	UNF	16	20	28	—	1¼
	1⁵⁄₁₆	1.3125	—	—	18	—	—	8	12	16	20	28	—	1⁵⁄₁₆
1⅜		1.3750	6	12	18	—	UNC	8	UNF	16	20	28	—	1⅜
	1⁷⁄₁₆	1.4375	—	—	18	—	6	8	12	16	20	28	—	1⁷⁄₁₆
1½		1.5000	6	12	18	—	UNC	8	UNF	16	20	—	—	1½
	1⁹⁄₁₆	1.5625	—	—	18	—	6	8	12	16	20	—	—	1⁹⁄₁₆
1⅝		1.6250	—	—	18	—	6	8	12	16	20	—	—	1⅝
	1¹¹⁄₁₆	1.6875	—	—	18	—	6	8	12	16	20	—	—	1¹¹⁄₁₆
1¾		1.7500	5	—	—	—	6	8	12	16	20	—	—	1¾
	1¹³⁄₁₆	1.8125	—	—	—	—	6	8	12	16	20	—	—	1¹³⁄₁₆
1⅞		1.8750	—	—	—	—	6	8	12	16	20	—	—	1⅞
	1¹⁵⁄₁₆	1.9375	—	—	—	—	6	8	12	16	20	—	—	1¹⁵⁄₁₆
2		2.0000	4½	—	—	—	6	8	12	16	20	—	—	2
	2⅛	2.1250	—	—	—	—	6	8	12	16	20	—	—	2⅛
2¼		2.2500	4½	—	—	—	6	8	12	16	20	—	—	2¼
	2⅜	2.3750	—	—	—	—	6	8	12	16	20	—	—	2⅜
2½		2.5000	4	—	—	UNC	6	8	12	16	20	—	—	2½
	2⅝	2.6250	—	—	—	4	6	8	12	16	20	—	—	2⅝
2¾		2.7500	4	—	—	UNC	6	8	12	16	20	—	—	2¾
	2⅞	2.8750	—	—	—	4	6	8	12	16	20	—	—	2⅞
3		3.0000	4	—	—	UNC	6	8	12	16	20	—	—	3
	3⅛	3.1250	—	—	—	4	6	8	12	16	—	—	—	3⅛
3¼		3.2500	4	—	—	UNC	6	8	12	16	—	—	—	3¼
	3⅜	3.3750	—	—	—	4	6	8	12	16	—	—	—	3⅜
3½		3.5000	4	—	—	UNC	6	8	12	16	—	—	—	3½
	3⅝	3.6250	—	—	—	4	6	8	12	16	—	—	—	3⅝
3¾		3.7500	4	—	—	UNC	6	8	12	16	—	—	—	3¾
	3⅞	3.8750	—	—	—	4	6	8	12	16	—	—	—	3⅞
4		4.0000	4	—	—	UNC	6	8	12	16	—	—	—	4
	4⅛	4.1250	—	—	—	4	6	8	12	16	—	—	—	4⅛
4¼		4.2500	—	—	—	4	6	8	12	16	—	—	—	4¼
	4⅜	4.3750	—	—	—	4	6	8	12	16	—	—	—	4⅜
4½		4.5000	—	—	—	4	6	8	12	16	—	—	—	4½
	4⅝	4.6250	—	—	—	4	6	8	12	16	—	—	—	4⅝
4¾		4.7500	—	—	—	4	6	8	12	16	—	—	—	4¾
	4⅞	4.8750	—	—	—	4	6	8	12	16	—	—	—	4⅞
5		5.0000	—	—	—	4	6	8	12	16	—	—	—	5
	5⅛	5.1250	—	—	—	4	6	8	12	16	—	—	—	5⅛
5¼		5.2500	—	—	—	4	6	8	12	16	—	—	—	5¼
	5⅜	5.3750	—	—	—	4	6	8	12	16	—	—	—	5⅜
5½		5.5000	—	—	—	4	6	8	12	16	—	—	—	5½
	5⅝	5.6250	—	—	—	4	6	8	12	16	—	—	—	5⅝
5¾		5.7500	—	—	—	4	6	8	12	16	—	—	—	5¾
	5⅞	5.8750	—	—	—	4	6	8	12	16	—	—	—	5⅞
6		6.0000	—	—	—	4	6	8	12	16	—	—	—	6

ISO Metric Standard Screw Thread Series

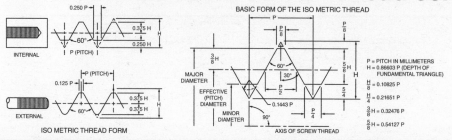

Nominal Size. Diam. (mm)			Series with Graded Pitches		Series with Constant Pitches												Nominal Size Diam. (mm)
Column[a]			Pitches (mm)														
1	2	3	Coarse	Fine	6	4	3	2	1.5	1.25	1	0.75	0.5	0.35	0.25	0.2	
0.25			0.075	—	—	—	—	—	—	—	—	—	—	—	—	—	0.25
0.3			0.08	—	—	—	—	—	—	—	—	—	—	—	—	—	0.3
	0.35		0.09	—	—	—	—	—	—	—	—	—	—	—	—	—	0.35
0.4			0.1	—	—	—	—	—	—	—	—	—	—	—	—	—	0.4
	0.45		0.1	—	—	—	—	—	—	—	—	—	—	—	—	—	0.45
0.5			0.125	—	—	—	—	—	—	—	—	—	—	—	—	—	0.5
	0.55		0.125	—	—	—	—	—	—	—	—	—	—	—	—	—	0.55
0.6			0.15	—	—	—	—	—	—	—	—	—	—	—	—	—	0.6
	0.7		0.175	—	—	—	—	—	—	—	—	—	—	—	—	—	0.7
0.8			0.2	—	—	—	—	—	—	—	—	—	—	—	—	—	0.8
	0.9		0.225	—	—	—	—	—	—	—	—	—	—	—	—	—	0.9
1			0.25	—	—	—	—	—	—	—	—	—	—	—	—	0.2	1
	1.1		0.25	—	—	—	—	—	—	—	—	—	—	—	—	0.2	1.1
1.2			0.25	—	—	—	—	—	—	—	—	—	—	—	—	0.2	1.2
	1.4		0.3	—	—	—	—	—	—	—	—	—	—	—	—	0.2	1.4
1.6			0.35	—	—	—	—	—	—	—	—	—	—	—	—	0.2	1.6
	1.8		0.35	—	—	—	—	—	—	—	—	—	—	—	—	0.2	1.8
2			0.4	—	—	—	—	—	—	—	—	—	—	—	0.25	—	2
	2.2		0.45	—	—	—	—	—	—	—	—	—	—	—	0.25	—	2.2
2.5			0.45	—	—	—	—	—	—	—	—	—	—	0.35	—	—	2.5
3			0.5	—	—	—	—	—	—	—	—	—	—	0.35	—	—	3
	3.5		0.6	—	—	—	—	—	—	—	—	—	—	0.35	—	—	3.5
4			0.7	—	—	—	—	—	—	—	—	—	0.5	—	—	—	4
	4.5		0.75	—	—	—	—	—	—	—	—	—	0.5	—	—	—	4.5
5			0.8	—	—	—	—	—	—	—	—	—	0.5	—	—	—	5
		5.5	—	—	—	—	—	—	—	—	—	—	0.5	—	—	—	5.5
6			1	—	—	—	—	—	—	—	—	0.75	—	—	—	—	6
		7	1	—	—	—	—	—	—	—	—	0.75	—	—	—	—	7
8			1.25	1	—	—	—	—	—	—	1	0.75	—	—	—	—	8
		9	1.25	—	—	—	—	—	—	—	1	0.75	—	—	—	—	9
10			1.5	1.25	—	—	—	—	—	1.25	1	0.75	—	—	—	—	10
		11	1.5	—	—	—	—	—	—	—	1	0.75	—	—	—	—	11
12			1.75	1.25	—	—	—	—	1.5	1.25	1	—	—	—	—	—	12
	14		2	1.5	—	—	—	—	1.5	1.25[b]	1	—	—	—	—	—	14
		15	—	—	—	—	—	—	1.5	—	1	—	—	—	—	—	15
16			2	1.5	—	—	—	—	1.5	—	1	—	—	—	—	—	16
		17	—	—	—	—	—	—	1.5	—	1	—	—	—	—	—	17
	18		2.5	1.5	—	—	—	2	1.5	—	1	—	—	—	—	—	18
20			2.5	1.5	—	—	—	2	1.5	—	1	—	—	—	—	—	20
	22		2.5	1.5	—	—	—	2	1.5	—	1	—	—	—	—	—	22
24			3	2	—	—	—	2	1.5	—	1	—	—	—	—	—	24
		25	—	—	—	—	—	2	1.5	—	1	—	—	—	—	—	25
		26	—	—	—	—	—	—	1.5	—	1	—	—	—	—	—	26
	27		3	2	—	—	—	2	1.5	—	1	—	—	—	—	—	27
		28	—	—	—	—	—	2	1.5	—	1	—	—	—	—	—	28
30			3.5	2	—	—	(3)	2	1.5	—	1	—	—	—	—	—	30
		32	—	—	—	—	—	2	1.5	—	—	—	—	—	—	—	32
	33		3.5	2	—	—	(3)	2	1.5	—	—	—	—	—	—	—	33
		35[c]	—	—	—	—	—	—	1.5	—	—	—	—	—	—	—	35[c]
36			4	3	—	—	—	2	1.5	—	—	—	—	—	—	—	36
		38	—	—	—	—	—	—	1.5	—	—	—	—	—	—	—	38
	39		4	3	—	—	—	2	1.5	—	—	—	—	—	—	—	39
		40	—	—	—	—	3	2	1.5	—	—	—	—	—	—	—	40
42			4.5	3	—	4	3	2	1.5	—	—	—	—	—	—	—	42
	45		4.5	3	—	4	3	2	1.5	—	—	—	—	—	—	—	45

a Thread diameter should be selected from columns 1, 2 or 3; with preference being given in that order.
b Pitch 1.25 mm in combination with diameter 14 mm has been included for spark plug applications.
c Diameter 35 mm has been included for bearing locknut applications.
The use of pitches shown in parentheses should be avoided wherever possible.
The pitches enclosed in the bold frame, together with corresponding nominal diameters in Column 1 and 2, are those combinations which have been established by the ISO Recommendations as a selected "coarse" and "fine" series for commercial fasteners. Sizes 0.25 mm through 1.4 mm are covered in ISO Recommendation R 68 and, except for the 0.25 mm size, in AN Standards ANSI B1.10.

(ANSI)

Machine Screw and Cap Screw Heads

	Size	A	B	C	D
Fillister Head	#8	.260	.141	.042	.060
	#10	.302	.164	.048	.072
	1/4	3/8	.205	.064	.087
	5/16	7/16	.242	.077	.102
	3/8	9/16	.300	.086	.125
	1/2	3/4	.394	.102	.168
	5/8	7/8	.500	.128	.215
	3/4	1	.590	.144	.258
	1	1 5/16	.774	.182	.352
Flat Head	#8	.320	.092	.043	.037
	#10	.372	.107	.048	.044
	1/4	1/2	.146	.064	.063
	5/16	5/8	.183	.072	.078
	3/8	3/4	.220	.081	.095
	1/2	7/8	.220	.102	.090
	5/8	1 1/8	.293	.128	.125
	3/4	1 3/8	.366	.144	.153
Round Head	#8	.297	.113	.044	.067
	#10	.346	.130	.048	.073
	1/4	7/16	.183	.064	.107
	5/16	9/16	.236	.072	.150
	3/8	5/8	.262	.081	.160
	1/2	13/16	.340	.102	.200
	5/8	1	.422	.128	.255
	3/4	1 1/4	.526	.144	.320
Hexagon Head	1/4	.494	.170	7/16	
	5/16	.564	.215	1/2	
	3/8	.635	.246	9/16	
	1/2	.846	.333	3/4	
	5/8	1.058	.411	15/16	
	3/4	1.270	.490	1 1/8	
	7/8	1.482	.566	1 5/16	
	1	1.693	.640	1 1/2	
Socket Head	#8	.265	.164	1/8	
	#10	5/16	.190	5/32	
	1/4	3/8	1/4	3/16	
	5/16	7/16	5/16	7/32	
	3/8	9/16	3/8	5/16	
	7/16	5/8	7/16	5/16	
	1/2	3/4	1/2	3/8	
	5/8	7/8	5/8	1/2	
	3/4	1	3/4	9/16	
	7/8	1 1/8	7/8	9/16	
	1	1 5/16	1	5/8	

Screw Thread Elements for Unified and National Form of Thread

Threads per Inch (n)	Pitch (p) $p = \dfrac{1}{n}$	Single Height Subtract from basic major diameter to get basic pitch diameter	Double Height Subtract from basic major diameter to get basic minor diameter	83 1/3% Double Height Subtract from basic major diameter to get minor diameter of ring gage	Basic Width of Crest and Root Flat $\dfrac{P}{8}$	Constant for Best Size Wire Also Single Height of 60° V-Thread	Diameter of Best size Wire
3	.333333	.216506	.43301	.36084	.0417	.28868	.19245
3 1/4	.307692	.199852	.39970	.33309	.0385	.26647	.17765
3 1/2	.285714	.185577	.37115	.30929	.0357	.24744	.16496
4	.250000	.162379	.32476	.27063	.0312	.21651	.14434
4 1/2	.222222	.144337	.28867	.24056	.0278	.19245	.12830
5	.200000	.129903	.25981	.21650	.0250	.17321	.11547
5 1/2	.181818	.118093	.23619	.19682	.0227	.15746	.10497
6	.166666	.108253	.21651	.18042	.0208	.14434	.09623
7	.142857	.092788	.18558	.15465	.0179	.12372	.08248
8	.125000	.081189	.16238	.13531	.0156	.10825	.07217
9	.111111	.072168	.14434	.12028	.0139	.09623	.06415
10	.100000	.064952	.12990	.10825	.0125	.08660	.05774
11	.090909	.059046	.11809	.09841	.0114	.07873	.05249
11 1/2	.086956	.056480	.11296	.09413	.0109	.07531	.05020
12	.083333	.054127	.10826	.09021	.0104	.07217	.04811
13	.076923	.049963	.09993	.08327	.0096	.06662	.04441
14	.071428	.046394	.09279	.07732	.0089	.06186	.04124
16	.062500	.040595	.08119	.06766	.0078	.05413	.03608
18	.055555	.036086	.07217	.06014	.0069	.04811	.03208
20	.050000	.032475	.06495	.05412	.0062	.04330	.02887
22	.045454	.029523	.05905	.04920	.0057	.03936	.02624
24	.041666	.027063	.05413	.04510	.0052	.03608	.02406
27	.037037	.024056	.04811	.04009	.0046	.03208	.02138
28	.035714	.023197	.04639	.03866	.0045	.03093	.02062
30	.033333	.021651	.04330	.03608	.0042	.02887	.01925
32	.031250	.020297	.04059	.03383	.0039	.02706	.01804
36	.027777	.018042	.03608	.03007	.0035	.02406	.01604
40	.025000	.016237	.03247	.02706	.0031	.02165	.01443
44	.022727	.014761	.02952	.02460	.0028	.01968	.01312
48	.020833	.013531	.02706	.02255	.0026	.01804	.01203
50	.020000	.012990	.02598	.02165	.0025	.01732	.01155
56	.017857	.011598	.02320	.01933	.0022	.01546	.01031
60	.016666	.010825	.02165	.01804	.0021	.01443	.00962
64	.015625	.010148	.02030	.01691	.0020	.01353	.00902
72	.013888	.009021	.01804	.01503	.0017	.01203	.00802
80	.012500	.008118	.01624	.01353	.0016	.01083	.00722
90	.011111	.007217	.01443	.01202	.0014	.00962	.00642
96	.010417	.006766	.01353	.01127	.0013	.00902	.00601
100	.010000	.006495	.01299	.01082	.0012	.00866	.00577
120	.008333	.005413	.01083	.00902	.0010	.00722	.00481

Note: Using the Best Size Wires, measurement over three wires minus Constant for Best Size Wire equals Pitch Diameter.

Dimensioning and Tolerancing Symbols

SYMBOL FOR:	ASME Y14.5	ISO
STRAIGHTNESS	—	—
FLATNESS	▱	▱
CIRCULARITY	○	○
CYLINDRICITY	⌀⧸	⌀⧸
PROFILE OF A LINE	⌒	⌒
PROFILE OF A SURFACE	⌓	⌓
ANGULARITY	∠	∠
PERPENDICULARITY	⊥	⊥
PARALLELISM	//	//
POSITION	⊕	⊕
CONCENTRICITY	◎	◎
SYMMETRY	≡	≡
CIRCULAR RUNOUT	↗	↗
TOTAL RUNOUT	↗↗	↗↗
AT MAXIMUM MATERIAL CONDITION	Ⓜ	Ⓜ
AT LEAST MATERIAL CONDITION	Ⓛ	Ⓛ
DIAMETER	⌀	⌀
BASIC DIMENSION	50	50
REFERENCE DIMENSION	(50)	(50)
DATUM FEATURE	▨Ⓐ	▨ or ▨Ⓐ
CONICAL TAPER	▷	▷
SLOPE	◁	◁
COUNTERBORE	⌴	NONE
SPOTFACE	SF	NONE
COUNTERSINK	⌵	NONE
DEPTH/DEEP	↓	NONE
SQUARE	□	□
NUMBER OF TIMES/PLACES	8X	8X
ARC LENGTH	⌒105	⌒105
RADIUS	R	R
SPHERICAL RADIUS	SR	SR
SPHERICAL DIAMETER	S⌀	S⌀
BETWEEN	←→	←→ (proposed)

Symbols for Materials in Section

Cast iron and malleable iron. Also for use of all materials.

Steel

Brass, bronze, and compositions

White metal, zinc, lead, babbitt, and alloys

Magnesium, aluminum, and aluminum alloys

Rubber, plastic, electrical insulation

Cork, felt, fabric, leather, and fiber

Titanium and refractory metals

Electric windings, electromagnets, resistance, etc.

Marble, slate, glass, porcelain, etc.

Water and other liquids

Across grain
With grain

Wood

Steel Numbering System

The Society of Automotive Engineers (SAE) has standardized a numbering system for steel identification. Each steel is assigned a four- to five-digit numerical name. The first two digits identify the key alloys added to the steel. (Alloys are other metals added to the steel to change its properties, such as strength or hardness.) The last two digits (shown as XX in the chart) identify the carbon percentage of the steel in hundredths of a percent. For example, 1040 steel is a plain carbon steel (no other alloys) with a 0.40% carbon content. In general, the higher the carbon content, the stronger, harder, and more brittle the steel.

Numerical Name	Key Alloys	Numerical Name	Key Alloys
10XX	Carbon only (plain carbon)	501XX	Chromium
11XX	Carbon only (free-cutting)	511XX	Chromium
13XX	Manganese	521XX	Chromium
23XX	Nickel	514XX	Chromium
25XX	Nickel	515XX	Chromium
31XX	Nickel-Chromium	61XX	Vanadium
33XX	Nickel-Chromium	81XX	Nickel-Chromium-Molybdenum
303XX	Nickel-Chromium	86XX	Nickel-Chromium-Molybdenum
40XX	Molybdenum	87XX	Nickel-Chromium-Molybdenum
41XX	Chromium-Molybdenum	88XX	Nickel-Chromium-Molybdenum
43XX	Nickel-Chromium-Molybdenum	92XX	Silicon-Manganese
44XX	Manganese-Molybdenum	93XX	Nickel-Chromium-Molybdenum
46XX	Nickel-Molybdenum	94XX	Nickel-Chromium-Molybdenum-Manganese
47XX	Nickel-Chromium-Molybdenum	98XX	Nickel-Chromium-Molybdenum
48XX	Nickel-Molybdenum	XXBXX	Boron
50XX	Chromium	XXLXX	Lead
51XX	Chromium		

Glossary

A

additive manufacturing: Technology that builds parts by adding material in a series of layers. (1)

alphabet of lines: Guidelines for specific types of lines outlined in the ASME Y14.2, *Line Conventions and Lettering* standard. (3)

American Society of Mechanical Engineers (ASME): An independent, not-for-profit organization that defines drawing standards for engineering drawings. (3)

angle: The amount of rotation or turn between two lines that converge at a vertex. The division of a circle into 360 degrees (°) defines the measurement of an angle. (10)

angle of projection block: Identifies whether the drawing is a first-angle or third-angle projection. (4)

angular dimensioning: The measurement of the angle of a line, a surface, or an origin from a given reference point. (7)

angularity: A geometric control applied to a surface or axis that is at a specified angle (other than 90°) to a datum plane or axis. A basic dimension is used, and no tolerance of degrees is needed. (15)

application block: Identifies a part's assemblies, systems, and subsystems. (4)

arc: Any curved edge with a constant radius and an angle of less than 360°. (8)

assembly: A collection of parts fitted together to form a machine or structure. (17)

assembly drawing: A drawing that uses two-dimensional or pictorial views to show how individual parts are connected to form an assembly. (17)

auxiliary plane: The viewing plane that is parallel to the inclined surface. (14)

auxiliary view: A view showing the true size and shape of an inclined surface or features that are not parallel to the six principal viewing planes. (14)

B

balloon: Circles containing numbers or letters used to identify parts on a drawing. (17)

basic dimension: A numeric value stating the theoretically exact size, shape, orientation, or position of a feature or datum. (15)

basic size: The specified theoretical value from which limits of size are applied. (7)

bevel: The angle one surface makes with another surface when they are not at right angles. Normally used to eliminate sharp corners. (10)

bilateral tolerance: A variance from a dimension in both directions: plus (+) and minus (–). (7)

blend radius: A curve th at is tangent to other lines or arcs. (8)

blind hole: A hole drilled to a specific depth that does not go completely through the part or material. (9)

bolt circle: A circular centerline used to locate holes in a circular pattern. Also known as a circle of centers. (9)

boring: The process of enlarging a hole to a close tolerance and fine finish. A bored hole is more accurate than a drilled hole, but less accurate than a reamed hole. (9)

boss: A raised cylindrical surface on a casting, which provides additional material on the part. A cylindrical pad. (12)

break line: A line used to separate a broken-out section from the otherwise normal view. (3, 13)

broken-out section: A sectional view contained within another view, separated by a break line. Used when only a small portion of the view must be shown in sectional form. (13)

C

CAGE code: The Commercial and Government Entity code identifies government projects by a five-digit government classification code. (4)

caliper: A measuring tool used to transfer distances to be read with a rule. (6)

centerline: A line that shows the location of the center point of a hole, an axis of a part, the center of an arc, or a path of motion. (3)

chain line: A line that notes a special treatment or specification about a specific surface of a part. (3)

chamfer: A beveled edge or angle applied to a hole, shaft, or an edge to remove sharp corners. (10)

circle: A closed curve consisting of an edge that loops 360° around a center point at a fixed distance. (8)

circular runout: A geometric control providing control of individual circular elements of a surface as the part is rotated about a datum axis. (15)

circularity: A geometric control applied to circular elements of round parts, regardless of the axis or datums. Also known as *roundness*. (15)

circumference: The distance around the closed curve of a circle starting and ending at the same point. (8)

computer-aided drafting (CAD): Software used to create digital part drawings, assemblies, and 3D models. (1)

concentricity: A geometric control applied to the elements of a surface to maintain equal midpoint distance of all elements on each side of a datum axis, regardless of feature shape or size. (15)

contour: A curved outline of an object other than a circle. (8)

counterboring: The process of enlarging the end of a hole cylindrically to a specified depth. Normally used to provide recess for a fastener head or a bearing seat. (9)

counterdrilling: A two-step process of drilling a conical hole to a specified depth that allows a fastener's head to sit at or below the part's surface. (9)

countersinking: The process of enlarging the end of a hole conically. Normally used to allow recess for a flat-head screw. (9)

cutting plane: An imaginary plane that divides an object to produce a sectional view. (13)

cutting-plane line: A line that indicates the location of a cutting path along a plane, as well as the viewing direction for sectional and removed views. (3, 13)

cylindricity: A geometric control applied to a cylindrical surface to maintain the form of the surface elements, regardless of the axis or a datum. (15)

D

datum: Exact point, axis, or plane serving as the origin from which location or geometric characteristics of features of a part are derived. (7, 15)

datum feature: An actual feature of a part used to establish a datum. (15)

datum feature symbol: A symbol including an identification letter and an equilateral triangle used to identify a datum feature. (15)

datum target: Specific point, line, or area of contact on a part specified as a datum. (15)

decimal: A fraction using increments of ten and written with a decimal point. (5)

decimal inch: The decimal equivalent of a fraction of an inch. (7)

decimal inch dimension: Measurements based on decimal inches. (7)

degree: The unit of measurement for an angle based on 360 divisions of a circle's circumference, represented by the symbol °. (7)

detail drawing: A drawing that provides all the information necessary for the production of a part, including dimensions, tolerances, and correct views. (1, 16)

detailed representation: Method of representing threads on drawing that shows an approximate true representation of a thread in a detailed, but simplified form. (11)

diameter: The distance of the center axis of a circle from outer edge to outer edge. (8)

dimension: A measurement of definite value of the distance between two given points on an object. (7)

dimension line: A line with arrowheads that spans the distance between extension lines and includes a numerical value. (3)

dimensioning: A method of representing measurements on a drawing using lines and numerical values. (3)

direct tolerancing: The method of specifying tolerances directly to dimensions that control location or size. (7)

dovetail slot: Type of slot used on machine tools to create an interlocking assembly to provide reciprocating motion. (12)

drawing number: Unique number assigned to a drawing for identification and archival purposes. (4)

drawing standards: Documented practices used to develop drawings. (3)

drill: A cylindrical shaped tool with a sharpened point and edges that cuts a specific sized hole. (9)

drilling: The process of cutting a hole in or through a surface with a drill. (9)

dual dimensioning: Dimensioning using both US Customary and SI metric units. (7)

E

exploded assembly drawing: A type of assembly drawing where the individual parts are placed in the correct positions, but spread apart from each other. (17)

extension line: A line that does not touch and extends away from the corners or surfaces of a part to indicate the points of measurement. (3)

external thread: Screw thread cut on an external surface. (11)

F

feature: Universal term applied to a portion of a part, such as a surface, hole, thread, or groove. (4, 15)

feature control frame: Within geometric dimensioning and tolerancing, the rectangular box containing the geometric control, tolerance amount, and datum references. (15)

fillet: A radius applied to the inside corner of a part. (8)

first-angle projection: A system that serves as the foundation for orthographic projection used primarily in countries *other* than the United States. See *third-angle projection*. (2)

flat: A depression on a shaft or shank providing a seat for a setscrew. (12)

flatness: A geometric control applied to a surface to maintain flatness of the surface elements. (15)

foreshortening: Reducing the depth of a surface to give it the illusion of projection. (2)

form tolerance: A geometric tolerance control used to define the form or shape of various geometric figures. (15)

fraction: A part of a whole number. (5)

fractional dimension: Measurements based on fractional numbers. (7)

full section: A sectional view in which the entire object has been cut. (13)

G

general notes: Notes providing general information relevant to the whole drawing. (4)

geometric dimensioning and tolerancing (GD&T): A system of specifying the allowable variation from exact shape or position of a part with respect to the actual function or relationship of part features. (15)

graduations: Small marks on a measuring instrument used to read a measurement. (6)

groove: A recess on the interior surface of a cylindrical piece (internal groove). Also, used interchangeably with neck (external groove). (12)

H

half section: A view in which one-half of an object is shown normally and the other half is shown as if cut for a sectional view. Normally used on symmetrical objects. (13)

hidden line: A line that shows edges or surfaces that are not visible when viewing a part from a specific view. (3)

I

internal thread: Screw thread cut on an internal surface. (11)

K

key: Small piece of metal used to secure a part to a shaft. Common keys include the flat, square, and Woodruff types. (12)

keyseat: An external groove machined along the length of a shaft providing a seat for a key. (12)

keyway: An internal groove machined along the length of a hole providing a slot for a key. (12)

L

lead: Distance a nut will travel during one complete rotation of the screw thread. (11)

leader line: A line that directs the reader to notes, symbols, item numbers, part numbers, dimensions, or specific operations vital to the machining process. (3)

least material condition (LMC): Condition using the limits resulting in the minimum amount of material remaining on the part. This is the upper limit for internal features (such as holes) and the lower limit for external features (such as shafts). (15)

limits: Maximum and minimum allowable size for a dimension. (7)

local notes: Notes providing additional information related to a specific part or feature. (4)

location tolerance: A geometric tolerance control that defines the allowable variation of a feature from the exact or true position shown on the drawing. (15)

M

major diameter: Largest diameter of the thread. (11)

material condition modifier: A modifier applied to a tolerance specification indicating the applicable material condition. (15)

maximum material condition (MMC): Condition using the limit resulting in the maximum amount of material remaining on the part. This is the lower limit for internal features (such as holes) and the upper limit for external features (such as shafts). (15)

metric dimensioning: A system of measurement based on the International System of Units (SI system). (7)

metric system: The system of weights and measures used in most other countries of the world. Compare with the US Customary system. Also known as the *SI system*. (6)

micrometer: A measuring tool that uses a fine pitch screw to take precise measurements. Also known as a "mike." (6)

minor diameter: Smallest diameter of the thread. (11)

multiview drawing: A drawing that provides a means of visualizing a three-dimensional object through the use of two-dimensional views. For most objects, three views are usually sufficient. (2, 16)

N

neck: A recess in the outer edge of a cylindrical workpiece used to allow surfaces of mating pieces to fit flush to each other. (12)

nominal pipe size: Set of standard pipe sizes related to the inside diameter of the pipe. Range from 1/16" to 24". (11)

nominal size: A general or stock size used for the identification of a part. (7)

O

offset section: A sectional view similar to a full section where the cutting-plane line diverts, or is offset, to include nearby features. (13)

orientation tolerance: A geometric tolerance used to control the angularity, perpendicularity, or parallelism of a feature. (15)

orthographic projection: A technique that uses perpendicular projectors to create various views of an object. (2)

P

pad: A raised surface on a casting identical to a boss, but in a shape other than cylindrical. (12)

parallelism: A geometric control applied to a surface or axis to maintain the feature parallel in reference to a datum surface or axis. (15)

part number: Unique number assigned to a part for identification purposes. (4)

partial auxiliary view: An auxiliary view showing only the inclined surface or features. (14)

parts list: A list of all individual pieces needed to create the part or assembly detailed in the drawing. Also called materials list or bill of materials. (4, 17)

perpendicularity: A geometric control applied to a surface or axis to maintain the feature perpendicular in reference to a datum surface or axis. (15)

phantom line: A line that can indicate movement of parts, repeated details, extra material on a part before machining, or filleted and rounded corners on a part. (3)

pipe thread: Thread specifically used for joining pipes and fittings. (11)

pitch: Distance between consecutive crests on a threaded part, measured parallel to the thread axis. The pitch may also be expressed as the number of threads per inch. (11)

pitch diameter: Diameter of an imaginary cylinder that would pass through the thread at a point halfway between the top (crest) and the bottom (root) of the thread. (11)

plus and minus tolerancing: A method of tolerancing that specifies a dimension's variation in a positive and negative direction. (7)

polar coordinate dimensioning: A method of locating a point, line, or surface with a linear distance and an angular measurement from a fixed point of two intersecting perpendicular planes. (7)

position: A geometric control used to define the permitted variation of a feature from the exact or true position indicated on the drawing. (15)

primary auxiliary view: An auxiliary view that has been projected from one of the six principal views. (14)

print reading: The act of interpreting the information shown on a drawing or print. (1)

profile of a line: Two-dimensional geometric control of the elements of an irregular surface. (15)

profile of a surface: Three-dimensional geometric control that extends throughout an irregular surface. (15)

profile tolerance: A geometric tolerance control used to control form, or a combination of size, form, and orientation. Specifies a line or surface profile within which all points or elements of the feature must lie. (15)

R

radius: The distance from the center point to the outer edge of a circle or arc. (8)

reamer: A straight or helical multi-fluted rotary cutting tool used to enlarge, smooth, and size a drilled hole by removing a small amount of material. (9)

reaming: The process of enlarging the interior of a hole to a specific size and finish. (9)

rectangular coordinate dimensioning: A system of dimensioning that uses distances of two or three intersecting planes referenced from a baseline or a datum. (7)

reference dimension: A dimension without tolerances that gives only basic measurement information. (7)

regardless of feature size (RFS): A tolerance zone that remains constant, regardless of the actual size of the feature. (15)

removed section: A sectional view similar to a revolved section but shown separately from the view in which the cutting-plane line is shown. (13)

revision: Any change made to an original drawing. (4)

revision history block: Record of changes made to a drawing. Also called a revision block or change block. (4)

revision status of sheets: A record of the revision status for each sheet of a multiple sheet drawing. Located next to the revision block or as a separate document. (4)

revolved section: A sectional view showing a cross section of the object. A revolved section is shown within another view. (13)

round: A radius applied to the outside corner of a part. (8)

runout tolerance: A geometric control used to define the relationships of one or more features of a part to its axis. Determined using a dial indicator and revolving the part about a designated datum axis. (15)

S

schematic representation: Method of representing threads on a drawing using alternating short and long lines drawn perpendicular to the center axis to represent the major and minor diameters of the thread. (11)

screw thread: A V-shaped groove machined on a shaft or in a hole that follows a helical (spiral) path. Also called a *thread*. (11)

secondary auxiliary view: An auxiliary view showing the true size and shape of an oblique surface. (14)

section: A "cut away" view of a part that shows hidden interior details of a primary view. (3)

section line: Cross-hatching used to represent material "cut" by the cutting plane in a sectional view. (3, 13)

sectional view: A view of an object as it would appear if cut by an imaginary plane. (13)

sheet format: A standard that controls the layout of information on a drawing. The industry standard for decimal inch drawings is ASME Y14.1 and ASME Y14.1M for metric. (4)

sheet size: The size of a drawing layout presented in digital form (soft copy) or on paper (hard copy). (4)

SI (International System of Units) system: The system of weights and measures used in most other countries of the world. Abbreviated SI from the French *Systeme International*. Compare with the US Customary system. Also known as the *metric system*. (6)

simplified representation: Method of representing threads on a drawing using hidden lines parallel to the center axis to represent straight and tapered 60° thread. (11)

size feature: In geometric dimensioning and tolerancing, a feature with a center plane or center axis. (15)

slot: A feature machined into a piece that serves as the female part of a connection to another piece. Common slots include T-slots and dovetail slots. (12)

specified tolerance: Any tolerance directly applied to a dimension. (7)

spotfacing: The process of providing a smooth, flat surface around a hole to accommodate a washer or bolt head. (9)

steel rule: A measuring tool with graduations in one of three systems: fractional inch, decimal inch, or metric. (6)

stitch line: A line that indicates a sewing process on a part or an assembly. (3)

straightness: A geometric control applied to an element, axis, or elements of a surface to maintain straightness of the elements, regardless of a datum. (15)

subassembly: An assembly that can be inserted into a larger assembly. (17)

surface texture: The desired surface condition of a manufactured part. (4)

symmetry: A geometric control applied to maintain the elements of a surface equidistant about a center plane, regardless of feature size. (15)

symmetry line: A line that shows the center axis of a part where both sides are symmetrical. (3)

T

tabular dimensioning: A system of dimensioning that uses letters or numbers referenced to a table instead of dimension lines. (7)

tangent: Touching a curved surface at a single point. (8)

tap: Cutting tool used to machine an internal thread in a hole. (11)

taper: A conical surface with a uniformly changing diameter along its length. (10)

telescoping gauge: A measuring tool used to transfer distances. Primarily used to measure internal features, such as holes and slots. (6)

third-angle projection: A system that serves as the foundation for orthographic projection used in the United States. See *first-angle projection*. (2)

thread: A V-shaped groove machined on a shaft or in a hole that follows a helical (spiral) path. Also called a *screw thread*. (11)

thread class: Definition of the fit or tolerance between an external thread and the mated internal thread. (11)

thread form: Description of the shape and characteristics of a thread. (11)

thread series: Description of a thread based on the number of threads per inch. Available in various series that match specific applications. (11)

three-dimensional (3D) model: A computer-based representation of a part or assembly, including the physical geometry and material characteristics. (1)

three-dimensional (3D) printer: A manufacturing device that uses an electronic file of a 3D model to create a part through the process of additive manufacturing. (1)

through hole: A hole drilled through the entire thickness of the material. (9)

title block: Boxed area normally located in the lower-right corner of a drawing where general information is provided. (4)

tolerance: The total amount that a dimension or feature is allowed to vary. The difference between the lowest allowable value and the highest allowable value (minimum and maximum limits). (4, 7, 15)

tolerance block: Boxed area on a drawing that indicates the general tolerance limits specified for the drawing. (4)

total runout: A geometric control providing a composite control of all surface elements as the part is rotated about a datum axis. (15)

true position: The theoretically exact location of the center plane or center axis of a size feature. (15)

true size and shape: The actual, not scaled, measurement and outline of an object. (2)

T-slot: Type of slot machined into machine tool tables for the purpose of fastening down workholding devices. (12)

U

unilateral tolerance: A variance from a dimension in only one direction, either plus (+) or minus (–). (7)

unspecified tolerance: A tolerance that applies to a dimension that does not have a specified tolerance. (7)

US Customary system: The system of weights and measures used in the United States. Compare with the SI (metric) system. Also known as the US Conventional system. (6)

V

vertex: The point at which two angle-forming lines converge. (10)

V-groove: A V-shaped groove, normally found on pulleys used with V-belts. (12)

viewing-plane line: A line that indicates the viewing direction for alternate drawing views. (3)

visible line: A line that defines the shape and surfaces of an object. Also known as object lines or outlines. (3)

W

whole number: Numbers without fractions or decimal points; such as 1, 2, 3, 4, etc. Also known as counting numbers. (5)

Z

zones: Smaller areas of a larger drawing identified by numbers and letters. Used for easier location of features. (4)

Index

V

vertex, 153
V-groove, 193
viewing-plane line, 28
visible line, 25
visualizing shapes, 11–24

W

weight block, 52
whole numbers
    addition and subtraction, 61
    division, 62–63
    multiplication, 61–62
Woodruff keyseats, 195

Z

zones, 54